Capital Account Liberation

Liberation

Methods and Applications

Systems Evaluation, Prediction, and Decision-Making Series

Series Editor

Yi Lin, PhD

Professor of Systems Science and Economics
School of Economics and Management
Nanjing University of Aeronautics and Astronautics

Capital Account Liberation: Methods and Applications
Ying Yirong and Jeffrey Yi-Lin Forrest
ISBN 978-1-4987-1226-2

Grey Game Theory and Its Applications in Economic Decision-Making
Zhigeng Fang, Sifeng Liu, Hongxing Shi, and Yi Lin
ISBN 978-1-4200-8739-0

Hybrid Rough Sets and Applications in Uncertain Decision-Making
Lirong Jian, Sifeng Liu, and Yi Lin
ISBN 978-1-4200-8748-2

Introduction to Theory of Control in Organizations
Vladimir N. Burkov, Mikhail Goubko, Nikolay Korgin, and Dmitry Novikov
ISBN 978-1-4987-1423-5

Investment and Employment Opportunities in China
Yi Lin and Tao Lixin
ISBN 978-1-4822-5207-1

Irregularities and Prediction of Major Disasters
Yi Lin
ISBN: 978-1-4200-8745-1

Measurement Data Modeling and Parameter Estimation
Zhengming Wang, Dongyun Yi, Xiaojun Duan, Jing Yao, and Defeng Gu
ISBN 978-1-4398-5378-8

Optimization of Regional Industrial Structures and Applications
Yaoguo Dang, Sifeng Liu, and Yuhong Wang
ISBN 978-1-4200-8747-5

Systems Evaluation: Methods, Models, and Applications
Sifeng Liu, Naiming Xie, Chaoqing Yuan, and Zhigeng Fang
ISBN 978-1-4200-8846-5

Systemic Yoyos: Some Impacts of the Second Dimension
Yi Lin
ISBN 978-1-4200-8820-5

Theory and Approaches of Unascertained Group Decision-Making
Jianjun Zhu
ISBN 978-1-4200-8750-5

Theory of Science and Technology Transfer and Applications
Sifeng Liu, Zhigeng Fang, Hongxing Shi, and Benhai Guo
ISBN 978-1-4200-8741-3

Capital Account Liberation

Methods and Applications

Ying Yirong • Jeffrey Yi-Lin Forrest

CRC Press
Taylor & Francis Group
Boca Raton London New York

CRC Press is an imprint of the
Taylor & Francis Group, an **informa** business

CRC Press
Taylor & Francis Group
6000 Broken Sound Parkway NW, Suite 300
Boca Raton, FL 33487-2742

First issued in paperback 2019

© 2015 by Taylor & Francis Group, LLC
CRC Press is an imprint of Taylor & Francis Group, an Informa business

No claim to original U.S. Government works

ISBN-13: 978-1-4987-1226-2 (hbk)
ISBN-13: 978-1-138-89456-3 (pbk)

Contents

Preface

Capital account liberalization has always been accompanied by a large number of international capital flows and can easily lead to volatility in the prices of financial assets. History has clearly shown the fact that major international financial crises, the related contagions, the frequent occurrence of the crises, and the depth, breadth, and speed of the contagions have been growing over time, and each of these issues has found clear connections with the flows of international hot money. The existing literature also indicates that capital account liberalization and financial instability, as well as financial crises and contagions, are closely related. However, capital account liberalization is the inevitable development trend of the world. At the present moment, developed countries have realized their currency convertibility, while some of the most developed economies have literally opened their capital accounts and had their currencies converted freely in open markets. Relatively speaking, the more developed a country is, the higher the degree of opening to the outside world, and, likewise, the higher the level of capital account opening and currency convertibility.

Capital account liberalization is of far-reaching significance, which has been mainly shown in three respects in terms of modern China. First, capital account liberation will contribute to China's economic integration with the global economy and to the promotion of the opening of China to the outside world. The implementation of capital account liberalization in China will be conducive to the implementation of the nation's macroscopic strategy of economic development and international trades, leading to the establishment of an open Chinese economy. Not only will it consist of extensive international exchanges of goods, services, and ideas, but it will also be beneficial for the continuous expansion of the international economy and cross-cultural, technical cooperation. Second, it will bring dynamic economic benefits to China's industries in general and the financial industry in particular because the pressure from foreign competition will force Chinese financial institutions to constantly improve their quality of service, to vary their modes of operation, and to best reduce their operating costs in order to adapt to the greater environment of the international financial market. Finally, capital account liberation can accelerate the establishment of the market economic system in China. The implementation of capital account liberalization is conducive to the rational formation mechanism of the renminbi (RMB) exchange rate; it will further promote the opening of the Chinese capital market to the outside world. Consequently, the standardized operation of Chinese capital market and its efficiency will be greatly improved.

Following the financial crisis of the recent years, the world economy has recovered slowly. The impact of the debt problems of some European countries has become increasingly visible. While the quantitatively loose monetary policy of the United States continues, the international hot money is flooding into the emerging economies, which have been developing faster than then developed countries, so that the external environment for the economic and financial development of most countries is full of uncertainty, increasing the possibility for a new financial crisis to break out. Along with the current trend of economic and financial globalization, when a new financial crisis breaks out, it is imperative to prevent the rapid spread of capital along the flow path provided by various capital account liberalizations. Not only will the rapid spread of capital lead to greater economic losses, it will also seriously disrupt the global economic order and create a breeding ground for the next round of global financial crises and infections. Therefore, this book focuses on the research of the contagious power, status, effects, and other relevant issues of financial crises along with more and more capital accounts being liberated. We attempt to establish a theoretical analysis framework that can be used to effectively analyze the contagion phenomenon of financial crises along with financial market volatility under capital account liberalization.

The management of international capital flows is a key issue facing the international economy. Along with the development of economic globalization, many countries have begun to relax their controls on their capital accounts. However, the financial crises in Latin American countries and Mexico and the exchange rate crises in the Southeast Asian countries have shown that there is a major risk associated with capital account liberalization. Therefore, how to fully understand the benefits and risks of capital account liberalization and how to take an open-door policy at the appropriate time in order to reduce the risk to the lowest possible level have become an important problem for many emerging market economies. However, the existing literature on this has been limited to the investigation of certain policies, especially on capital controls. In fact, the management of international capital flows is a problem of systems science and systems engineering. Besides capital controls, it compromises problems of macroeconomics and finance. Therefore, it is impossible to employ just a single policy to effectively manage a sudden surge of capital inflows. The main reasons for this impossibility include the following.

First, although there are theories on the management of international capital flows, they are merely on the influence mechanism of international capital flows and the understanding of associated risks. Here, the influence mechanism includes that of exchange rates, asset pricing, wealth effects, the reputation, and the policy effect. International capital inflows bring to the host country not only the economic benefits but also the risks of the macroeconomic and financial stability. If the risks are not managed appropriately, the economy will become more vulnerable. Once an external shock appears or an unexpected change occurs, this vulnerability will be materialized, leading possibly to a systemic financial crisis. Second, the situation of the Asian countries has become more specific after the 1997 financial crisis. When the international capital inflows are massive, the host country may be sterilized in the short term, and in the long term, the country will be forced into such a regime that allows the exchange rate to be flexible, to enhance its macroeconomic policies and perfect its financial system. But as the cost of sterilization increases, it is inevitable for the country to face the dilemma of whether to keep the interest rate or the exchange rate stable. However, the effects of reforming the exchange rate and perfecting the financial system take a long time to unfold. So the host country may well choose to use capital control measures in the short and medium terms to make up for the shortfall of the two measures. Finally, this book analyzes the challenge and the future of the management of international capital flows after the recent global financial crisis. This financial crisis has had a great impact on the import and export trade of Asia and affects the economic growth of the region. Because of the uncertainty in the external economic conditions, the management of international capital flows in Asia confronts many unprecedented challenges.

This book includes six chapters together with an introduction and a conclusion. The main contents are organized as follows.

In Chapter 1, we establish a complete mathematical analysis framework for the study of the problem of capital account liberalization. Specifically, a few important models developed for the study of capital account liberalization, which must be solved by using different methods, are introduced. These models deal with such problems as follows:

- Capital control and monetary independence
- Long-term optimal capacity levels of investment projects
- Capital controls
- Capital flows and the real exchange rate
- The optimal model about capital flows
- The limit cycle theory in the studies of population

For instance, to estimate the impact of capital controls on macroeconomic variables, we introduce a vector error correction model that includes the index of industrial production as constructed by the national statistics agency (Colombia: National Administrative Department of Statistics [DANE]), a multilateral real exchange rate as constructed by the Central Bank of Colombia (Banco de la República), total capital inflows as reported by the Colombian Central Bank, an important index of terms of trade as constructed by the Central Bank, and the global emerging markets bond index spread as constructed by JP Morgan to control changes in global financial conditions.

In Chapter 2, we research the influence of capital account liberalization on the stability of the financial market by greatly expanding the scope of ordinary differential equation theory to the analysis of local stabilities.

In this chapter, we investigate optimal trade-offs between growth and instability of open economies based on their policy preferences about the risky growth opportunities offered by international capital flows. Policy choices of many countries often force these countries to pay close attention to the potential economic instability because of their financial vulnerability when seeking high rates of growth via risky capital flows. This chapter reveals the c-effect of capital flows. This concept indicates that in a widely open economy with lower risk aversion but without a sound financial sector, a great instability must be endured in exchange for a high rate of growth. This chapter also establishes the b-effect, which implies that high instability is an inevitable price paid for having rapid growth if the host country permits wide opening of its capital market without first strengthening its financial sector.

In Chapter 3, the combined effect of multiple factors, especially the dynamic effect, is investigated by employing the theory of partial differential equations. On the basis of option pricing, we carefully analyze the problem of currency substitution. With capital account liberalization, investors will naturally choose different currency assets according to the corresponding returns, costs, and preferences of risk, and the possibility of currency substitution will be strengthened because of the free movement of capital. Extraordinarily active currency substitution can be easily transformed into financial crises, and the large context of the phenomenon of currency substitution combined with increased international capital flows will lead to an inevitable contagion of financial crises. Introducing the concept and formula of relative risk preference coefficient under the condition of the exponential utility function, we analyze the need to balance foreign currencies and the domestic currency in terms of the optimal portfolio choice; we establish four propositions and discover that there are four tendencies for investors to hold the optimal ratio of foreign currency assets.

In Chapter 4, although the theory of limit cycles is a complex theory, where some problems such as Hilbert's 16th problem have not been solved as of this writing, we creatively apply it to the study of problems related to capital account liberalization and discuss the contagion of financial crises among the financial markets of different countries. We build a transmission-based model that can identify the contagion between the financial markets of two countries. We then identify the *regulatory power*, the *investor confidence cohesion*, and the *immunity* of an affected country as the main reasons for causing the nonlinear fluctuations in the stock returns of the two countries after the crisis. By using limit cycle theory, we conclude that the financial contagion goes through three stages shortly after the outbreak of the crisis: the instability and obstruction of an initial and weak infection followed by a limited and controlled oscillation of the infection and then a short-term uncontrolled strong infection.

In Chapter 5, many problems related to capital account liberalization are formulated as optimization models, showing the fact that much broader economic issues can be solved by employing optimization methods. After comparing five classic economic cases, we solve a few special cases of the Verdier equation and provide the financial interpretations of the corresponding cases.

In this chapter, we focus on the discussion of the Venison optimization model, whose objective function takes an integrated form. By analyzing the capital accumulation model, we come to the conclusion that an optimal control problem with a lagging term can be converted into an optimal model without any lagging term, namely, the Venison optimal model. We also discuss applications of the Venison optimization model in three typical markets represented as bullish, bearish, and equilibrium market. We obtain the constraining condition and the optimal solution of the objective function of the integral form for the three markets, namely, the optimal overseas investment strategy of qualified domestic institutional investor (QDII) after capital account liberalization. Overall, we conduct the analysis on the benefit and risk of QDII and also probe into its current state and future development.

Additionally, in this chapter we introduce the Verdier optimal model to research the capital flow by discussing the origin and analytic solutions of the model with different constraining conditions and conducting the optimization analysis of several special cases. Consequently, we successfully derive the optimal solution of free capital flows. At the end, we analyze the current stage of capital account liberalization in China as a case study and come to the conclusion that the basic prerequisites of openness have already been met. And we also look into the future on the basis of all our analyses presented and recognize that there will be no priority order in the series of reforms to be taken in the future; namely, interest rate liberalization, exchange rate reformation, the internationalization of RMB, and the liberalization of capital account can be promoted coordinately at the same time.

In Chapter 6, by using shareholder variance and risk (SVAR) models and the impulse response analysis, we compare the contagion effect of financial markets between nations with a relatively high degree of openness and those with a moderate degree of openness. There is a major difference between the global financial contagion triggered by the U.S. subprime mortgage crisis and past financial contagions. That is a nonsystemic risk of a single country or local area turning into a risk of the global financial system along the tracks of *globalization, integration,* and *freedom.* The contagion is more complex, the infection intensity is greater, and different nations experience different states of infection corresponding to their individual levels of capital account liberalization.

Ying Yirong
Jeffrey Yi-Lin Forrest

MATLAB® is a registered trademark of The MathWorks, Inc. For product information, please contact:

The MathWorks, Inc.
3 Apple Hill Drive
Natick, MA 01760-2098 USA
Tel: 508-647-7000
Fax: 508-647-7001
E-mail: info@mathworks.com
Web: www.mathworks.com

Acknowledgments

First, we express our appreciation and thanks to the following people, whose well-developed, important works have helped us to enrich the presentation of this book:

- Professor Laurie J. Batesain, Department of Economics, Bryant University
- Professors Salvatore Capasso and Tullio Jappelli, University of Naples
- Professor Philip DeCiccaa, NBER and Department of Economics, McMaster University
- Professors Alberto Porto and Agustin Lodola, Universidad Nacional de La Plata
- Professor Marcel Schröder, Arndt-Corden Department of Economics, Australian National University

Second, many of the results and conclusions contained in this book have been previously authored by us and published in various sources, and we are grateful to the copyright owners for permitting us to use the materials. They include the International Association for Cybernetics (Namur, Belgium), Gordon and Breach Science Publishers (Yverdon, Switzerland, and New York), Hemisphere (New York), the International Federation for Systems Research (Vienna, Austria), the International Institute for General Systems Studies, Inc. (Grove City, Pennsylvania), Kluwer Academic and Plenum Publishers (Dordrecht [the Netherlands] and New York), MCB University Press (Bradford, United Kingdom), Pergamon Journals, Ltd. (Oxford), Springer and Taylor & Francis Group (London), World Scientific Press (Singapore and New Jersey), Wroclaw Technical University Press (Wroclaw, Poland), Northwest University Press (China), and Shanghai University Press (China).

Third, there are countlessly many colleagues who have contributed to our career successes over the years. Although it is impossible for us to list them all, we especially thank

- Professor Fan Jianping, University of North Carolina (United States), from whom the first author had greatly benefited from their fruitful discussions in the areas of econophysics and the technology of visualization

- Professor Wu Congsheng, University of Bridgeport (United States), from discussions with whom the first author had formulated many new ideas in the area of financial engineering during his visit to the United States
- Professor Yao Feng, Kagawa University (Japan), with whom the first author had the honor to explore how to employ statistical methods to analyze some of the challenging problems in the frontier of finance
- Professor Qin Xuezhi, Dalian University of Technology (China), with whom the first author has had rewarding collaborations for many years
- Professor Ben Fitzpatrick, Auburn University (United States), the second author's PhD supervisor, for his teaching and academic influence
- Professor Shutang Wang, Northwest University (China), the second author's MS supervisor, for his career motivation
- Professor George Klir, New York State University at Binghamton (United States), from whom the second author found the direction in his career
- Professors Mihajlo D. Mesarovic, Case Western Reserve University (United States), and Yasuhiko Takaraha, Tokyo Institute of Technology, from whom the second author was affirmed his chosen endeavor in his academic career
- Professor Gary Becker, University of Chicago (United States), a Nobel laureate in economics; his rotten kid theorem had initially brought the second author deeply into economics, finance, and corporate governance

Fourth, we acknowledge the financial support given by the National Natural Science Foundation of China (71171128) and the "College Teachers Visiting Plan Abroad" of the Foundation of Shanghai Education Committee (B.60-A133-13-002).

After several years of dedicated work on this manuscript, we are delighted to see the final product take its shape. We hope that it will be a joy for anyone to read through the pages of this book as either a reference or a source of inspiration for future research. If you have any comment or suggestion, please contact us at yingyirong@sina.com (for Ying Yirong) and at jeffrey.forrest@sru.edu (for Jeffrey Yi-Lin Forrest).

Authors

 **Ying Yirong** is professor of finance and is associate chair of the Department of Finance, College of Economics, Shanghai University, Shanghai, China. He earned his BSc in mathematics in 1982 from the Mathematics Department of Northwest University (China) and his PhD in mathematics in 2000 from the Mathematics Department of Xidian University. In 2002, Dr. Yirong did one year of postdoctoral study at the Institute of Contemporary Finance, Shanghai Jiao-Tong University.

Professor Yirong has taught many different courses in the areas of economics and finance, such as econometrics, financial economics, financial physics, applied statistics, financial engineering, economic cybernetics, and low carbon economy.

His research interests include financial engineering, financial mathematics, securities pricing, and risk management. He has published four monographs:

- *The Significance, Methodology and Application about Infinite*. Xi'an: Press of Northwest University, 1996
- *Capital Account Liberation in the Post Financial Crisis Era*. Shanghai: Press of Shanghai University, 2012
- *Econometrics*. Shanghai: Press of Shanghai University, 2012
- *Applied Statistics*. Shanghai: Press of Shanghai University, 2013

Additionally, Dr. Yirong has published more than 50 research papers in various professional journals, such as *Kybernetes, Advances in Systems Science and Applications, Annals of Differential Equations, Journal of Management Science, Journal of System Management, Journal of Chang'an University, Chinese Journal of Engineering Mathematics, International Journal of Research in Business and Technology,* and *International Journal of Engineering, Mathematics, and Computer Sciences.*

Currently, Dr. Yirong is a director of the Shanghai Financial Engineering Association; the Systems Science, Management Science and System Dynamics Association; and the Chinese Enterprise Operations Research Association.

He has won numerous research excellence awards. His works have been financially funded by the National Natural Science Foundation (China), the Ministry of Education (China), and various agencies in Shanghai.

Jeffrey Yi-Lin Forrest holds all his educational degrees (BSc, MS, and PhD) in pure mathematics, respectively, from Northwest University (China), Auburn University (United States) and Carnegie Mellon University (United States), where he has one-year postdoctoral experience in statistics. Currently, he is a guest or specially appointed professor in economics, finance, systems science, and mathematics at several major universities in China, including Huazhong University of Science and Technology, the National University of Defense Technology, and Nanjing University of Aeronautics and Astronautics, and a tenured professor of mathematics at the Pennsylvania State System of Higher Education (Slippery Rock campus). Since 1993, Dr. Forrest has been serving as the president of the International Institute for General Systems Studies, Inc. Along with various professional endeavors he has organized, Dr. Forrest has had the honor to mobilize scholars from more than 80 countries representing more than 50 different scientific disciplines.

Over the years, Dr. Forrest served on the editorial boards of 11 professional journals, including *Kybernetes: The International Journal of Systems & Cybernetics, Cybernetics and Management Science, Journal of Systems Science and Complexity, International Journal of General Systems*, and *Advances in Systems Science and Applications*. He is the editor-in-chief of three book series: *Systems Evaluation, Prediction and Decision-Making*, published by CRC Press (New York, United States) an imprint of Taylor & Francis Group since 2008, *Communications in Cybernetics, Systems Science and Engineering—Monographs*, and *Communications in Cybernetics, Systems Science and Engineering—Proceedings*, published by CRC Press (Balkema, the Netherlands), an imprint of Taylor & Francis Group since 2011.

Some of Dr. Forrest's research was funded by the United Nations, the State of Pennsylvania, the National Science Foundation of China, and the German National Research Center for Information Architecture and Software Technology.

Dr. Forrest's professional career started in 1984 when his first paper was published. His research interests are mainly in the area of systems research and applications in a wide-ranging number of disciplines of the traditional science, such as mathematical modeling, foundations of mathematics, data analysis, theory and methods of predictions of disastrous natural events, economics and finance, management science, and philosophy of science. As of 2013, he had published more than 300 research papers and 40 monographs and edited special topic volumes by such prestigious publishers as Springer, Wiley, World Scientific, Kluwer Academic, and Academic Press. Throughout his career, Dr. Forrest's scientific achievements have been recognized by various professional organizations and academic publishers.

Chapter 1

From Theory to Visualization: General Analysis Framework in Finance

1.1 Mathematical Applications in Finance

The recent global economic crisis has been a hard-to-find ground for validating methods of modeling the economy, because a sharp break has been seen in the established trends that existed in all economic processes. The economy of the *golden billion* could physically grow, but it does not want to. In this case, the virtual economy turned out to be more stable than the real one: because of their overabundance, food, fuel, metals, and gold lost much more value than services and information did. Even in the financial sphere, what crashed was what had been considered the most reliable investment for as long as anyone could remember: mortgages, which are loans given to real people secured by real collateral, while, for example, loans provided for promoting websites with advertisements are still being repaid. The farther away from basic sectors something is in the technological chain, the less it has suffered from the crisis.

Today it is still hard to tell where all of this will end, especially if we take into account the problem of capital account liberalization. In the long run, it may lead to a more efficient allocation of resources, while a growing number of authors are now prepared to acknowledge its destabilizing short-run and medium-run consequences. This is because capital account liberalization could create an economic environment in which banks may engage in morally hazardous behaviors.

However, this end might still be a long way off from the current situation; it is only because the existing system of economic mechanisms works only when there is a prospect of economic growth. So in the foreseeable future, either growth will nevertheless resume or a whole series of unsuccessful attempts to restore it will follow. And until new mechanisms adapted to zero growth appear, processes of recession and restoration of growth are all that can be seriously modeled.

The current economic subsystem of the society, which manages the production, distribution, and consumption of resources, goods, and services, has to solve very difficult problems. We are talking about the production of billions of types of goods and their distribution among billions of individuals and corporations. For this reason, the economy as a managing system is always relatively decentralized. In our opinion, precisely, the catastrophic disparity between the means of management and supervision and the rapidly growing complexity of economic connections were the main reasons for the collapse of the central planning concept. Thus, in modeling an economy, we are dealing with a complex system. Complex systems (such as an economy) are distinguished not by the fact that they consist of a very large number of elements, but primarily by their uniqueness and, most important, their capability of endogenous qualitative changes. Therefore, when we study a complex system, we are actually observing a unique trajectory that does not reproduce itself with statistical reliability and does not show all of the system's possibilities. In doing so, we go beyond the bounds of the empirical method, the basis of success in the natural sciences. It is not surprising that successes in modeling complex systems have been considerably more modest thus far than successes in modeling, say, technical systems. Experiments and mass observation are not possible with complex systems. Therefore, no universal model has yet been created for any complex system from which all others would follow as particular cases in the same way as, say, models of electronic devices follow from a theoretical model of electrodynamics. We have to deal with a large number of models of one system, each of which uses its own conceptual language and examines the system from its own perspective, ignoring quantities that are in no way insignificant. While models of physical systems should explain the results of experiments that have been done and predict the results of planned experiments, models of complex systems are primarily called upon to take the place of experiments. There is not just a multitude of models of the economy, but a multitude of methodological approaches to modeling it. With respect to the objects of modeling, there are models of specific firms or markets, micro-models (models of typical firms or markets), and macro-models (models of entire economic systems). It would seem that micro-models should be derived from specific typifications, and macro-models from micro-models by aggregation. But in practice, this has been very rarely done. According to the construction procedure, applied macro-models can be broken down, somewhat arbitrarily, into the following types.

Econometric models are distinguished by their attempt to rely primarily on observed correlations between time series of economic indicators, rather than on hypotheses about a system of causal relationships in an economy. If it turns out that a series of values of some quantity can be described using other quantities with a small and independent error, then some pattern is considered to have been found in the economy. When there are enough of such functions to determine all of the quantities in terms of their past values, a system of correlations is obtained that is formally capable of predicting future values of the indicators. This is an econometric model. Such models sometimes reach enormous dimensions: they contain hundreds of thousands of variables and correlations. Real-life practice shows that an econometric model is not capable of predicting a break in a trend: when tuned to growth, it will predict growth. Econometric models are also not very suitable for analytic calculations. That is, they do not produce answers to the question, what would happen if a different policy were applied.

Theoretical models are constructed as a formalization of any authoritative descriptive economic theory. Contemporary economics textbooks are full of simple models of this type, and they are frequently used to justify economic policy. Close analysis shows, however, that in all substantive theories, the concepts are somewhat hard to pin down, changing from one chapter to the next, from one subject to another. As a result, with the transition from explaining a theory's basic premises to applied problems, we get not a single model corresponding to the theory, say, of John Maynard Keynes, but thousands of Keynesian models that are incompatible with each other. *Computable general equilibrium (CGE) models* have become the main tool used for economic forecasting in international practice since the 1990s. It has become clear that, in order to adequately describe a contemporary economy, it is not enough to take into account only technological constraints (balance models), extrapolate previous trends (econometric models), or rectilinearly impose external constraints (system dynamics models). CGE models date back to Kenneth Arrow and Gerard Debreu's dynamic version of a model of general economic equilibrium. General equilibrium models, especially dynamic ones, are complex because they contain a great many optimization problems of nonlinear behaviors. In CGE models, these dynamic connections are usually replaced by some phenomenological assumptions. Here, it is important to make a comment regarding the term *equilibrium*. In contemporary science, it is used in three originally different meanings: dynamic equilibrium (the balance of forces acting on a system), static equilibrium (the balance of probabilities of transitions between states of a system), and economic equilibrium or, more generally, Nash theoretical game equilibrium (the balance of interests of parties to a conflict). There is a connection between these concepts, to be sure, but no one has yet managed to precisely trace it in a reasonably general form. Therefore, the usual mixing of these concepts for superficially scientific discourse should be considered an impermissible vulgarization. Somehow, economic equilibrium in the general sense does not in any way imply static or simple dynamics.

Balance models appeared as a method of supporting economic planning. The main part of these models is a system of material balances for some set of products that together encompasses the whole economy. In the 1950s–1970s, planning procedures worldwide were based on *Leontief models*. Recently, balance models have often been supplemented with financial balance systems. But balance models do not describe market mechanisms. The strong point of balance models is that they consist of the most reliable balance correlations in economics and are based on data collected for the model. Their shortcoming is that the balance language does not reflect the relationships of economic agents (EAs) and characteristics of their behaviors. Therefore, balance models are often incapable of capturing the actual problems that an economic development encounters.

Simulation models were originated in attempts to apply the methods used for modeling technical systems to the economy. The basic technique of simulation modeling is to break down a system into units corresponding to significant processes or objects. The description of the system consists of descriptions of individual units. It is hard to apply this method to modeling an economy because, in contrast to a technical system, which is created from individual parts, an economy emerges in a process of self-organization, and its division into parts is not at all unique. Global dynamics simulation models are the most popular. However, the arbitrary nature of the premises and the break with economic theory provoked serious criticism of these models, and their gloomy forecasts did not prove to be correct. The system dynamics method has persisted, however. System dynamics models, which describe the established practice of decision making, are now used in almost all large corporations. It is hard to judge how useful they actually are because they always have a large illustrative and advertising component, and it is often difficult to distinguish them from computer games.

1.1.1 Synergetic Approach

A trend that is opposite of all described earlier, in a sense, has drawn considerable interest recently: construct comparatively simple models on the basis of analogies with well-studied physical and biological processes. To this end, qualitative phenomena inherent in dynamic systems of a certain mathematical type serve as the basic analogy. These models are called synergetic or, more recently, econophysics models. They do indeed sometimes reveal unexpected effects and connections, but for now, it seems to us, they are not sufficiently reliable or practical.

The conceptual framework of systems analysis of a developing economy (SADE) was created in 1975 by academician Aleksandr Petrov at the Computer Center of the Academy of Sciences of the Soviet Union. This line of research synthesized a methodology for the mathematical modeling of complex systems and development in natural sciences, on one hand, and the achievements of advanced economic theory, on the other. Systems analysis models of a developing economy are similar in meaning to CGE models, but they focus more on the specifics of established economic relationships. These studies began 15 years before the first CGE models appeared.

A model that reproduced the basic qualitative features of the evolution of a planned economy was constructed in 1988. Therefore, by the time economic transformations began in the Soviet Union, and then in Russia, an approach to the analysis of the economic changes that were occurring had already been developed. In particular, 2 years before the 1992 reform, its short-term consequences were predicted. Models were constructed for the economy in the period of high inflation in 1992–1995, for the economy in the period of *financial stabilization* in 1995–1998, and for estimating the prospects of economic development after the crisis of 1998, which was based on a system of hypotheses about the nature of economic relationships that were formed in the relevant period in Russia. These models made it possible to understand the internal logic of development of economic processes, which had been hidden behind the visible, often seemingly paradoxical pattern of economic phenomena, which did not fit in the familiar theoretical schemes. Experience in using the models showed that they serve as a reliable tool for analyzing macroeconomic patterns and forecasting the consequences of macroeconomic decisions, provided that the existing relationships are maintained. It can be said that a *chronicle* of Russian economic reforms was produced, written in the language of mathematical models.

Here we list three classical mathematical applications in finance as follows:

Example 1.1: Arrow's Impossibility Theorem

Arrow's Impossibility Theorem is a classical example by using accurate mathematical description. In the 1970–1980s, it was established that every Arrow's social welfare function was associated with a hierarchy of decisive coalitions topped with a dictator.

A natural approach to the problem was to *measure* the size of the decisive coalitions and to characterize them as majorities or minorities. This idea goes back, respectively, to *families of qualitative majorities* and *families of majorities*, meaning their impact on social choice as if of actual majorities. Having considered Arrow's model with arbitrary measures, Tangian (2010) showed that these coalitions were in a sense *quantitative majorities*, because their average size surpassed 50% of voters. This conclusion put in question the prohibition of decisive hierarchies that represented a majority rather than a minority. The same applies to dictators as their equivalents.

The next step was a numerical evaluation of the representative capacity of Arrow's social welfare functions/decisive hierarchies/dictators. Tangian (2010) reported that the quantitative evaluation enabled one to find the *best* Arrow's social welfare functions/decisive hierarchies/dictators in every specific realization of the model. The best ones were always rather representatives, whereas nonrepresentative ones could be missing completely. All of these results show that Arrow's condition, which prohibits dictators/decisive hierarchies, is stronger than commonly supposed, excluding *good* dictators together with *bad* ones. The general prohibition can be hardly justified in typical circumstances, where it is *too intolerant*. Practical decision making needs some more room for compromises.

In 1951, Arrow studied the relation between individual preferences for given decision-making options (candidates for president, political issues, competing projects, etc.) and a *social* preference. His model includes a society of *n individuals* $I = \{1, 2, ..., n\}$, a set of *m alternatives* $X = \{x, y, z, ...\}$, and five axioms, which turn out to be inconsistent.

The model operates on individual *preferences*, which are schedules of alternatives ordered by priority.

A *preference* $\succ$ is a binary relation on a set X of alternatives, which is asymmetric ($x \succ y \Rightarrow y \not\succ x$). Two alternatives are *equivalent* or *indifferent* if none is preferred: $x \sim y \Leftrightarrow x \not\succ y$ and $y \not\succ z$.

Axiom 1.1 (*Number of alternatives*). There are at least three alternatives: $|X| \geq 3$.

Axiom 1.2 (*Universality*). For every *preference profile*, that is, a combination of n individual preferences $(\succ_1, ..., \succ_n)$, there exists a social preference $\succ$ denoted also by $\sigma(\succ_1, ..., \succ_n)$.

Axiom 1.3 (*Unanimity*). An alternative preferred by all individuals is also preferred by the society: $x \succ iy$ for all $i \Rightarrow x \succ y$.

Axiom 1.4 (*Independence of irrelevant alternatives*). The social preference on two alternatives is determined exclusively by individual preferences on these alternatives and is independent of individual preferences on other alternatives. In other words, if individual preferences on two alternatives remain the same under two profiles, then the social preference on these alternatives also remains the same under these profiles:

$$(\succ_1|_{xy}, ..., \succ_n|_{xy}) = (\succ_1'|_{xy}, ..., \succ_n'|_{xy}) \Rightarrow \succ|_{xy} = \succ'|_{xy}$$

Axiom 1.5 (*No dictator*). There is no dictator, that is, no individual i whose strict preference always holds for the society: $x \succ iy \Rightarrow x \succ y$.

Arrow's Impossibility Theorem: *Every mapping* $\sigma: \succ_1, ..., \succ_n \rightarrow \succ$, *which satisfies Axioms 1.1–1.4 (said to be Arrow's social welfare function) is dictatorial. Consequently, Axioms 1.1–1.5 are inconsistent.*

According to Tangian (2010), modeling a large society needs an infinite set of individuals and a coalition algebra that does not contain individuals as its elements. The distinction of a *large* society is that only sufficiently large groups are considered influential, whereas single individuals are negligible. It relates to Arrow's dictators as well, but if they are negligible, they are no longer dictators.

Since the same coalition algebra up to an isomorphism can be realized on different sets of individuals, the set of individuals is not retained under the group of isomorphic model. Just this effect of inclusion/exclusion of dictators was discovered in the infinite model. It is not observed in the finite model, where individuals are at the same time minimal coalitions (atoms) and are therefore the model invariants. Thus, the infinite model turned out to be the most illuminating.

The non-invariant behavior of Arrow's dictators under isomorphisms evokes questions. From the mathematical viewpoint, the elements that appear and disappear under isomorphisms of the model are *incorrect*, or secondary, because they are irrelevant to its fundamental structures. By the same reasons, such elements cannot determine the model axioms. Consequently, Arrow's prohibition of dictators must be transferred to their invariant substitutes, hierarchies of decisive coalitions. Therefore, we shall analyze the properties of dictators with regard to the associated decisive coalitions, which are invariants of the model and carriers of its properties. Fishburn has built his proof with a reference to two-valued measures on the set of individuals, and Kirmann–Sondermann's proof was based on the equivalent notion of ultrafilter. The very idea of a measure in the given context implied comparing decisive coalitions with majorities and minorities. Furthermore, it prompts to measure decisive coalitions in a more accurate way to evaluate their representative capacity and thereby to evaluate the representative capacity of Arrow's dictators.

Example 1.2: Markowitz Portfolio Selection

Without mathematical Lagrange, it is impossible to describe and measure the various financial risks.

In 1952, Markowitz published the article *Portfolio Selection*, which is seen as the origin of modern portfolio theory. Portfolio models are tools intended to help portfolio managers determine the weights of the assets within a fund or portfolio. Markowitz's ideas have had a great impact on portfolio theory and have withstood the test of time. However, in practical portfolio management, the use of Markowitz's model has not had the same impact as it has had in academia. Many fund and portfolio managers consider the composition of the portfolio given by the Markowitz model as unintuitive. The practical problems in using the Markowitz model motivated Black and Litterman to develop a new model in the early 1990s (Black and Litterman, 1992). The model, often referred to as the Black–Litterman model (hereafter the BL model), builds on the basis of Markowitz's model and aims at handling some of its practical challenges. While the optimization in the Markowitz model begins from the null portfolio, the optimization in the BL model begins from, what Black and Litterman refer to as, the equilibrium portfolio (often assessed as the benchmark weights of the assets in the portfolio). *Bets* or deviations from the equilibrium portfolio are then taken on assets to which the investor has assigned views. The manager assigns a level of confidence to each view indicating how sure he/she is of that particular view. The level of confidence affects how much the weight of that particular asset in the BL portfolio differs from the weights of the equilibrium portfolio.

Markowitz focuses on a portfolio as a whole, instead of an individual security selection when identifying an optimal portfolio (Markowitz, 1952). Previously, little research concerning the mathematical relations within portfolios of assets had been carried out. Markowitz began from John Burr Williams's *Theory of Investment Value*. Williams claimed that the value of a security should be the same as the net present value of future dividends. Since the future dividends of most securities are unknown, Markowitz claimed that the value of a security should be the net present value of *expected* future returns. Markowitz claims that it is not enough to consider the characteristics of individual assets when forming a portfolio of financial securities. Investors should take into account the co-movements represented by covariance of assets. If investors take covariance into consideration when forming portfolios, Markowitz argues that they can construct portfolios that generate higher expected return at the same level of risk or a lower level of risk with the same level of expected return than portfolios ignoring the co-movements of asset returns. Risk, in Markowitz' model (as well as in many other quantitative financial models), is assessed as the variance of the portfolio. The variance of a portfolio in turn depends on the variance of the assets in the portfolio and on the covariance between the assets.

The practical problems in using the model prompted Black and Litterman to continue the development of portfolio modeling (Black and Litterman, 1992). Markowitz showed that investors under certain assumptions can, *theoretically*, build portfolios that maximize expected return given a specified level of risk, or minimize the risk given a level of expected return. The model is primarily a normative model. The objective for Markowitz has not been to explain how people select portfolios, but how they should select portfolios.

Even before 1952, diversification was a well-accepted strategy to lower the risk of a portfolio without lowering the expected return. But until then, no thorough theoretical foundation existed to validate the concept of diversification. Markowitz's model has remained to date the cornerstone of modern portfolio theory.

According to Markowitz, inputs needed to create optimal portfolios include expected returns for every asset, variances for all assets, and covariance between all of the assets handled by the model. In his model, investors are assumed to want as high an expected return as possible, but at as low a risk as possible. This seems quite reasonable. There may be many other factors that investors would like to consider, but this model focuses on risk and return. The following problem needs to be solved in order to derive the set of attainable portfolios (derived from the expected return and the covariance matrix estimated by the investor) that an investor can reach:

$$\begin{cases} \min\limits_{w} w^T \Sigma w \\ w^T \bar{r} = \bar{r}_p \end{cases} \tag{1.1}$$

$$\begin{cases} \max\limits_{w} w^T \bar{r} \\ w^T \Sigma w = \sigma_p^2 \end{cases} \tag{1.2}$$

Recently, various portfolio models were put forward, such as the fuzzy multi-objective programming model. Assume that investors choose risk-free securities and n kinds of risky securities into the portfolio. According to Markowitz's portfolio investment theory, r_0, r_i ($i=1,2,...,n$) represent respectively the expected rate of return and risk-free securities held by the ith risky securities during the period, by using w_0, w_i ($i=1,2,...,n$) to represent the shares of risk-free securities and the ith securities, and σ_{ij} represents the covariance of i and j kinds of securities held during the period. In case no put options are allowed, the expected return rate R and risk σ^2 are

$$R = r_0 w_0 + \sum_{i=1}^{n} r_i w_i \left(\varepsilon_0 + \sum_{i=1}^{n} \varepsilon_1 = 1, \varepsilon_0 \geq 0, \varepsilon_1 \geq 0 \right)$$

$$\sigma^2 = \sum_{i=1}^{n}\sum_{j=1}^{n} w_i w_j \sigma_{ij}$$

Taking into account the benefits and risks of ambiguity, the symbols r_i and σ_{ij} should be seen as fuzzy numbers. The fuzzy multi-objective programming model for portfolio investment can then be created as follows:

$$\max \overline{R} = r_0 w_0 + \sum_{i=1}^{n} r_i w_i$$

$$\min \sigma^2 = \sum_{i=1}^{n}\sum_{j=1}^{n} w_i w_j \sigma_{ij}$$

where

$$w_0 \geq 0, \quad w_1 \geq 0, \quad w_0 + \sum_{i=1}^{n} w_i = 1, i = 1,2,...,n$$

The practical application of a basic method for solving multi-objective programming is the evaluation function. The basic idea can be briefly described as follows: first, find each single-objective optimization problem of the multi-objective optimization problem of the optimal solution. Second, through the single-objective optimization problems, find the optimal solution by means of either geometric or intuitive application background or structure, known as the evaluation function. Third, construct the optimal value and the satisfactory solution of the original multi-objective programming by solving the evaluation function. The ideal point method is one method of evaluation function.

In short, the most important challenges in using the Markowitz model are as follows:

1. The Markowitz model optimizers maximize errors. Since there are no correct and exact estimates of either expected returns or variances and covariance, these estimates are subject to estimation errors. Markowitz optimizers overweight securities with high expected return and negative correlation, and underweight those with low expected returns and positive correlation. However, the argument appears somewhat contradictory. The reason for investors to estimate a high expected return on assets should be that they believe this asset is prone to high returns. It then seems reasonable that the manager would appreciate that the model overweighs this asset in the portfolio (taking covariance into consideration).

2. Mean–variance models are often unstable, meaning that small changes in inputs might dramatically change the portfolio. The models are especially unstable in relation to the expected return input. One small change in expected return on one asset might generate a radically different portfolio. That drastic change mainly depends on an ill-conditioned covariance matrix. One can exemplify ill-conditioned covariance matrices by those estimated with *insufficient historical data*.

3. The Markowitz model does not account for the market capitalization weights of assets. This means that if assets with a low level of capitalization have high expected returns and are negatively correlated with other assets in the portfolio, the model can suggest a high portfolio weight. This is actually a problem, especially when adding a shorting constraint. The model then often suggests very high weights in assets with low levels of capitalization. The Markowitz mean–variance model does not differentiate between different levels of uncertainty associated with the estimated inputs to the model.

Example 1.3: Black–Scholes Formula

Black–Scholes (B–S) formula is an outstanding example with suitable mathematical tool and perfect mathematical description. Exact explicit solution of the lognormal stochastic volatility option model has remained an open problem for two decades. In derivative pricing, the starting point is usually the specification of a model for the price process of the underlying asset. Such models tend to be of an ad hoc nature. For example, in the Black–Scholes–Merton theory, the underlying asset has a geometric Brownian motion as its price process. More generally, the economy is often modeled by a probability space equipped with the filtration generated by a multidimensional Brownian motion, and it is assumed that asset prices are adapted to this filtration. This is the so-called standard model within which a great deal of financial engineering has been carried out.

The derivation of the closed formula assumes a B–S model as follows:

$$dS_t = S_t(r\,dt + \sigma\,dW_t) \tag{1.3}$$

with constant interest rate r and constant volatility. However, the latter can be easily modified to allow for time-dependent volatility. Furthermore, the arguments consider and are valid only for European-style options. The price of European-style call option would thus be a function of the stock price and time. Let us now state the celebrated B–S formula for the sake of completeness of our presentation:

$$C(S_t, t) = S_t N(d_1) + Ke^{-r(T-r)} N(d_2) \tag{1.4}$$

$$d_{1,2} = \frac{\log(S_t/K) + (r \pm (\sigma^2/2))(T-t)}{\sigma\sqrt{T-t}}$$

As usual, K and T are the strike price and the maturity date, respectively. Veiga and Wystup (2009) presented a closed pricing formula for European-style options under the B–S model as well as formulas for its partial derivatives. The formulas are developed by making use of Taylor series expansions and a proposition that relates expectations of partial derivatives with partial derivatives themselves. The closed formulas are attained by assuming that dividends are paid in any state of the world. The results are readily expandable to time-dependent volatility models. As it is going to be used extensively throughout this section, we take here the opportunity to also present the general formula for the ith derivative with respect to the first variable S_t of Formula (1.4) as developed by Carr.

Theorem 1.1 (Carr). Let $C(S_t,t)$ and all its derivatives be continuous functions in its first variable. Then,

$$E_t^Q\left[\partial_1^i C(S_{t_k},t_k)\right]=\exp\left[-\left(r+\frac{i\sigma^2}{2}\right)(t_k-t)\right]\partial_1^i\left(S_t e^{-i\sigma^2(t_k-t)},t\right) \qquad (1.5)$$

For completeness, readers can check the numerical results in Wu and Wang (2007), covering calls and puts, together with results on their partial derivatives. The closed formulas presented there allow a fast calculation of prices or implied volatilities when compared with other valuation procedures that rely on numerical methods.

1.2 Mathematical Models in Finance

There are many mathematical models in finance. In this section, we introduce a few newest mathematical models about capital account liberation.

Model 1.1 The long-term optimal capacity level for investment project

Consider an investment project that produces a single commodity. The project's operation yields payoff at a rate that depends on the project's installed capacity level and on an underlying economic indicator such as the output commodity's price or demand, which we model by an ergodic, one-dimensional Ito diffusion. The project's capacity level can be increased dynamically over time. The objective is to determine a capacity expansion strategy that maximizes the ergodic or long-term average payoff resulting from the project's management. We prove that it is optimal to increase the project's capacity level to a certain value and then take no further actions. The optimal capacity level depends on both the long-term average and the volatility of the underlying diffusion.

Let us consider an investment project that operates within a random economic environment, the state of which is modeled by the following one-dimensional Ito diffusion:

$$dX_t = b(X_t)d_t + \sigma(X_t)dW_t, \qquad X_0 = x > 0 \qquad (1.6)$$

where
b, $\sigma:(0,\infty)\to R$ are given functions with R being the set of all real numbers
W is a one-dimensional Brownian motion

In practice, we can think of such an investment project as a unit that can produce a single commodity. In this context, the state process X can be used to model an economic indicator such as the commodity's demand or the commodity's price.

With reference to the general theory of one-dimensional diffusions, we impose the following standard assumption that is sufficient for (1.6) to define a diffusion that is unique in the sense of probability laws.

Assumption 1.1 The deterministic functions b, σ:$(0,\infty) \to R$ satisfy the following conditions: $\sigma^2(x) > 0$, for all $x \in (0,\infty)$, and there exists $\varepsilon > 0$ such that

$$\int_{x-\varepsilon}^{x+\varepsilon} \frac{1+|b(s)|}{\sigma^2(s)} \, ds < \infty \quad \text{for all } x \in (0,\infty)$$

This assumption also ensures that the scale function p and the speed measure $\tilde{m}$ are given by

$$p(1) = 0, \quad p'(x) = \exp\left(-2\int_1^x \frac{b(s)}{\sigma^2(s)}\right), \text{ for } x \in (0,+\infty)$$

In this section, we consider only the special case that the solution to (1.6) is nonexplosive and recurrent. It means that it arises when the state process X is modeled by the stochastic differential equation:

$$dX_t = \kappa(\theta - X_t) \, dt + \sigma\sqrt{X_t} \, dW_t$$

where k, θ, $\sigma > 0$ are constants satisfying $2k\theta > \sigma^2$. This diffusion is identical to the short rate process in the Cox–Ingersoll–Ross interest rate model and is widely adopted as a model for commodity prices. Verifying that this diffusion satisfies earlier assumptions is a standard exercise. Also, it is straightforward to verify that the normalized speed measure of X is given by

$$m(dx) = \Gamma^{-1}\left(\frac{2k\theta}{\sigma^2}\right) x^{(2k\theta/\sigma^2)-1} \exp\left(\frac{2k}{\sigma^2}\left[\theta\ln\left(\frac{2k}{\sigma^2}\right) - x\right]\right) dx$$

where
  Γ is the gamma function
  $\Gamma^{-1}(\cdot) = 1/\Gamma(\cdot)$

We also assume that the running payoff function h is given by

$$h(x, y) = x^\alpha y^\beta - cy, \ c \in (0,\infty)$$

where α, $\beta \in (0, 1)$ and $c \in (0, \infty)$ are constants. The term $x^\alpha y^\beta$ here identifies with the so-called Cobb–Douglas production function, while the term cy provides a measure for the cost of capital utilization.

With reference to the optimization theory and methods, the project's optimal capacity level $\bar{y}$ is the maximum of the project's initial capacity y, and the solution to the algebraic equation

$$\left(\int_0^\infty x^\alpha m(dx) \right) y^\beta - cy = 0$$

In light of the calculation

$$\int_0^\infty x^\alpha m(dx) = \Gamma\left(\frac{2k\theta}{\sigma^2} + \alpha \right) \Gamma^{-1}\left(\frac{2k\theta}{\sigma^2} \right)\left(\frac{\sigma^2}{2k} \right)^\alpha$$

it follows that the optimal capacity level of the investment project is given by $(z = 2\kappa/\sigma^2)$

$$\bar{y} = \max\left\{ y, \left[\frac{\beta\Gamma(z\theta + \alpha)}{c\Gamma(z\theta)z^\alpha} \right]^{1/(1-\beta)} \right\}$$

Model 1.2 Capital control and monetary independence

One of the reasons for governments to employ capital controls is to obtain some degree of monetary independence. Recent capital control proxies are used in order to determine the date of capital account liberalization for a panel of Western European and emerging countries. Results show that capital controls have a very limited effect on observed deviations from interest parities, even when accounting for the political risk associated with capital controls. In this part of the chapter, we take an alternative route to assess the effect of capital controls on *monetary freedom*. Straetmans et al. (2013) investigated that to what extent capital controls contribute to deviations from the (covered and uncovered) interest parity conditions for foreign exchange. Given the potential of capital controls to limit arbitrage and speculation, exchange rate parity conditions constitute a natural testing framework for the *monetary freedom* hypothesis: the well-known covered (CIP) and uncovered interest parity (UIP) relations relate cross-border interest differentials to current and future (expected) price formation in foreign exchange markets in the following way:

$$(f - s)_t = (i - i^*)_t \quad \text{and} \quad E_t S_{t+1} - S_t = (i - i^*)_t \tag{1.7}$$

with E_t being the rational expectations operator, S_t and f_t the natural logarithms of the nominal bilateral spot and 1-month forward exchange rate expressed in domestic currency per unit of foreign currency, and *i* and *i** domestic and foreign interest rates on monthly deposits, respectively.

Straetmans et al. (2013) considered regression equations of the framework outlined in Equation 1.7, augmented with proxies for capital controls and political risk premiums. Augmenting the parity regressions with those variables enables the authors to determine the contributions of these variables to observed deviations from CIP and UIP. These authors find some evidence that capital controls can effectively distort the covered interest arbitrage condition. However, capital controls seem unable to explain even a fraction of the observed forward discount bias in the UIP relation. This also implies that capital controls are ineffective in creating more monetary freedom for domestic monetary authorities. For some countries, the installment of capital controls even leads to an erosion of monetary freedom. Surprisingly, these results are not fundamentally altered when controlling for political risk premia is employed.

Model 1.3 Capital controls, capital flows, and the real exchange rate

To estimate the impact of capital controls on macroeconomic variables, we introduce a vector error correction (VEC) model that includes an index of industrial production constructed by the national statistics agency—DANE, a multilateral real exchange rate constructed by the Central Bank of Colombia—Banco de la República, total capital inflows as reported by the Colombian Central Bank, an important index of terms of trade constructed by the Central Bank, and the global EMBI spread constructed by JP Morgan to control for global financial conditions. Concha et al. (2011) estimated the following model:

$$\Delta y_t = c + \alpha\beta'y_{t-1} + \sum_{i=1}^{p-1}\Gamma_i\Delta y_{t-i} + \varphi D_t + \varepsilon_t \tag{1.8}$$

$$y_t = (EMBI_t, TOT_t, Tx_t, CF_t, RER_t, IP_t) \tag{1.9}$$

where
  EMBI is the spread of emerging markets sovereign bond returns with respect to
    US treasuries (in logs)
  TOT is the terms of trade index (in logs)
  Tx is the tax-equivalent measure of capital controls
  CF is the total net capital flows
  RER is the multilateral real exchange rate index (in logs)
  IP is the industrial production index (in logs)
  D a dummy variable capturing changes in the *power* of capital controls as
    described earlier

These models are estimated using weekly data covering the same time period. Results suggest that the capital controls used since 1998 have been ineffective in reducing capital flows and the trend of the Colombian peso to appreciate. In addition, there is no evidence suggesting a change in the composition of capital flows induced by capital controls. This work found some evidence in favor of capital controls in reducing nominal exchange rate volatility at high frequencies.

Model 1.4 An optimization model about capital flows

Physical and human capital alone cannot explain why capital flows with low intensity and flows to middle-income countries. Silva (2010) presented an optimization model to show how managerial ability, the ability to run risky projects, can increase the total factor productivity and explain the pattern of capital flows as follows:

$$\max_{z, k_S, k_T} = F(z,x)y(k_S) + \Phi(z,x)y_H(k_T) + \Theta(z,x)y_L(k_T) \qquad (1.10)$$

$$s.t. \quad F(z,x)k_S + [1 - F(z,x)]k_T = k \qquad (1.11)$$

where

$$\Phi(z,x) = \int_z^1 \pi(a)dF(a,x), \quad \Theta(z,x) = \int_z^1 [1 - \pi(a)]dF(a,x) \qquad (1.12)$$

Given z, $\Phi(z, x)$ is the measure of successful risky projects, and $\Theta(z, x)$ is the measure of unsuccessful risky projects.

It is obvious that the optimal quantity of capital in the safe project is independent of the level of ability. The model implies that countries with more high-ability managers use more risky projects and have higher productivity. In this model, Silva (2010) defined proxies for managerial ability with data on physical and human capital, schooling, and entrepreneurship. Consistent with the pattern of capital flows, the model predicts similar returns to capital across countries and higher returns in middle-income countries.

Model 1.5 Verdier equation

What drives capital inflows in the long run? This model illustrates how capital movements conform surprisingly well to the predictions of a neoclassical model with credit constraints (Verdier, 2008). The most intriguing prediction of this class of models is that, contrary to a pure neoclassical model, domestic savings should act as a complement rather than a substitute to capital inflows. Nevertheless, this class of models maintains the neoclassical prediction that capital should flow to countries where it is most scarce. Using foreign debt data from 1970 to 1998, qualitative and quantitative evidence are found that support these predictions. Then, the equilibrium with $dt = kt$ is defined by the following two equations:

$$(1+n)(1+g)z_{t+1} = s_y \left[ad_t^\rho + (1-a)z_t^\rho - (r+\delta)d_t + (1-\delta)z_t \right] \qquad (1.13)$$

$$a\eta \left(\frac{d_t}{v_t} \right)^\rho \frac{y_t}{d_t} = r + \delta \qquad (1.14)$$

We shall analyze a few special cases of this model in Chapter 5 later.

Model 1.6 The limit cycle theory in the applications of population

Kukla and Płatkowski (2013) put forward population games in which players are motivated in their choice of strategies by payoffs, popularity, and transcendent factors. The dynamics of the players is governed by balance conservation equations along with attractiveness-driven strategy choices. It has the following form:

$$\frac{dx}{dt} = c_1(1-x)x^{1-\alpha}v_1^{1-\beta} - x\left[c_2 y^{1-\alpha}v_2^{1-\beta} + c_3(1-x-y)^{1-\alpha}v_3^{1-\beta} \right] \qquad (1.15)$$

$$\frac{dy}{dt} = c_2(1-y)y^{1-\alpha}v_2^{1-\beta} - y\left[c_1 x^{1-\alpha}v_1^{1-\beta} + c_3(1-x-y)^{1-\alpha}v_3^{1-\beta} \right] \qquad (1.16)$$

where $z = 1 - x - y$, $c_i = f_i^{1-\beta} \in R^+$ are the transcendent factors, and

$$v_1 = Rmx + Smy + Rmz$$

$$v_2 = Tmx + Pmy + [T + P(m-1)]z$$

$$v_3 = Rmx + [S + P(m-1)]y + Rmz$$

Kukla and Płatkowski demonstrated that a strategy that increases the attractiveness does not necessarily lead to the increase in its frequency in the population in the long run. In particular, they also discover the existence of limit cycles in the iterated prisoner's dilemma and the rock–paper–scissors games in such populations.

1.3 Mathematical Principles in Finance

It seems that there are no common principles in the practice of modeling an economy due to disagreements in economic theories and the complexity of specific models. However, mathematical methods still supply some basic principles for us. In this section, we try to explain this fact from the modeling of the Russian economy in conditions of the global crisis and Japanese economy in the crisis period.

1.3.1 Financial Balances Equation

The intricate system of material balances requires aggregation of information about countless different kinds of material goods. This is primarily necessary not for an economist, but for anyone living in this system. It is not surprising that means of aggregation appeared a very long time ago, and a description of it is included in most models. The aggregation and transfer of information in the economy are accomplished via money. In any economy that is at all developed, for each systematically repeated exchange flow, there is a corresponding payment counterflow:

Flow of payments between agents for transfer of the good
= price of the good × amount of the good transferred between agents

in which it is assumed that the price does not depend on which pair of agents makes the exchange. Multiplying material balances by prices and adding them up for groups of agents and products, we get a description of flows of aggregated goods between macro agents. In particular, summation for goods and services, but without including resources, and for agents located in the territory of some country produces the country's basic macroeconomic balance:

$$GDP = consumption + accumulation + exports - imports$$

The agents' money supplies (balances) are an additive quantity. In contemporary conditions, when there is no natural creation of money, these supplies satisfy the following balance equation:

$$\Delta M = A + B + C + D$$

where
ΔM is the change in the agent's supply of money
A is the balance of payments for goods from each of the other agents*
B is the balance of money transfers from each of the agents
C is the total increase in liabilities to each of the other agents
D is the total increase in receivables from each of the other agents

---

* Including financial services, the balance of interest received and paid.

To reconcile accounting calculations, which avoid negative quantities, with the algebra of balance equations, it is necessary to interpret debts (liabilities) as negative supplies. Since all liabilities are someone's demands, by adding up the financial balances for all agents, we find that the sum of the agents' supplies of money does not increase with time (the flows of money are closed). From this fact, it follows that the supplies of money must be negative for some agents (issuers), and the rest of the agents' money is the issuers' liability. In particular, cash is the liability of central banks. And this is not just a matter of formal bookkeeping. Issuing cash is not an independent source of money. One can get a currency credit from a Russian bank and transfer it to a friend in Argentina. This is entirely legal. Noncash dollars start to wander around the world, and the U.S. Federal Reserve will find out about them only when someone wants to cash them, and then the Federal Reserve will be forced to print dollar bills. All new money in today's world comes from the process of creating money through credit, a simultaneous increase in assets (positive supplies, demands) and liabilities (negative supplies, debts). Therefore, one ought not to naively expect that when one financial market collapses, money will move out of it into another financial market, or try to find out who *found* the billions that someone lost in the crash. In a crisis, assets shrink with liabilities, and everyone loses: some lose the benefit from purchases on credit, and others lose the hope that debt will be repaid. The possibility of financial bubbles is part of the very nature of modern money. Nevertheless, when material balances are unsuitable in the presence of nonadditive goods, only monetary flows remain reliably measurable quantities. We should not ignore supposedly *inflated* financial parameters, but learn to use them properly. In terms of physical analogies, money is a very distinctive material. If we return to the basic function of currency—to aggregate information about different kinds of material goods—then we have to admit that there is no close analogy to such aggregation and transmission of information neither in physical nor in biological or technical systems. In particular, the popular comparison of a financial system to the circulatory system is inappropriate: the blood transmits information by a change in its quality (hormonal composition), while money does so only by its quantity. A budget indicates an agent's current capabilities, while credit indicates future ones.

1.3.2 A Model in Canonical Form

The next narrower principle consists in considering agents not simply as nodes of material and financial flows, but as entities that make decisions regarding the magnitude of flows within their competence. Material and financial balances, along with the natural conditions of nonnegativity of flows and supplies, serve as internal constraints on an agent's possible choices. Technological limitations on the possibility of transforming some goods into others function as other such constraints. The main task of economics as a management system is to determine transmission (exchange) flows regarding the magnitude of which counterparties must make a joint decision. The idea of economic equilibrium is that each agent offers his own *plan* for the magnitude of this flow (or supply or demand in economic language). This plan is conditional: it depends on the values of particular information variables (prices, interest rates, and exchange rates) that give the agent information about the condition of the whole system. The plans allowed by established economic relationships have *institutional* (external) constraints that contain information variables. (The correlation mentioned earlier between flows of money and goods at a given price is the simplest example of such a constraint.) Agents' plans are coordinated in the process of their interaction so that the balance correlations included in the model are fulfilled throughout the entire system. In the general case, we get the canonical form of a model (see Figure 1.1). The correlations of a model in the canonical form are broken down into units describing the behavior of EAs and units describing the agents' interaction (AI). In the general case, interaction does not mean that the counteragents' plans are carried out, but sometimes simply amounts to an exchange of information between them. All SADE models can be written in the canonical form, including a model of how a planned economy functions, and so can all CGE models, *theoretical* models, and most simulation ones (but far from all econometric, balance, or synergetic models).

The main task for creating a model in the canonical form is to select the aggregated additive quantities for which the complete system of material and financial balances is written and to represent the institutional constraints describing economic relationships in terms of these quantities. It is precisely in the form and set of these constraints that the difference between competitive and monopoly markets, joint-stock and equity property, and the possibility of using money surrogates and shadow transaction channels is expressed. Another difference between CGE and SADE models is their systems of institutional constraints: while the former takes institutional constraints mostly from textbooks, in SADE models, we try to reflect the specific nature of the economy that is being modeled.

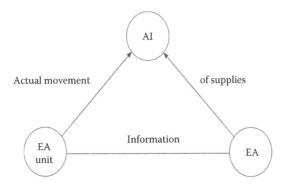

Figure 1.1   Diagram of model in canonical form.

1.3.3 Rationality of Behavior

The constraints of a particular model usually leave the agents a fairly wide margin of freedom in their choice of plans. In simulation models, agents' actions are described by direct assignment of the rules of decision making. Such an approach may be justified in describing an organizational routine within a corporation. But in a whole economic system that is capable of self-organization, such a direct decision for everyone seems overly self-confident, and it seems very hard to modify it without mistakes. Therefore, in economic theories and in CGE and SADE models, the assumption of rational behavior is postulated for most agents. That is, within the constraints, all agents choose such strategies that maximize some parameters (utility, profit, capitalization, etc.). This premise constantly raises doubts among specialists and is sharply rejected by nonspecialists. They say that economic theories are impractical because they ignore the human factor. However, people are not atoms. Why dream up motives for them and principles by which they act when you can just ask them why they do what they do and not something else? Such surveys are constantly being conducted, but they do not give the whole picture, while economists' view from the sidelines enables them to note certain patterns. The behavior of individual people really is capriciously determined by a great many factors that are not taken into account by economic theories (a person got married, got sick, was promoted, lost money at the casino, became unhinged, etc.). It seems that each of us is more complex than the whole economy. Therefore, all attempts that we know of to base an economic theory on the laws of psychology have been complete failures, and in economics, we can understand something only because just a small part of the richness of a person's internal world is manifested in it. In economics, people act under the conditions of impersonal threats (judicial or administrative punishment, being fired or ruined, moral censure, etc.). As a result, their behavior is standardized, and individual differences are leveled out. There remain at least two possible ways to define a rational agent. The existing theory resorts to *representative agents*, each of whom is characterized by assigned invariant interests. In microeconomic studies, attempts are sometimes made to identify a difference of interests, but macroeconomic models almost always consider one representative agent of each type (consumer, producer, merchant, etc.). We believe that the idea of absolutely autonomous representative individuals contradicts the fact that people and organizations interact with each other. Competition and imitation take place within large groups of agents that perform similar roles in the economy. As a result, the collective behavior of such a group turns out to be simpler and more consistent than the behavior of any of its members, and can be described as a simple endeavor to maximize consumption, profit, wealth, and the like. That can be confirmed by direct measurements. For instance, using a time series of sets of goods purchased by some group of consumers and the series of prices at which these goods were purchased, it is possible to constructively determine whether there is a utility function such that the observed purchases maximize this utility at the observed prices, with the corresponding budget constraints. The results of these checks are as follows:

1. An individual family does not have a utility function or even a single-valued demand function that depends on prices and incomes.
2. A uniform social strata (the population of a suburb of New York) as a whole does not have a utility function, although it does have a demand function.
3. The population of a whole country (even an open one such as the Netherlands or an unstable one such as Hungary during the transition from socialism to capitalism) as a whole does have a utility function, which describes the observed annual demand over a period of approximately 10 years across a range of 20–300 products.
4. The totality of buyers at a large store does have a utility function, which describes the weekly demand over several years across a range of 2000 products.

Unfortunately, there are no results for Russia as a whole due to the lack of suitable trade statistics. But the final consumption reflected in the intersectoral balances for 1995–2003 is rationalizable, if financial services are excluded. This surprising result can be explained as follows: the utility function does not reside in the heads of consumers but originates in the process of interaction of irrational consumers and merchants manipulating prices to their own advantage. Based on this reasoning and observations, it is reasonable to assign interests to macro agents. That is, we consider these interests as variational principles distinguishing the observed behavior of macro agents from all alternatives permitted by the model's constraints. This premise also has a retroactive effect. When we are dealing with an influential agent with a unified will, for example, the state, there is no need to try to describe its actions by the principle of optimality. It is better to ask what it intends to do, that is, to describe its behavior by scenarios of possible actions.

1.3.4 Rational Expectations Principle

The principles cited earlier are sufficient to construct a static model but not a dynamic one. In dynamic models, the agent plans his actions for the future and, hence, has to forecast future changes in the situation (information variables). This creates a paradox: we construct a model so as to forecast the situation, but to construct the model, we have to know how agents forecast the situation. The principle of rational expectations is a radical solution for this paradox. The simplest formulation of this principle is that model agents use for their forecasts the same model that we construct. At first, it seems surprising that anything nontrivial can be obtained from this principle. But it is actually possible because agents have different planned variables and goals. Although the principle of rational expectations raises reasonable doubts (since it implies that the model agents *know everything in advance*), we take the risk of applying it to modeling the real Russian economy. As a result, we achieved greater success than with the phenomenological simplification of the description of agents' behavior that characterizes CGE models and early SADE models. In the determinant case under consideration here, the principle of rational expectations leads to an inter-temporal economic equilibrium model. In such a model, each agent, based on his own goals, capabilities, and forecasts, determines his supply and demand for products, resources, and financial instruments at the current and all future points in time, and then the forecasts (which are the same for everyone) are determined from the condition that supply and demand must match at the current and all future points in time. Inter-temporal equilibrium models have long been known, but until now, they were used exclusively for studying certain theoretical questions on stationary regimes of rather abstract models of an economy.

An industrial and economic consumption (IEC) model becomes particularly simple and natural with uniform (scale-invariant) descriptions of production and consumption. The model is uniform if the absolute size of the system that is being modeled is unimportant for it: if the input additive quantities (flows and supplies of products, resources, and financial instruments, including their initial values) are doubled, the output additive quantities also double. Assumptions of uniformity (constant return to scale) are present, to some extent, in all popular models of an economy. It is sufficient to note that the majority of well-known economic patterns (e.g., Robert Solow's famous golden rule) were derived from the analysis of balanced growth in simple models, and balanced growth is realized only on the condition of uniformity. Uniformity of the real economy raises certain doubts: natural resources are limited, the growth rate of production does not coincide with the growth rate of the population, the need for some specific good is saturated—all of this contradicts the idea of uniformity and balanced growth. But over the 200 years of its existence, the industrial economy has managed to continue its exponential growth, overcoming (so far) all external constraints: of land at the beginning of the nineteenth century, labor at the beginning of the twentieth century, and energy resources in the 1980s. And forecasts based on the limitation of possibilities (from Malthus to Forrester inclusive) have been far off the mark. In regard to economic processes, as long as we are interested only in rates and proportions, the assumption of uniformity will be justified.

In a uniform IEC model, the description of agents' interests is standardized: both producers and consumers can be thought to strive to maximize their properly understood capitalization. An important parameter of the *true* estimate of an agent's net assets comes into play here, which, for the sake of brevity, we will simply call an agent's capital. Formally, it is an estimate of net assets in objectively determined prices (dual variables to balance constraints in conditions of optimality of the agent's behavior). For liquid instruments and supplies, the objectively determined estimates coincide with market prices, while for those that are not completely liquid, they prove to be less than the market prices for assets and greater than the market prices for liabilities. Capital satisfies a simple equation:

Growth in capital = balance sheet profit − distributed profit × (1 + effective tax)

Dividends serve as the distributed profit for a producer, and consumer spending does so for a consumer. The effective tax is computed in the model and includes actual taxes and rates of indirect loss associated, say, with the need to keep supplies of money to cover the flow of payments. Balance sheet profit can be expressed in two forms:

$$\text{Balance sheet profit} = A_1 + A_2 + A_3 + A_4 + A_5 \qquad (1.17)$$

where

A_1 is the value added of production
A_2 is the value added of commercial and financial intermediation
A_3 is the balance of payments for resources
A_4 is the profit from revaluation of liquid assets
A_5 is the depreciation of not completely liquid assets

For a consumer, the value added consists of the interest rate on savings. We emphasize that this expression is not postulated but derived with the rate of depreciation being determined from the model so as to reduce the estimate in market prices to an estimate in objectively determined prices. We interpret this result as follows: accounting includes profit from revaluation as income, which is not received from anyone, and as spending depreciation that is not paid to anyone. This is done to bring the estimate of net assets obtained by balancing actual spending and income into line with the *true* estimate that gives the amount of an agent's capital:

$$\text{Balance sheet profit} = B_1 \times B_2 \qquad (1.18)$$

where
B_1 is the agent's capital
B_2 is the internal rate of return

Here the variable internal rate of return no longer depends on the absolute values of flows and supplies but is a nontrivial combination, unique for each agent, of current and expected parameters of the real rate of return of individual financial instruments and individual production processes in which the agent participates. In a uniform IEC model, an agent's capital coincides with the amount of expected distributed profit discounted at the internal rate of return, and also with the maximum amount of the agent's capitalization. Thus, in the IEC context, all three estimates of a firm's value used in economic analysis (according to the value of net assets, expected discounted profit [net present value], and market capitalization) coincide, if the prices of the assets, the discount rate, and the exchange rate are computed from conditions of optimality of the agent's behavior. Finally, it was found that the amount of an agent's capital could be used for a new solution of the old problem of terminal conditions at the end of a planning interval. We suggest setting a single terminal condition of growth of capital by a certain factor. With this condition, on the optimum trajectory, an agent's capital will be positive, which replaces the no Ponzi game condition, which usually has to be set in a very complex form. It is remarkable, but all of these properties of an agent's capital in the model do not require any additional assumptions. They stem from conditions of optimality and independence from scale (just as in physics, the conservation of energy stems from the principle of least action and the independence of a physical system's motion from calendar time). In terms of the framework of an IEC model, it is precisely this internal rate of return (current and expected) that a firm communicates to its owner, who, using the equation for the firm's capital, determines the growth rate of distributed profit (which is, generally speaking, variable) that he wants.

In inter-temporal equilibrium models, an agent plans his own actions for the future, and therefore, it would seem, the optimal actions should depend heavily on the agent's motives and his knowledge of the future. However, strange as it may seem, for problems of optimizing economic processes, this is not at all the case. Economic processes are characterized by the mainline effect, which is unknown in engineering or physics: the influence of future goals and external impacts on a current optimal decision diminishes exponentially as the future becomes more remote from the present. In other words, an economic system has universally optimal trajectories that provide an acceptable result with different scenarios of the remote future. No wonder several Nobel prizes have been awarded for studying this remarkable property. In all applied models constructed according to the principles described earlier, with values of the constant parameters with which the model reproduces the statistics, the mainline effect is so strongly manifested that the influence of the future virtually disappears completely in one calculation step. From a mathematical point of view, this means that an agent's behavior is described by a dynamic system. From a substantive point of view, it turns out that, although in a rational expectations model we permit an agent to know the future, with proper values of the parameters, institutional constraints narrow his possibilities so that to make the optimal decision, it is sufficient to know the current situation. However, it is nevertheless necessary to state the problem and derive optimality conditions for it, because otherwise it would be impossible to guess the specific type of expression for, say, the internal rate of return.

1.3.5 Model of the Russian Economy in the Crisis Period

An applied IEC model is derived from a perfect competition model by adding to the description of the agents' constraints and possibilities reflecting the current economic relationships as well as introducing additional agents—for example, the banking system. A description of the additional agents is facilitated by the fact that in an IEC model, the agents' interests are standardized. In addition, it is necessary to take into account the actions of individual agents (the Ministry of Finance, the Central Bank, the outside world), whose behavior is described by a scenario. If we combine the optimality conditions for all macro agents (the EA unit in Figure 1.1) with the scenario description of the actions of individual agents, and supplement these with descriptions of their interactions (the AI unit in Figure 1.1), which contain, first of all, the balances of financial and material flows between agents as well as certain flow of information, then one can get a macro model of inter-temporal equilibrium with capital management (IEC). The specifics of the economy being studied are expressed, first, in a set of aggregate additive quantities (material and financial assets), in the language of which the agents' behavior and interaction is described, and, second, in a set and the form of institutional constraints. In such a model, of course, equilibrium will no longer be efficient, but the remaining properties are preserved, for the most part. For individual agents, however, uniformity may be disrupted—for example, because of an agent's monopoly position in the market, in the presence of an economy of production scale, or because of constraints such as quotas. Additional terms representing indirect incomes or lost profit appear on the right side of the equation for capital. The integrals of these terms over future time, discounted according to the rate of return, can be interpreted as an objectively determined estimate of an agent's intangible assets.

A realistic macro model of this type amounts to a system that initially contains more than 100 nonlinear finite and differential equations, and since the model requires that optimal management problems be solved for the agents, it cannot (in contrast to an ordinary simulation or CGE model) be calculated step by step, starting from the original condition. Very intricate algorithms have to be used for the calculations in an IEC model. At the same time, there are not many adjustment parameters in the model: 10–20, as opposed to hundreds in econometric, simulation, or CGE macro models. Therefore, when an IEC model does a poor job of reproducing the statistics, the form of the initial correlations has to be changed, and the calculations have to be done all over again. It is absolutely impossible to do such work, writing the model down on paper and converting the notation to programming language, in a reasonable time without errors and distortions of the underlying assumptions. The only way to work efficiently with an IEC model is by using the equation knitted operation for modified order degree (EKOMOD) system, which supports mathematical modeling and is implemented in a Maple computer algebra environment. EKOMOD verifies that the model is correct, can automatically repeat calculations if the original equations are changed, implements the calculation algorithm directly in literal form of the model's system of correlations, and writes out the model in a form suitable for publication in all stages. The first IEC model describing the development of the Russian economy was constructed for the tax administration in 2003–2004 to estimate the amount of shadow transaction and tax potential on the basis of data from outside the tax administration. This model was modified in 2006. Only later did it turn out that it correctly described, both in time and in volume, the transition from capital outflow to inflow in Russia in 2006. It must be taken into account here that the model *knew* nothing about the organization during this period of state corporations whose foreign loans accounted for the bulk of capital inflow. This result showed that the IEC model is capable of describing a break in trends, which meant that it could be successfully used during a crisis period.

The latest version of the Russian economic model describes the real sector, which produces products for the domestic market and exports and consumes domestic products and imports, and the financial sector. The financial flows accompanying the production, distribution, and consumption of products are described as the circulation of seven financial instruments: cash, balances in bank accounts, balances in correspondent accounts in the Central Bank, bank loans, bank deposits, deposits/credits in the Central Bank, and foreign currency. Products, labor, and financial instruments form the set of additive quantities for which a complete system of balances is constructed, with the flows of financial instruments being divided into legal and shadow ones. Economic development, expressed by the movement of additive quantities, is described as the result of the activity of nine macro agents:

1. A producer, representing the totality of nonfinancial commercial companies
2. A bank, representing the totality of financial commercial companies
3. The population, representing individuals—consumers and employees
4. An owner, representing individuals and corporations that manage the movement of capital between sectors of the national economy and outside the country
5. A merchant, as a pure middleman between consumers, the producer, the exporter, and the importer
6. The government, whose activity is represented explicitly in the model by an aggregated description of Ministry of Finance activity and implicitly by setting various parameters of economic policy (tax rate, government spending, reserve requirements, etc.)
7. The Central Bank, which is represented in the model by its functions as the issuer of the national currency, holder of currency reserves, a settlement center, and a creditor of commercial banks
8. An exporter
9. An importer

The most important challenging change was the transition from a single-product model describing all material flows as part of real GDP to a three-product model in which the components of the real macroeconomic balance are expressed as various nonlinear combinations of three products: domestic (including services), exported, and imported. In the model, if the route of disaggregating the basic macroeconomic balance is selected, then it would turn over to the merchant agent. The merchant splits (nonlinearly) the producer's product (real GDP) into domestic and exported, and then combines (once again, nonlinearly) the consumer and investment product from domestic and imported. The use of three products enabled us to significantly improve the description of foreign trade in the model. While in previous versions the amount of exports was assigned exogenously and had to be forecasted independently of the model, in the latest version, exports are an endogenous quantity. What has to be assigned independently of the model now is the dynamics of import and export (international) prices, the dynamics of government spending, policy regarding changes in the ruble exchange rate in relation to foreign currency, and employment.

1.3.6 Model of the Japanese Economy in the Crisis Period

In 1997, the collapse of mega-banks in Japan increased regulatory pressure, market scrutiny, and the distress of the financial system, thereby causing a crisis in the domestic financial sector (Sawada, 2011). The financial crisis in Japan is often referred to as a typical example of a *credit crunch*, which is conventionally defined as a sharp decline in bank loans caused by supply factors such as risk-based capital standards imposed on banks. Casual evidence shows that the negative impact of the credit crunch in Japan was serious at the household level. The responses of firms, households, and the government appeared to have played important roles in coping with the macroeconomic crisis. Hence, unless a thorough empirical study is undertaken, the overall social impact of the credit crunch is not necessarily clear.

The nonlinear skill can help us to bridge the gap by estimating a consumption Euler equation that is augmented by endogenously imposed credit constraints using a switching regression. The methodology permits us to derive a density function of the probabilities of binding credit constraints for each year, which makes it possible to quantify the seriousness of the credit crunch at the household level.

Under the financial crisis, households faced the problem of reconciling the realized income shortfall with a desirable level of stable consumption. Households have devised several methods, such as self-insurance and mutual insurance, to protect their levels of consumption against the exposit risks of negative income shocks. In this section, we suppose that households may be constrained from borrowing for a variety of reasons such as asymmetric information between lenders and borrowers. The existence of credit constraints has important negative impacts on the risk-coping ability of households. For example, faced with a shortfall in the real income, credit-constrained households may be forced to reduce consumption expenditure since credit cannot be used as insurance. A credit crunch could magnify this negative impact of credit constraints. In order to formalize the role of credit in smoothing consumption, it may be a viable way to construct a model that provides optimal consumer behavior under uncertain income and potential credit constraints. Suppose a householder's decision maker has a concave instantaneous utility $U(.)$ of the household consumption C. The household then has to choose a value of C, with a subjective discount rate b such that the discounted lifetime utility is maximized, which is subject to inter-temporal budget constraints with the interest rate r. Generally, when the household income is stochastic, analytical solutions to this problem cannot be derived. However, in order to obtain an optimum solution, one can derive a set of first-order necessary conditions by forming a value function and a Bellman equation. Let X represent the Lagrange multiplier associated with credit constraint $C < A + y + z$, where A is the household asset at the beginning of the period, y is the stochastic household income, and the maximum supply of credit possible for this household is represented by z. On combining the envelope condition derived from the first-order conditions, we obtain a consumption Euler equation without credit constraints:

$$U'(C_{it}) = E_t \left[U'(C_{it+1}) \left(\frac{1+r_{it}}{1+\delta} \right) \right] + \lambda_{it} \tag{1.19}$$

Equation 1.19 is the augmented Euler equation. Given other variables, an increase in the current income of a credit-constrained household leads to a decline in the marginal utility of current consumption, thereby causing the Lagrange multiplier to decline. This theoretical property provides us with a basis for testing the validity of the theoretical framework.

It is easy to obtain a nonlinear recursive equation for variable C by taking a second-order Taylor approximation of Equation 1.19. After replacing the expected consumption growth with the observed consumption growth as well as the expectation error, Equation 1.19 becomes

$$\hat{C}_{it+1} = X_{it}\beta + f(\lambda_{it}) + v_{it+1}$$

where
  f is an increasing function that takes zero if X becomes 0
  v_{it} indicates a well-behaved stochastic error term

In order to control for the changes in preferences and household characteristics, variables such as household size, age of the respondent, and age squared are included in X. The term X is an implicit consumption price of binding credit constraints, which is equal to the increase in the expected lifetime utility that would result if the current constraint were relaxed by 1 unit.

Now, let C^* represent the optimal consumption in the absence of a current credit constraint or, simply, in the canonical model of the life cycle permanent income hypothesis. $C^* = C$ if the credit constraint is not binding, and $C^* > C$ if the credit constraint is binding. Next, we define the gap between optimal consumption under the perfect credit accessibility and cash in hand without credit constraints, that is, $H = C^* - (A + y + z)$. Furthermore, following Jappelli et al. (2005), we assume that the conditional expectation of optimal consumption C^* can be approximated by a quadratic function. Hence, the reduced form of the optimal consumption C^* can be expressed as a linear function of observables, such as current income, wealth, and age, as well as the quadratic terms of these variables. The maximum amount of borrowing is also assumed to be a linear function of the same variables.

After defining a matrix as a set of these variables, one can derive the econometric model of the augmented Euler equation (1.19) with a set of endogenous credit constraints. Such augmented Euler equation can be estimated by empirical approach introduced in Chapter 6.

1.4 Mathematical Estimations in Finance

Properties, such as infinite divisibility, skew, excess kurtosis and suitable model, and the heavy-tailed distribution characteristic of real finance asset yield, make the normal-Laplace (NL) distribution a good candidate model for option pricing. However, a few constraints of the NL parameter and the density function's lack of an enclosed expression may cause failures to the traditional moment estimation and maximum likelihood estimation. Therefore, we put forward an elementary enclosed expression for the density function of the NL distribution under some special conditions in order to remedy the aforementioned problem.

1.4.1 Introduction

The NL distribution is first introduced by Reed and Jorgensen (2004). Meanwhile, they put forward the probability density function (pdf) of the NL distribution, the cumulative distribution function (cdf), and moment-generating function (MGF). They also briefly introduced the advantages of NL distribution in modeling in a number of areas, including economics (income and profit distribution), finance (stock prices), geography (human settlements population distribution), physics (particle size model distribution), and geology (oil) field analysis. Although they focused on the introduction and presentation of NL distribution, they did not sufficiently discuss its applications in detail in these areas. Reed used the NL distribution in option pricing analysis and introduced the generalized normal-Laplace (GNL) distribution. Properties such as infinite divisibility, skew, excess kurtosis, and the ability to better model the heavy-tailed distribution characteristic of real finance asset yield make the GNL distribution a good candidate model in option pricing. Simos and Tsionas (2010) considered that the generalized NL distribution is a useful law for modeling asymmetric data exhibiting excess kurtosis. They constructed goodness-of-fit tests for this distribution, which utilize the corresponding MGF and its empirical counterpart. Tong (2013) further extended the application of NL distribution in option pricing. He made the distribution a condition of heavy-tailed distribution and analyzed a bivariate GARCH option pricing model.

Chen and He (2006) proved that the probability distribution of the enterprise scale is normal, and the probability distribution of growth rate is Laplace, which is a natural phenomenon and not affected by its own attributes and external environmental factors. Due to the fact that the asymmetric Laplace distribution (ALD) has an explicit expression, it facilitates the calculation of the digital features and parameter estimation. Hence, for the investors of stock index futures, using ALD to calculate the VAR and CVAR of the stock index futures yield would be a better choice. Zeng and Liu (2012) used ALD to fit Shanghai and Shenzhen stock daily and weekly yield data. Their results showed that an ALD can reflect the spikes, thick tail, and skew features of Shanghai and Shenzhen stock daily and weekly yield data better than the normal distribution.

In terms of the model parameter estimation, since Reed and Jorgensen (2004) put constraints to GNL parameter, and the density function's lack of an enclosed expression may lead to failures of the traditional moment estimation and maximum likelihood estimation, the quadratic distance estimator method has been used in the GNL model parameter estimation.

The analysis of Tobit model with non-normal error distribution is extended to the case of ALD. Since the ALD pdf is known to be continuous but not differentiable, the usual mode-finding algorithms such as maximum likelihood can be difficult and result in the inconsistent parameter estimates. Wichitaksorn and Tsurumi (2013) used a survey dataset on the wage earnings of Thai male workers and compared the Tobit model with normal and ALD errors through the model's marginal likelihood; the results reveal that the model with the ALD error is preferred. Barbiero (2014) proposed an alternative discrete skew Laplace distribution by using the general approach of discretizing a continuous distribution while retaining its survival function. He explored the properties of the distribution and compared the distribution to a Laplace distribution on integers recently proposed in the literature.

Morris and McNicholas (2013) introduced a dimension reduction method for model-based clustering via a finite mixture of shifted ALDs. It is based on the existing work within the Gaussian paradigm and relies on the identification of a reduced subspace. Their clustering approach is illustrated on both simulated and real data, where it performs favorably compared to its Gaussian analogue.

Reed and Jorgensen (2004) first introduced the NL distribution, then Jose et al. (2008) discussed a first-order autoregressive process with NL stationary marginal distribution and relevant properties. The process gives a combination of Gaussian as well as non-Gaussian time series models for the first time and is free from the zero-defect problem. They also discussed the applications in modeling data from various contexts. Recently, Harandi and Alamatsaz (2013) put forward a new class of skew distributions containing both unimodal and bimodal distributions with flexible hazard rate behavior. They investigated some distributional properties of this class and presented a characterization, a generating method, and parameter estimation. Finally, they also examined the model by using real datasets. However, a few constraints on NL parameter and density function that lacks an enclosed expression may lead to the traditional failure of moment estimation and maximum likelihood estimation. In this section, we put forward an elementary enclosed expression of the density function of NL distribution for some special cases and try to remedy the aforementioned lack.

1.4.2 Estimation

Reed and Jorgensen (2004) first introduced a new distribution, the NL distribution. Then, Simos and Tsionas (2010) tried to test for the GNL distribution with applications. Wichitaksorn and Tsurumi (2013) showed its usefulness for the estimation of Tobit model with non-normal error. If Φ and ϕ are respectively the cdf and the pdf of a standard normal distribution, and the Mills ratio is

$$R(x) = \frac{1 - \Phi(x)}{\phi(x)} \tag{1.20}$$

then the pdf of the NL distribution can be expressed as

$$F(x) = \Phi\left(\frac{x-\nu}{\tau}\right) - \phi\left(\frac{x-\nu}{\tau}\right) \frac{\beta R\left(\alpha\tau - \frac{x-\nu}{\tau}\right) - \alpha R\left(\beta\tau + \frac{x-\nu}{\tau}\right)}{\alpha + \beta} \tag{1.21}$$

We shall refer to this as the NL distribution and use $x \sim NL(\alpha, \beta, \nu, \tau^2)$ to indicate that X follows this distribution. Let $\nu \in R$ stand for the location parameter, $\tau > 0$ stand for the scale parameter, and $\alpha > 0$ and $\beta > 0$ determine the degree of obesity of the right and left ends of the tail. And, the greater the values of α and β are, the thinner their tails will be. In particular, when $\alpha \to \infty$, $\beta \to \infty$, the NL distribution would degenerate into a normal distribution. The pdf of the NL distribution is given by

$$f(x) = \frac{\alpha\beta}{\alpha + \beta} \phi\left(\frac{x-\nu}{\tau}\right) \left(R\left(\alpha\tau - \frac{x-\nu}{\tau}\right) + R\left(\beta\tau + \frac{x-\nu}{\tau}\right)\right) \tag{1.22}$$

It is well known that the MGF of X is the product of the MGFs s of its normal and Laplace components. Precisely, we can express $X = W + Z$, where Z follows a $N(\nu, \tau^2)$ distribution and W a skewed Laplace distribution with pdf

$$f_w(x) = \begin{cases} \dfrac{\alpha\beta}{\alpha + \beta} e^{\beta x}, & x \geq 0, \\[3mm] \dfrac{\alpha\beta}{\alpha + \beta} e^{-\alpha x}, & x > 0, \end{cases}$$

Notice that

$$\Phi(x) = \frac{1}{\sqrt{2\pi}} \cdot \int_{-\infty}^{x} e^{-t^2/2} dt$$

We can rewrite the Mills ratio as follows:

$$R(x) = e^{x^2/2} \left[\frac{1}{2} - \int_{-\infty}^{x} e^{-t^2/2} \, dt \right] \tag{1.23}$$

Denote the regions

$$S_x = \{(s,t) \,|\, 0 \le s \le x, 0 \le t \le x\} \quad \text{and} \quad D_x = \{(s,t) \,|\, s^2 + t^2 \le x^2, s \ge 0, t \ge 0\}$$

It is ready to see that for arbitrary $x > 0$, we have the relationship: $D_x \subset S_x \subset D_{\sqrt{2}x}$.

In light of the properties of two-dimensional integrals, the following inequality holds true:

$$\iint_{D_x} e^{-(s^2 + t^2)/2} \, ds dt \le \iint_{S_x} e^{-(s^2 + t^2)/2} \, ds dt \le \iint_{D_{\sqrt{2}x}} e^{-(s^2 + t^2)/2} \, ds dt \tag{1.24}$$

After some simple calculations, we obtain

$$\iint_{S_x} e^{-(s^2 + t^2)/2} \, ds dt = \int_{0}^{x} e^{-s^2/2} \, ds \int_{0}^{x} e^{-t^2/2} \, dt = \left(\int_{0}^{x} e^{-t^2/2} \, dt \right)^2$$

$$\iint_{D_x} e^{-(s^2 + t^2)/2} \, ds dt = \int_{0}^{\frac{\pi}{2}} d\theta \int_{0}^{x} e^{-r^2/2} r \, dr = \frac{\pi}{2} (1 - e^{-x^2/2})$$

$$\iint_{D_{\sqrt{2}x}} e^{-(s^2 + t^2)/2} \, ds dt = \int_{0}^{\frac{\pi}{2}} d\theta \int_{0}^{\sqrt{2}x} e^{-r^2/2} r \, dr = \frac{\pi}{2} (1 - e^{-x^2})$$

Therefore, inequality (1.24) is equivalent to the following:

$$\sqrt{\frac{\pi}{2} (1 - e^{-x^2/2})} \le \int_{-\infty}^{x} e^{-t^2/2} \, dt \le \sqrt{\frac{\pi}{2} (1 - e^{-x^2})}$$

According to the intermediate value theorem, there exists a constant k that is only related to x such that

$$\int_{-\infty}^{x} e^{-t^2/2} \, dt = \sqrt{\frac{\pi}{2} (1 - e^{-kx^2})}$$

Therefore, we derive the following elementary expression for the Mills ratio:

$$R(x) = e^{x^2/2} \left[\frac{1}{2} - \sqrt{\frac{\pi}{2}(1 - e^{-kx^2})} \right] \tag{1.25}$$

By substituting the function $R(x)$ in (1.25) into (1.22), we obtain an element representation for the density function of the NL distribution as follows:

$$f(x) = \frac{\alpha\beta}{\alpha + \beta} \cdot \frac{e^{-\frac{1}{2}\left(\frac{x-\nu}{\tau}\right)^2}}{\sqrt{2\pi}} \cdot \left\{ e^{\frac{1}{2}\left(\alpha\tau - \frac{x-\nu}{\tau}\right)^2} \left[\frac{1}{2} - \sqrt{\frac{\pi}{2}\left(1 - e^{-k_1\left(\alpha\tau - \frac{x-\nu}{\tau}\right)^2}\right)} \right] \right.$$

$$\left. + e^{\frac{1}{2}\left(\beta\tau + \frac{x-\nu}{\tau}\right)^2} \left[\frac{1}{2} - \sqrt{\frac{\pi}{2}\left(1 - e^{-k_2\left(\beta\tau + \frac{x-\nu}{\tau}\right)^2}\right)} \right] \right\} \tag{1.26}$$

where k_1 is only related to $\alpha\tau - (x-\nu)/\tau$, k_2 is only related to $\beta\tau + (x-\nu)/\tau$, and $(1/2) < k_1 < 1$, $(1/2) < k_2 < 1$.

1.4.3 Conclusions

It is ready to see that the right-hand side of representation (1.21) is a simple compound of elementary functions. In some special regions in the space of the parameters, we have a few simple elementary expressions for the density function of the NL distribution.

Theorem 1.2 If $\beta\tau + (x-\nu)/\tau$ is sufficiently small, then we have the following approximate representation:

$$f(x) \approx K_1 \cdot \exp\left(\frac{1}{2}\alpha^2\tau^2 - \alpha(x - \nu) \right)$$

where

$$K_1 = \frac{\alpha\beta}{\alpha + \beta} \cdot \left[\frac{1}{\sqrt{2\pi}} - \sqrt{1 - \exp[-k_1(\alpha + \beta)^2\tau^2]} \right]$$

Theorem 1.3 If $\alpha\tau - (x-v)/\tau$ is sufficiently small, then we have the following approximate expression:

$$f(x) \approx K_2 \cdot \exp\left(\frac{1}{2}\beta^2\tau^2 + \beta(x-v)\right)$$

where

$$K_2 = \frac{\alpha\beta}{\alpha+\beta} \cdot \left[\frac{1}{\sqrt{2\pi}} - \sqrt{1 - \exp[-k_2(\alpha+\beta)^2\tau^2]}\right]$$

Theorem 1.4 If τ is sufficiently small, then we have the following approximate representation:

$$f(x) \approx K_3 \cdot \left[\exp(-\alpha(x-v)) + \exp(\beta(x-v))\right]$$

where

$$K_3 = \frac{1}{2\sqrt{2\pi}} \cdot \frac{\alpha\beta}{\alpha+\beta}$$

Theorem 1.5 If τ is sufficiently small and $\alpha = \beta$, then we have the following approximate expression:

$$f(x) \approx K_4 \cdot \cosh(-\alpha(x-v))$$

where

$$K_4 = \frac{\alpha}{2\sqrt{2\pi}}$$

1.5 Mathematical Approximations in Finance

Based on the monetary and fiscal policy rules that maximize the welfare, in a real business cycle model augmented with sticky prices, we develop two different second-order accurate estimations to the Schmitt-Grohé (S-G) model. Our estimations illustrate that the welfare costs are monotonically increasing or monotonically decreasing in different regions of the parameter γ.

1.5.1 Introduction

Since the 1980s, going along with the increasing economic and financial globalization, some countries have picked up their steps toward capital account liberation. From Latin America to Asia and to Europe, many developing countries one after another have relaxed their restrictions for capital flows. Capital account liberation has been one focus in the process of the economic and financial globalization. On one hand, capital account liberation can attract a great amount of international capital for developing countries. On the other hand, international capital flows have also brought serious turbulence for those countries, even leading to the breakout of finance crises. These crises can proliferate continuously in the form of contagion. Their impacts go way beyond the fields of economy and finance. They severely threaten the national security, and as the degree of liberation of the capital account strengthens, the seriousness of the problem of financial security is also increasing with time. American subprime crisis had affected the whole world. It had especially brought economic disasters to some countries. All of these are more or less related to the over-liberation of the relevant capital accounts. Although capital account liberation can improve the allocation efficiency of capital through promoting the relevant factors of products, it is still not known whether or not large amounts of capital are truly integrated into the development of the national economy. Does international capital flow bring hidden trouble to the finance stability of the host country? What the optimal degree of capital account liberation for China should be? All these and other related problems need to be investigated urgently.

The methodology of optimization has been one of the perfect mathematical approaches and has found its generally successful applications in the field of nature science, such as physics (Ding et al., 2013; Gulich and Zunino, 2014; Zou and Li, 2014) and applied economics (Pastor, 1990; Tornell and Velasco, 1992; Guitian, 1998; Eichengreen and Leblang, 2002; Johnson and Mitton, 2003; Doraisami, 2004; Alfaro and Hammel, 2007; Blair et al., 2001; Klein and Olivei, 2008; Liu, 2012).

Recently, Kitano (2011) computed the unconditional lifetime utility levels for different degrees of capital controls. He also compared them with those under perfect capital mobility by adopting the method developed by Schmitt-Grohé and Uribe (2006). And then, he defined the welfare associated with the time-invariant equilibrium as implied by the Ramsey policy conditional on a particular state of the economy in period 0 as

$$V_0^r \equiv E_0 \sum_{t=0}^{\infty} \beta^t U\left(c_t^r, h_t^r\right) \tag{1.27}$$

where c_t^r and h_t^r denote the contingent plans for consumption and hours under the Ramsey policy. U is a period utility index assumed to be strictly increasing in its first argument, strictly decreasing in its second argument, and strictly concave.

Kitano defined λ as the welfare benefit of adopting policy regime a rather than policy regime b. Formally, λ is implicitly defined by

$$EV_0^a \equiv E_0 \sum_{t=0}^{\infty} \beta^t U((1-\lambda)c_t^a, h_t^a) \qquad (1.28)$$

However, in reality, the choice of exchange rate regime would matter for the optimal degree of capital control. And λ is one of important parameters that affect the optimal degree of capital control. It means that a fraction of regime b's consumption process compensates a household to be as well-off under regime b as under regime a. So, Kitano expressed λ as follows:

$$\lambda = \left\{ \frac{(1-\gamma)EV_0^a + (1-\beta)^{-1}}{(1-\gamma)EV_0^b + (1-\beta)^{-1}} \right\}^{1/\omega(1-\gamma)} - 1 \qquad (1.29)$$

In this section, we develop a few accurate estimations for the S-G model.

1.5.2 Main Results

For simplicity, let us denote

$$a = EV_0^a, \qquad b = EV_0^b, \qquad \text{and} \qquad c = (1-\beta)^{-1}$$

The following estimations can be obtained by calculating the first-order and second-order derivatives.

Theorem 1.6 If $|1-\gamma| < 0.5$, then there exists a second-order estimation:

$$\lambda \approx m_0 + m_1(1-\gamma) + m_2(1-\gamma)^2$$

where

$$m_0 = \exp\left(\frac{a-b}{\omega c}\right) - 1, \qquad m_1 = \frac{b^2 - a^2}{2\omega c^2} \exp\left(\frac{a-b}{\omega c}\right)$$

$$m_2 = \frac{8(a^3 - b^3) + 3(a^2 - b^2)^2}{24\omega c^3} \exp\left(\frac{a-b}{\omega c}\right)$$

Proof. According to Taylor formula, we have

$$m_0 = \lim_{\gamma \to 1} it \, \lambda = \lim_{\lambda \to 1} \left\{ \frac{a(1-\gamma) + c}{b(1-\gamma) + c} \right\}^{1/(\omega(1-\gamma))} - 1 = \exp\left(\frac{a-b}{\omega c}\right) - 1$$

$$m_1 = \frac{d\lambda}{d\gamma}\bigg|_{\gamma=1} = \lim_{\gamma\to1}\left\{\left\{\frac{a(1-\gamma)+c}{b(1-\gamma)+c}\right\}^{1/(\omega(1-\gamma))} \cdot \frac{d}{d\gamma}\frac{\ln(a(1-\gamma)+c)-\ln(b(1-\gamma)+c)}{\omega(1-\gamma)}\right\}$$

$$= \frac{b^2-a^2}{2\omega c^2}\exp\left(\frac{a-b}{\omega c}\right)$$

$$m_2 = \frac{d^2\lambda}{d\gamma^2}\bigg|_{\gamma=1} = \lim_{\gamma\to1}\left\{\frac{a(1-\gamma)+c}{b(1-\gamma)+c}\right\}^{1/(\omega(1-\gamma))}\cdot\left[\frac{d}{d\gamma}\frac{\ln(a(1-\gamma)+c)-\ln(b(1-\gamma)+c)}{\omega(1-\gamma)}\right]^2$$

$$+ \lim_{\gamma\to1}\left\{\frac{a(1-\gamma)+c}{b(1-\gamma)+c}\right\}^{1/(\omega(1-\gamma))}\frac{d^2}{d\gamma^2}\frac{\ln(a(1-\gamma)+c)-\ln(b(1-\gamma)+c)}{\omega(1-\gamma)}$$

$$= \frac{8(a^3-b^3)+3(a^2-b^2)^2}{24\omega c^3}\exp\left(\frac{a-b}{\omega c}\right)$$

Theorem 1.7 If $|\gamma| < 0.5$, then there exists a second-order estimation

$$\lambda \approx n_0 + n_1\gamma + n_2\gamma^2$$

where

$$n_0 = \left(\frac{a+c}{b+c}\right)^{1/\omega} - 1, \quad n_1 = \left(\frac{a+c}{b+c}\right)^{1/\omega}\left(\ln\frac{a+c}{b+c} + \frac{a}{a+c} - \frac{b}{b+c}\right)$$

$$n_2 = \left(\frac{a+c}{b+c}\right)^{1/\omega}\left(\ln\frac{a+c}{b+c} + \frac{a}{a+c} - \frac{b}{b+c} + \frac{a^2}{2(a+c)^2} - \frac{b^2}{2(b+c)^2}\right)$$

$$+ \frac{1}{2}\left(\frac{a+c}{b+c}\right)^{1/\omega}\left(\ln\frac{a+c}{b+c} + \frac{a}{a+c} - \frac{b}{b+c}\right)^2$$

Proof. According to Taylor formula, we have

$$n_0 = \lim_{\gamma\to0}\lambda = \left(\frac{a+c}{b+c}\right)^{1/\omega} - 1$$

$$n_1 = \frac{d\lambda}{d\gamma}\bigg|_{\gamma=0} = \left(\frac{a+c}{b+c}\right)^{1/\omega}\left(\ln\frac{a+c}{b+c} + \frac{a}{a+c} - \frac{b}{b+c}\right)$$

$$n_2 = \frac{d^2\lambda}{d\gamma^2}\bigg|_{\gamma=0} = \left(\frac{a+c}{b+c}\right)^{1/\omega}\left(\ln\frac{a+c}{b+c} + \frac{a}{a+c} - \frac{b}{b+c} + \frac{a^2}{2(a+c)^2} - \frac{b^2}{2(b+c)^2}\right)$$

$$+ \frac{1}{2}\left(\frac{a+c}{b+c}\right)^{1/\omega}\left(\ln\frac{a+c}{b+c} + \frac{a}{a+c} - \frac{b}{b+c}\right)^2$$

Theorem 1.8 If $1-\gamma$ is sufficiently small, then an asymptotic expression exists

$$\lambda \approx \frac{a-b}{\omega c + \omega(a+b)(1-\gamma)}$$

Proof. Expression (1.29) can be rewritten as follows:

$$\omega(1-\gamma)\ln(\lambda+1) = \ln\left[\frac{a}{c}(1-\gamma)+1\right] - \ln\left[\frac{b}{c}(1-\gamma)+1\right] \tag{1.30}$$

By selecting the midpoint between point $[(a/c)(1-\gamma)+1]$ and point $[(b/c)(1-\gamma)+1]$, we can transform two logarithmic functions in the right-hand side of (1.30) into a series by using Taylor formula. That is, we have

$$\ln\left[\frac{a}{c}(1-\gamma)+1\right] = \ln\left[\frac{a+b}{2c}(1-\gamma)+1\right] + \frac{1}{(1-\gamma)((a+b)/2c)+1}\cdot\left[(1-\gamma)\left(\frac{a-b}{2c}\right)\right]$$

$$- \frac{1}{2\left[(1-\gamma)((a+b)/2c)+1\right]^2}\cdot\left[(1-\gamma)\left(\frac{a-b}{2c}\right)\right]^2 + o((1-\gamma)^2)$$

$$\tag{1.31}$$

$$\ln\left[\frac{b}{c}(1-\gamma)+1\right] = \ln\left[\frac{a+b}{2c}(1-\gamma)+1\right] + \frac{1}{(1-\gamma)((a+b)/2c)+1}\cdot\left[(1-\gamma)\left(\frac{b-a}{2c}\right)\right]$$

$$- \frac{1}{2\left[(1-\gamma)((a+b)/2c)+1\right]^2}\cdot\left[(1-\gamma)\left(\frac{b-a}{2c}\right)\right]^2 + o((1-\gamma)^2)$$

$$\tag{1.32}$$

By subtracting Equation 1.32 from Equation 1.31, we obtain

$$\omega(1-\gamma)\ln(\lambda+1) = \frac{(1-\gamma)((a-b)/c)}{(1-\gamma)((a+b)/c)+1} + o((1-\gamma)^2) \tag{1.33}$$

Equation 1.33 can be rewritten as

$$\ln(\lambda+1) = \frac{a-b}{\omega c + \omega(a+b)(1-\gamma)} + o(1-\gamma)$$

After applying Taylor formula once again, we obtain the result.

Theorem 1.9 If $1-\gamma$ is sufficiently small, then there exists an asymptotic expression

$$\lambda \approx \frac{a-b}{\omega c + \omega(a+b)(1-\gamma)} + \frac{1}{2} \cdot \frac{(a-b)^2}{[\omega c + \omega(a+b)(1-\gamma)]^2} \cdot (1-\gamma)^2$$

$$+ \frac{1}{48} \cdot \frac{\omega c + (a-b) + \omega(a+b)(1-\gamma)}{[\omega c + \omega(a+b)(1-\gamma)]^4} \cdot (1-\gamma)^2$$

Proof. In the same way as in Theorem 1.8, we transform the two logarithmic functions in the right-hand side of (1.30) into a series at the midpoint between point $[(a/c)(1-\gamma) + 1]$ and point $[(b/c)(1-\gamma) + 1]$. That is, we have

$$\ln\left[\frac{a}{c}(1-\gamma)+1\right] = \ln\left[\frac{a+b}{2c}(1-\gamma)+1\right] + \frac{1}{(1-\gamma)((a+b)/2c)+1}\cdot\left[(1-\gamma)\left(\frac{a-b}{2c}\right)\right]$$

$$- \frac{1}{2[(1-\gamma)((a+b)/2c)+1]^2}\cdot\left[(1-\gamma)\left(\frac{a-b}{2c}\right)\right]^2$$

$$+ \frac{1}{48[(1-\gamma)((a+b)/2c)+1]^3}\cdot\left[(1-\gamma)\left(\frac{a-b}{2c}\right)\right]^3 + o(1-\gamma)^3$$

$$(1.34)$$

$$\ln\left[\frac{b}{c}(1-\gamma)+1\right] = \ln\left[\frac{a+b}{2c}(1-\gamma)+1\right] + \frac{1}{(1-\gamma)((a+b)/2c)+1}\cdot\left[(1-\gamma)\left(\frac{b-a}{2c}\right)\right]$$

$$- \frac{1}{2[(1-\gamma)((a+b)/2c)+1]^2}\cdot\left[(1-\gamma)\left(\frac{b-a}{2c}\right)\right]^2$$

$$+ \frac{1}{48[(1-\gamma)((a+b)/2c)+1]^3}\cdot\left[(1-\gamma)\left(\frac{b-a}{2c}\right)\right]^3 + o(1-\gamma)^3$$

$$(1.35)$$

By subtracting Equation 1.35 from Equation 1.34, we obtain

$$\ln(\lambda + 1) = \frac{a - b}{\omega c + (a + b)(1 - \gamma)} + \frac{1}{48} \cdot \frac{(a - b)^3}{\left[\omega c + (a + b)(1 - \gamma)\right]^3} \cdot (1 - \gamma)^2 + o(1 - \gamma)^3$$

By applying Taylor formula once again, we have proved that the result of Theorem 1.9 is true.

1.5.3 Conclusions

It is well known that the optimal degree of capital control for a country that adopts a fixed exchange rate is stricter than that for a country with a flexible exchange rate. It would be worthwhile to extend the S-G estimation to help with the study of the policy combination of exchange rate flexibility and capital control. It would be interesting to introduce different types of capital inflows, such as foreign direct investment and short-term borrowing into the S-G model and examine the effects on economic growth of different tax rates for the different types of capital inflow. Theorems 1.6 and 1.7 illustrate that the welfare costs of various γ can change in different ways. More specifically, in a region of $\gamma = 1$, if the operator of mathematical expectations conditionally responding to policy regime a is greater than that responding to policy regime b, then the welfare cost is monotonically decreasing in γ. If the operator of mathematical expectations conditionally responding to policy regime a is smaller than that responding to policy regime b, then the welfare cost is monotonically increasing in γ. In the region of $\gamma = 0$, if the operator of mathematical expectations conditionally responding to policy regime a is greater than that responding to policy regime b, then the welfare cost is monotonically increasing in γ. If the operator of mathematical expectations conditionally responding to policy regime a is smaller than that responding to policy regime b, then the welfare cost is monotonically decreasing in γ.

1.6 Game Theory in Finance

Game theory has been widely used in finance, and the main body of the literature can be briefly defined as a decision-making behavior as well as a decision-making equilibrium problem.

Liu (2009) realized that the act of illegal capital flows does not follow the rational assumptions of the neoclassical economics and the theory of expected utility. So the author used evolutionary game theory to analyze the behavior of two groups involved in an illegal capital transaction and the resultant equilibrium.

In this section, based on Liu (2009), we further consider the asymmetric evolutionary game process of capital flight involving two groups by adding the factor of matching rate in the dynamic equation so that the course of the game is much closer to an actual situation.

1.6.1 Model Assumptions

In the classical game theory, it is generally assumed that the participants are fully rational and have the capability to respond with the optimal strategy. But perfect rationality in a real-life decision making is difficult to reach. So this assumption is unreasonable in terms of practical applications. The evolutionary game theory developed in the 1970s makes up for this deficiency. It applies the bounded rationality assumption to study the strategy adjustment, evolution, and the final equilibrium process of the group members in the long term.

In this section, the key participants involved in capital flight are divided into two groups, namely, the perpetrators of capital flight (group X) and the helpers that assist the perpetrators to carry out the capital flight (group Y). They play the game by focusing on whether or not to carry on with capital flight. The strategies in the game are as follows:

- *Willing to implement and expect the counterpart to help with the capital flight* (referred to as the conspiracy strategy)
- *Don't want to implement the capital flight* (referred to as non-conspiracy strategy)

These two groups have the same strategy choices (conspiracy and non-conspiracy). Assume that X and Y group members can get returns b_1 and b_2 in normal activities. If both groups choose the non-conspiracy strategy, the return is still b_1 and b_2. If both groups select the conspiracy strategy at the same time and successfully conducted the capital flight, the members of group X can get the additional revenue θ_1, and the members of group Y can get the *help fee* of θ_1. If group X asks group Y for assistance, and the group Y refuses the request because of its concern with the related risk, then group X will suffer some losses c_1, which can be simply understood as the loss of earlier expense used for the planned capital flight, including information gathering, human resources, and the punishment, if reported to regulatory authorities. If group Y takes the initiative to find group X to help them because of their own interests while group X refuses to help with the transaction because they feel that the risk is too big or simply do not want to implement the capital flight, then group Y suffers the loss of c_2. This loss can be simply understood as that the members increased the risk of punishment by regulatory authorities. We can also assume that there is such a relationship as $\theta_1 > c_1$, $\theta_2 > c_2$ between additional gains and losses, because it is the high returns that cause the two groups to implement the capital flight regardless of the involved risk.

1.6.2 Evolutionary Game Model

Based on the previous assumptions, we can provide Table 1.1:

Assume that the proportion of selecting conspiracy strategy in group X is x, and the proportion of selecting conspiracy strategy in group Y is y. Then, the individual expected return of selecting the conspiracy strategy in group X is

$$E_{1c} = y(b_1 + \theta_1) + (1-y)(b_1 - c_1)$$

and the individual expected return of selecting the non-conspiracy strategy in group X is

$$E_{1f} = yb_1 + (1-y)b_1$$

So, the average return for group X is

$$\overline{E_1} = xE_{1c} + (1-x)E_{1f}$$

The individual expected return of selecting the conspiracy strategy in group Y is

$$E_{2c} = x(b_2 + \theta_2) + (1-x)(b_2 - c_2)$$

and the individual expected return of selecting the non-conspiracy strategy in group Y is

$$E_{2f} = xb_2 + (1-x)b_2$$

So, the average return in group Y is

$$\overline{E_2} = yE_{2c} + (1-y)E_{2f}$$

The replicated dynamic equation of choosing the conspiracy strategy in group X is

$$\frac{dx}{dt} = a_{11}x(E_{1c} - \overline{E_1}) + a_{12}y(E_{2f} - \overline{E_2})$$

$$= a_{11}x(1-x)[y(\theta_1 + c_1) - c_1] + a_{12}y^2[c_2 - x(\theta_2 + c_2)] \qquad (1.36)$$

Table 1.1 Payoff Matrix of the Two-Strategy Capital Flight

Participants		Group Y	
Group X	Strategy	Conspiracy	Non-Conspiracy
	Conspiracy	$b_1 + \theta_1, b_2 + \theta_2$	$b_1 - c_1, b_2$
	Non-conspiracy	$b_1, b_2 - c_2$	b_1, b_2

and the replicated dynamic equation of choosing the conspiracy strategy in group Y is

$$\frac{dy}{dt} = a_{21}x(E_{1f} - \overline{E_1}) + a_{22}y(E_{2c} - \overline{E_2})$$

$$= a_{21}x^2[c_1 - y(\theta_1 + c_1)] + a_{22}y(1-y)[x(\theta_2 + c_2) - c_2] \tag{1.37}$$

where a_{ij} ($i, j = 1, 2$; 1 represents group X, 2 represents group Y) stands for the matching rate between group i that selects the conspiracy strategy and group j that selects the conspiracy strategy. Let us consider the general situation where $a_{ij} \neq 0$.

In order to obtain the equilibrium point, let the right-hand sides of (1.36) and (1.37) be 0. Then, we can obtain the following algebraic equations:

$$a_{11}x(1-x)[y(\theta_1 + c_1) - c_1] = a_{12}y^2[x(\theta_2 + c_2) - c_2] \tag{1.38}$$

$$a_{21}x^2[y(\theta_1 + c_1) - c_1] = a_{22}y(1-y)[x(\theta_2 + c_2) - c_2] \tag{1.39}$$

In terms of the solution of Equations 1.38 and 1.39, it is clear that there are two equilibrium points as follows:

$$E_1(0,0) \quad E_2\left(\frac{c_2}{c_2 + \theta_2}, \frac{c_1}{c_1 + \theta_1}\right)$$

When $x \neq 0$, $y \neq 1$, by dividing (1.38) by (1.39), we can get

$$\frac{a_{11}(1-x)}{a_{21}x} = \frac{a_{12}y}{a_{22}(1-y)} \tag{1.40}$$

By letting $k = (a_{12}a_{21})/(a_{11}a_{22})$, we can rewrite (1.40) as follows:

$$x = \frac{1-y}{1 + (k-1)y} \tag{1.41}$$

Substituting (1.41) into the right-hand side of (1.37) leads to

$$a_{21}(1-y)[c_1 - y(\theta_1 + c_1)] + a_{22}y[(k-1)y + 1][(1-y)\theta_2 - c_2ky] = 0$$

So, we have the following cubic algebraic equation:

$$A_1y^3 + B_1y^2 + C_1y + D_1 = 0 \tag{1.42}$$

where

$$A_1 = a_{22}(\theta_2 + c_2 k)(1 - k)$$
$$B_1 = a_{22}[k(\theta_2 - c_2) - 2\theta_2] + a_{21}(\theta_1 + c_1)$$
$$C_1 = a_{22}\theta_2 - a_{21}(\theta_1 + 2c_1)$$
$$D_1 = a_{21}c_1$$

For simplicity, we will only analyze the situation of $k = 1$, that is, $a_{11}a_{22} = a_{12}a_{21}$. To this end, Equation 1.42 can be simplified into the following quadratic equation:

$$B_2 y^2 + C_2 y + D_2 = 0$$

where

$$B_2 = a_{21}(\theta_1 + c_1) - a_{22}(\theta_2 + c_2)$$
$$C_2 = a_{22}\theta_2 - a_{21}(\theta_1 + 2c_1)$$
$$D_2 = a_{21}c_1$$

To solve this quadratic equation for the case when $\Delta = C_2^2 - 4B_2 D_2$, let us denote $m = a_{22}(\theta_2 + c_2) - a_{21}(\theta_1 + c_1)$. Thus, $\Delta = (m - a_{22}c_2 - a_{21}c_1)^2 + 4a_{21}c_1 m$. Now, there are three cases as follows:

1. When $\Delta < 0$, the real roots of the equation do not exist, then models (1.36) and (1.37) have two equilibrium points.
2. When $\Delta = 0$, the equation has a real root. So models (1.36) and (1.37) have three equilibrium points.
3. When $\Delta > 0$, the equation has two real roots. So models (1.36) and (1.37) have four equilibrium points.

Let us do a specific analysis in the following.

1.6.3 Stability Analysis

The stability of the equilibrium point(s) can be determined by the Jacobian matrix, which can be obtained from (1.36) and (1.37) as follows:

$$J = \begin{pmatrix} a_{11}(1-2x)[y(\theta_1 + c_1) - c_1] & a_{11}x(1-x)(\theta_1 + c_1) \\ -a_{12}y^2(\theta_2 + c_2) & +2a_{12}y[c_2 - x(\theta_2 + c_2)] \\ a_{22}y(1-y)(\theta_2 + c_2) & a_{22}(1-2y)[x(\theta_2 + c_2) - c_2] \\ +2a_{21}x[c_1 - y(\theta_1 + c_1)] & -a_{21}x^2(\theta_1 + c_1) \end{pmatrix}$$

Case 1: $\Delta < 0$. In this case, there are only two equilibrium points.

First, we analyze the stability of the point $E_1(0,0)$. In this case, we have

$$J_1 = J\big|_{E_1} = \begin{pmatrix} -a_{11}c_1 & 0 \\ 0 & -a_{22}c_2 \end{pmatrix}$$

$$|\lambda I - J_1| = \begin{vmatrix} \lambda + a_{11}c_1 & 0 \\ 0 & \lambda + a_{22}c_2 \end{vmatrix}$$

$$= \lambda^2 + (a_{11}c_1 + a_{22}c_2)\lambda + a_{11}a_{22}c_1c_2 = 0$$

Because $trJ_1 = -(a_{11}c_1 + a_{22}c_2) < 0$, $a_{11}a_{22}c_1c_2 > 0$, $(trJ_1)^2 - 4|J_1| \geq 0$, we get the following result.

Theorem 1.10 The equilibrium point E_1 is a stable node.

Next, we analyze the stability of the point $E_2\left(\dfrac{c_2}{c_2 + \theta_2}, \dfrac{c_1}{c_1 + \theta_1}\right)$. In this case, we have

$$J_2 = J\big|_{E_2} = \begin{pmatrix} -\dfrac{a_{12}(c_2 + \theta_2)c_1^2}{(c_1 + \theta_1)^2} & \dfrac{a_{11}c_2\theta_2(c_1 + \theta_1)}{(c_2 + \theta_2)^2} \\ \dfrac{a_{22}c_1\theta_1(c_2 + \theta_2)}{(c_1 + \theta_1)^2} & -\dfrac{a_{21}(c_1 + \theta_1)c_2^2}{(c_2 + \theta_2)^2} \end{pmatrix}$$

$$|\lambda I - J_2| = \begin{vmatrix} \lambda + \dfrac{a_{12}(c_2 + \theta_2)c_1^2}{(c_1 + \theta_1)^2} & -\dfrac{a_{11}c_2\theta_2(c_1 + \theta_1)}{(c_2 + \theta_2)^2} \\ -\dfrac{a_{22}c_1\theta_1(c_2 + \theta_2)}{(c_1 + \theta_1)^2} & \lambda + \dfrac{a_{21}(c_1 + \theta_1)c_2^2}{(c_2 + \theta_2)^2} \end{vmatrix}$$

$$= \lambda^2 + \left[\dfrac{a_{12}(c_2 + \theta_2)c_1^2}{(c_1 + \theta_1)^2} + \dfrac{a_{21}(c_1 + \theta_1)c_2^2}{(c_2 + \theta_2)^2}\right]\lambda + \dfrac{(a_{12}a_{21}c_1c_2 - a_{11}a_{22}\theta_1\theta_2)c_1c_2}{(c_1 + \theta_1)(c_2 + \theta_2)}$$

Because

$$\dfrac{(a_{12}a_{21}c_1c_2 - a_{11}a_{22}\theta_1\theta_2)c_1c_2}{(c_1 + \theta_1)(c_2 + \theta_2)} = \dfrac{a_{11}a_{22}(c_1c_2 - \theta_1\theta_2)c_1c_2}{(c_1 + \theta_1)(c_2 + \theta_2)} < 0$$

we obtain the following result.

Theorem 1.11 The equilibrium point E_2 is an unstable saddle point.

Case 2: $\Delta = 0$. Other than the two equilibrium points E_1 and E_2 as discussed earlier, there is an additional equilibrium point E_3. From the analysis of case 1, the stability of E_1, E_2 is clear as stated earlier. And it is straightforward to obtain the following coordinates:

$$y_3 = \frac{m - a_{22}c_2 - a_{21}c_1}{2m}, \quad x_3 = 1 - y_3 = \frac{m + a_{22}c_2 + a_{21}c_1}{2m}$$

Let $J_3 = J|_{E_3} = \begin{pmatrix} h_{11} & h_{12} \\ h_{21} & h_{22} \end{pmatrix}$, where

$$h_{11} = -\frac{1}{4m^2}\Big\{2a_{11}(a_{22}c_2 + a_{21}c_1)\big[(m - a_{22}c_2 - a_{21}c_1)(\theta_1 + c_1) - 2mc_1\big]$$

$$+ a_{12}(m - a_{22}c_2 - a_{21}c_1)^2(\theta_2 + c_2)\Big\}$$

$$h_{12} = \frac{1}{4m^2}\Big\{a_{11}\big[m^2 - (a_{22}c_2 + a_{21}c_1)^2\big](\theta_1 + c_1)$$

$$+ 2a_{12}(m - a_{22}c_2 - a_{21}c_1)[2mc_2 - (\theta_2 + c_2)(m + a_{22}c_2 + a_{21}c_1)]\Big\}$$

$$h_{21} = \frac{1}{4m^2}\Big\{a_{22}\big[m^2 - (a_{22}c_2 + a_{21}c_1)^2\big](\theta_2 + c_2)$$

$$+ 2a_{21}(m + a_{22}c_2 + a_{21}c_1)\big[2mc_1 - (\theta_1 + c_1)(m - a_{22}c_2 - a_{21}c_1)\big]\Big\}$$

$$h_{22} = \frac{1}{4m^2}\Big\{2a_{22}(a_{22}c_2 + a_{21}c_1)\big[(m + a_{22}c_2 + a_{21}c_1)(\theta_2 + c_2) - 2mc_2\big]$$

$$- a_{21}(m + a_{22}c_2 + a_{21}c_1)^2(\theta_1 + c_1)\Big\}$$

$$|\lambda I - J_3| = \begin{pmatrix} \lambda - h_{11} & -h_{12} \\ -h_{21} & \lambda - h_{22} \end{pmatrix} = \lambda^2 - (h_{11} + h_{22})\lambda + (h_{11}h_{22} - h_{12}h_{21})$$

Because

$$h_{11} + h_{22} = -\frac{1}{4m^2}\Big\{[a_{21}(\theta_1 + c_1) + a_{12}(\theta_2 + c_2)]m^2 + 2(a_{22}c_2 + a_{21}c_1)$$

$$\times [a_{22}(\theta_2 - c_2) + a_{12}(\theta_2 + c_2) - a_{11}(\theta_1 - c_1) - a_{21}(\theta_1 + c_1)]m$$

$$+ (a_{22}c_2 + a_{21}c_1)^2[(2a_{22} - a_{12})(\theta_2 + c_2) + (2a_{11} - a_{21})(\theta_1 + c_1)]\Big\}$$

$h_{11}h_{22} - h_{12}h_{21}$

$$= \frac{1}{8m^3} \left\{ \left[a_{11}a_{21}\left(\theta_1^2 - c_1^2\right) + a_{12}a_{22}\left(\theta_2^2 - c_2^2\right) - 2a_{11}a_{22}(\theta_1 - c_1)(\theta_2 - c_2)\right]m^3 \right.$$

$$+ (a_{22}c_2 + a_{21}c_1)\left[a_{11}a_{21}(\theta_1 + c_1)(\theta_1 - 3c_1) - a_{12}a_{22}(\theta_2 + c_2)(\theta_2 - 3c_2)\right.$$

$$- 4a_{11}a_{22}(c_2\theta_1 - c_1\theta_2)\right]m^2 - (a_{22}c_2 + a_{21}c_1)^2\left[a_{11}a_{21}(\theta_1 + c_1)(\theta_1 + 3c_1)\right.$$

$$+ a_{12}a_{22}(\theta_2 + c_2)(\theta_2 + 3c_2) - 2a_{11}a_{22}(\theta_1 + c_1)(\theta_2 + c_2)\right]m - (a_{22}c_2 + a_{21}c_1)^3$$

$$\left. \times \left[a_{11}a_{21}(\theta_1 + c_1)^2 - a_{12}a_{22}(\theta_2 + c_2)^2\right]\right\}$$

we have the following results.

Theorem 1.12 If $|J_3| < 0$, then the equilibrium point E_3 is a unstable saddle point.

Theorem 1.13 If $|J_3| > 0$, $trJ_3 > 0$, then the equilibrium point E_3 is either an unstable node or a focus. If $|J_3| > 0$, $trJ_3 < 0$, then the equilibrium point E_3 is either a stable node or a focus. If $|J_3| > 0$, $trJ_3 = 0$, then the equilibrium point E_3 is a stable central point.

Case 3: $\Delta > 0$. Other than the two equilibrium points E_1 and E_2 as studied earlier, there are two additional equilibrium points E_4, E_5. The stability of E_1, E_2 is clear as analyzed earlier. For the additional equilibrium points E_4, E_5, it is ready to obtain the following coordinates:

$$y_4 = \frac{m - a_{22}c_2 - a_{21}c_1 + \sqrt{\Delta}}{2m}, \quad x_4 = 1 - y_4 = \frac{m + a_{22}c_2 + a_{21}c_1 - \sqrt{\Delta}}{2m}$$

$$y_5 = \frac{m - a_{22}c_2 - a_{21}c_1 - \sqrt{\Delta}}{2m}, \quad x_5 = 1 - y_5 = \frac{m + a_{22}c_2 + a_{21}c_1 + \sqrt{\Delta}}{2m}$$

Next, let us analyze the stability of E_4, E_5. Let $J_4 = J|_{E_4} = \begin{pmatrix} \eta_{11} & \eta_{12} \\ \eta_{21} & \eta_{22} \end{pmatrix}$, of which

$$\eta_{11} = -\frac{1}{4m^2}\left\{2a_{11}\left(a_{22}c_2 + a_{21}c_1 - \sqrt{\Delta}\right)\left[\left(m - a_{22}c_2 - a_{21}c_1 + \sqrt{\Delta}\right)(\theta_1 + c_1) - 2mc_1\right]\right.$$

$$\left. + a_{12}\left(m - a_{22}c_2 - a_{21}c_1 + \sqrt{\Delta}\right)^2 (\theta_2 + c_2)\right\}$$

$$\eta_{12} = \frac{1}{4m^2}\left\{a_{11}\left[m^2 - \left(a_{22}c_2 + a_{21}c_1 - \sqrt{\Delta}\right)^2\right](\theta_1 + c_1)\right.$$

$$\left. + 2a_{12}\left(m - a_{22}c_2 - a_{21}c_1 + \sqrt{\Delta}\right)\left[2mc_2 - (\theta_2 + c_2)\left(m + a_{22}c_2 + a_{21}c_1 - \sqrt{\Delta}\right)\right]\right\}$$

$$r_{21} = \frac{1}{4m^2}\left\{ a_{22}\left[m^2 - \left(a_{22}c_2 + a_{21}c_1 - \sqrt{\Delta} \right)^2 \right](\theta_2 + c_2) \right.$$

$$\left. + 2a_{21}\left(m + a_{22}c_2 + a_{21}c_1 - \sqrt{\Delta} \right)\left[2mc_1 - (\theta_1 + c_1)\left(m - a_{22}c_2 - a_{21}c_1 + \sqrt{\Delta} \right) \right] \right\}$$

$$r_{22} = \frac{1}{4m^2}\left\{ 2a_{22}\left(a_{22}c_2 + a_{21}c_1 - \sqrt{\Delta} \right)\left[\left(m + a_{22}c_2 + a_{21}c_1 - \sqrt{\Delta} \right)(\theta_2 + c_2) - 2mc_2 \right] \right.$$

$$\left. - a_{21}\left(m + a_{22}c_2 + a_{21}c_1 - \sqrt{\Delta} \right)^2 (\theta_1 + c_1) \right\}$$

Then, we have

$$|\lambda I - J_4| = \begin{pmatrix} \lambda - r_{11} & -r_{12} \\ -r_{21} & \lambda - r_{22} \end{pmatrix} = \lambda^2 - (r_{11} + r_{22})\lambda + (r_{11}r_{22} - r_{12}r_{21})$$

$$r_{11} + r_{22} = -\frac{1}{4m^2}\left\{ \left[a_{21}(\theta_1 + c_1) + a_{12}(\theta_2 + c_2) \right]m^2 + 2\left(a_{22}c_2 + a_{21}c_1 - \sqrt{\Delta} \right) \right.$$

$$\times \left[a_{22}(\theta_2 - c_2) + a_{12}(\theta_2 + c_2) - a_{11}(\theta_1 - c_1) - a_{21}(\theta_1 + c_1) \right]m$$

$$\left. + \left(a_{22}c_2 + a_{21}c_1 - \sqrt{\Delta} \right)^2 \left[(2a_{22} - a_{12})(\theta_2 + c_2) + (2a_{11} - a_{21})(\theta_1 + c_1) \right] \right\}$$

$$r_{11}r_{22} - r_{12}r_{21}$$

$$= \frac{1}{8m^3}\left\{ \left[a_{11}a_{21}\left(\theta_1^2 - c_1^2 \right) + a_{12}a_{22}\left(\theta_2^2 - c_2^2 \right) - 2a_{11}a_{22}(\theta_1 - c_1)(\theta_2 - c_2) \right]m^3 \right.$$

$$+ \left(a_{22}c_2 + a_{21}c_1 - \sqrt{\Delta} \right)\left[a_{11}a_{21}(\theta_1 + c_1)(\theta_1 - 3c_1) - a_{12}a_{22}(\theta_2 + c_2)(\theta_2 - 3c_2) \right.$$

$$\left. - 4a_{11}a_{22}(c_2\theta_1 - c_1\theta_2) \right]m^2 - \left(a_{22}c_2 + a_{21}c_1 - \sqrt{\Delta} \right)^2 \left[a_{11}a_{21}(\theta_1 + c_1)(\theta_1 + 3c_1) \right.$$

$$\left. + a_{12}a_{22}(\theta_2 + c_2)(\theta_2 + 3c_2) - 2a_{11}a_{22}(\theta_1 + c_1)(\theta_2 + c_2) \right]m$$

$$\left. - \left(a_{22}c_2 + a_{21}c_1 - \sqrt{\Delta} \right)^3 \left[a_{11}a_{21}(\theta_1 + c_1)^2 - a_{12}a_{22}(\theta_2 + c_2)^2 \right] \right\}$$

So, we have the following results.

Theorem 1.14 If $|J_4| < 0$, then the equilibrium point E_4 is an unstable saddle point.

Theorem 1.15 If $|J_4| > 0$, $trJ_4 > 0$, then the equilibrium point E_4 is either an unstable node or focus. If $|J_4| > 0$, $trJ_4 < 0$, then the equilibrium point E_4 is either a stable node or focus. If $|J_4| > 0$, $trJ_4 = 0$, then the equilibrium point E_4 is a stable central point.

Assume $J_5 = J\big|_{E_5} = \begin{pmatrix} n_{11} & n_{12} \\ n_{21} & n_{22} \end{pmatrix}$, where

$$n_{11} = -\frac{1}{4m^2}\left\{ 2a_{11}\left(a_{22}c_2 + a_{21}c_1 + \sqrt{\Delta}\right)\left[\left(m - a_{22}c_2 - a_{21}c_1 - \sqrt{\Delta}\right)(\theta_1 + c_1) - 2mc_1\right]\right.$$
$$\left. + a_{12}\left(m - a_{22}c_2 - a_{21}c_1 - \sqrt{\Delta}\right)^2 (\theta_2 + c_2)\right\}$$

$$n_{12} = \frac{1}{4m^2}\left\{ a_{11}\left[m^2 - \left(a_{22}c_2 + a_{21}c_1 + \sqrt{\Delta}\right)^2\right](\theta_1 + c_1)\right.$$
$$\left. + 2a_{12}\left(m - a_{22}c_2 - a_{21}c_1 - \sqrt{\Delta}\right)\left[2mc_2 - (\theta_2 + c_2)\left(m + a_{22}c_2 + a_{21}c_1 + \sqrt{\Delta}\right)\right]\right\}$$

$$n_{21} = \frac{1}{4m^2}\left\{ a_{22}\left[m^2 - \left(a_{22}c_2 + a_{21}c_1 + \sqrt{\Delta}\right)^2\right](\theta_2 + c_2)\right.$$
$$\left. + 2a_{21}\left(m + a_{22}c_2 + a_{21}c_1 + \sqrt{\Delta}\right)\left[2mc_1 - (\theta_1 + c_1)\left(m - a_{22}c_2 - a_{21}c_1 - \sqrt{\Delta}\right)\right]\right\}$$

$$n_{22} = \frac{1}{4m^2}\left\{ 2a_{22}\left(a_{22}c_2 + a_{21}c_1 + \sqrt{\Delta}\right)\left[\left(m + a_{22}c_2 + a_{21}c_1 + \sqrt{\Delta}\right)(\theta_2 + c_2) - 2mc_2\right]\right.$$
$$\left. - a_{21}\left(m + a_{22}c_2 + a_{21}c_1 + \sqrt{\Delta}\right)^2 (\theta_1 + c_1)\right\}$$

$$n_{11}n_{22} - n_{12}n_{21}$$

$$= \frac{1}{8m^3}\left\{\left[a_{11}a_{21}\left(\theta_1^2 - c_1^2\right) + a_{12}a_{22}\left(\theta_2^2 - c_2^2\right) - 2a_{11}a_{22}(\theta_1 - c_1)(\theta_2 - c_2)\right]m^3\right.$$

$$+ \left(a_{22}c_2 + a_{21}c_1 + \sqrt{\Delta}\right)\left[a_{11}a_{21}(\theta_1 + c_1)(\theta_1 - 3c_1) - a_{12}a_{22}(\theta_2 + c_2)(\theta_2 - 3c_2)\right.$$

$$\left. - 4a_{11}a_{22}(c_2\theta_1 - c_1\theta_2)\right]m^2 - \left(a_{22}c_2 + a_{21}c_1 + \sqrt{\Delta}\right)^2\left[a_{11}a_{21}(\theta_1 + c_1)(\theta_1 + 3c_1)\right.$$

$$+ a_{12}a_{22}(\theta_2 + c_2)(\theta_2 + 3c_2) - 2a_{11}a_{22}(\theta_1 + c_1)(\theta_2 + c_2)\right]m$$

$$\left. - \left(a_{22}c_2 + a_{21}c_1 + \sqrt{\Delta}\right)^3\left[a_{11}a_{21}(\theta_1 + c_1)^2 - a_{12}a_{22}(\theta_2 + c_2)^2\right]\right\}$$

$$n_{11} + n_{22} = -\frac{1}{4m^2} \left\{ \left[a_{21} \left(\theta_1 + c_1 \right) + a_{12} \left(\theta_2 + c_2 \right) \right] m^2 + 2 \left(a_{22} c_2 + a_{21} c_1 + \sqrt{\Delta} \right) \right.$$

$$\times \left[a_{22} \left(\theta_2 - c_2 \right) + a_{12} \left(\theta_2 + c_2 \right) - a_{11} \left(\theta_1 - c_1 \right) - a_{21} \left(\theta_1 + c_1 \right) \right] m$$

$$\left. + \left(a_{22} c_2 + a_{21} c_1 + \sqrt{\Delta} \right)^2 \left[\left(2 a_{22} - a_{12} \right) \left(\theta_2 + c_2 \right) + \left(2 a_{11} - a_{21} \right) \left(\theta_1 + c_1 \right) \right] \right\}$$

Then, we have the following results.

Theorem 1.16 If $|J_5| < 0$, then the equilibrium point E_5 is an unstable saddle point.

Theorem 1.17 If $|J_5| > 0$, $trJ_5 > 0$, then the equilibrium point E_5 is either an unstable node or a focus. If $|J_5| > 0$, $trJ_5 < 0$, then the equilibrium point E_5 is either a stable node or a focus. If $|J_5| > 0$, $trJ_5 = 0$, then the equilibrium point E_5 is a stable central point.

1.6.4 Model Summary

In this section, we establish an evolutionary game model between the perpetrators and assistants of capital flight by adding a matching rate and by analyzing its dynamic evolution.

The main consideration in the evolutionary game model is the general case where the matching rate is not equal to 0. When extreme situations appear, such as when $a_{11} = a_{22} = 1$, $a_{12} = a_{21} = 0$, one can easily find the following five equilibrium points, respectively:

$$E_1'(0,0),\ E_2'(1,0),\ E_3'(0,1),\ E_4'(1,1)\ \text{and}\ E_5'\left(\frac{c_2}{\theta_2 + c_2} \right)$$

where E_1', E_4' are stable nodes. As for whether the players tend to employ the strategy of *conspiracy* (i.e., E_1') or *non-conspiracy* (i.e., E_4'), it depends on the intensity of punitive measures. In addition, among the four strategy combinations, although three (excluding the *conspiracy* strategy) have shown self-restriction of capital flight, only the strategy combination when both parties choose the non-conspiracy strategy can exist stably in the long term. Even in a single game, as long as one party does not agree with the capital flight, this behavior does not ultimately occur. But this situation does not exist in a long time. In addition, under normal circumstances, there may be two, three, or four equilibrium points in the system, and when the matching rate satisfies certain conditions, these equilibrium points may become unstable, causing serious negative impacts. Therefore, the punitive measures on capital flight should continue to strengthen so as to reinforce the equilibrium points to evolve toward the stability of the origin point in order for the measures to play a role in suppression of capital flight.

The results of the game theory indicate that in order to suppress the occurrence of capital flight, the punishment measure should be increased; the system of financial regulations should be improved, among others. In the long run, the domestic macroeconomic regulation and control should be strengthened in order to maintain the confidence of the public, to increase internal coordination and international cooperation, and to combat illegal capital flows.

1.7 Visualization Technology in Finance

1.7.1 Introduction

The tools of visualization are of great significance for developing countries to study the problem of abnormal cross-border capital flows under the condition of capital account liberalization in terms of maintaining their financial stability and economic development.

The aim of this section is to introduce the visualization technology, such as the modern mapping techniques, to the finance community. Mapping techniques provide means for representing high-dimensional data with low-dimensional displays. We will focus on a methodology known as the self-organizing financial stability map (SOFSM) based upon data and dimensionality reduction that can be used for describing the state of financial stability and visualizing potential sources of systemic risks. Besides its visualization capabilities, the SOFSM can be used as a model of early warning that can be calibrated according to policy makers' preferences between missing systemic financial crises and issuing false alarms. A few applications of the SOFSM to the recent global financial crisis show that it performs on par with a statistical benchmark model and correctly called the crises that started in 2007 in the United States and in the euro area (Minsky, 1982; Kindleberger, 1996; Sarlin and Marghescu, 2011).

The recent global financial crisis has demonstrated the importance of understanding sources of domestic and global vulnerabilities that may lead to a systemic financial crisis. Early identification of financial stress would allow policy makers to introduce policy actions to decrease or prevent further buildup of vulnerabilities or otherwise to enhance the shock absorption capacity of the financial system. Finding the individual sources of vulnerability and risk is of central importance since that allows targeted actions for repairing specific cracks in the financial system.

A large number of indicators are often required to accurately assess the sources of financial instability (Sarlin and Peltonen, 2013). As with statistical tables, standard two- and three-dimensional visualizations have, of course, their limitations for high dimensions, not to mention the challenge of including a temporal or cross-sectional dimension or assessing multiple countries over time. Data reduction (or clustering) provides summaries of data by compressing information, while dimensionality reduction (or projection) provides low-dimensional representations of similarity relations in data. The self-organizing map (SOM) (Kohonen, 1982, 2001) holds promise for the task by combining the aims of data and dimensionality reduction. It is capable of providing an easily interpretable nonlinear description of the multidimensional data distribution on a two-dimensional plane without losing sight of individual indicators. The two-dimensional output of the SOM makes it particularly useful for visualizations, or summarizations, of large amounts of information.

The main aim of this section is to introduce and promote the awareness of mapping techniques in the field of finance and the policy-making community in general and financial stability surveillance in particular. The use of mapping techniques in financial stability surveillance is illustrated by laying out a methodology based upon data and dimensionality reduction for mapping the state of financial stability and visualizing potential sources of systemic risks.

In this section, we introduce the following five elements that are necessary for constructing the SOFSM:

1. Data and dimensionality reduction based upon the SOM
2. Identification of systemic financial crises
3. Macro-financial indicators of vulnerabilities and risks
4. A model evaluation framework for assessing performance
5. A model training framework

1.7.2 Self-Organizing Maps

Before providing a more formal description of the SOM algorithm, we focus on the intuition behind the basic SOM (Forte et al., 2002). We use the standard Euclidean batch SOM algorithm. That is, data are processed in batches rather than sequentially, and distances are compared using the Euclidean metric. The SOM grid consists of a user-specified number of reference vectors m_i (where $i = 1,2,...,M$) (i.e., mean profiles or units), which represent the same dimensions (number of variables) as the actual dataset Ω. Generally, the SOM algorithm operates according to the following steps:

Step 1. Initialize the reference vector values using the two principal components.
Step 2. Compare all data vectors x_j with all reference vectors m_i to find for each data value the nearest reference vector m_b (i.e., best-matching unit, BMU).
Step 3. Update each reference vector m_i to averages of the attracted data, including diminishing weight data located in a specified neighborhood.
Step 4. Repeat steps 2 and 3 a specified number of times.

In more formal terms, the vector quantization capability of the SOM performs the data reduction into reference vectors m_i. It models from the continuous space Ω with a pdf $f(x)$ to the grid of units, whose location depends on the neighborhood structure of the data Ω. While further details and a broad overview of the SOM are given in Kohonen (1982, 2001), we provide here a formal, but brief, description of the batch SOM algorithm. Important advantages of the batch algorithm are the reduction of computational cost and reproducible results (given the same initialization). The training process starts with initialization of the reference vector set to the direction of the two principal components of the input data. The principal component initialization not only further reduces computational cost and enables reproducible results, but also is shown to be important for convergence when using the batch SOM (Forte et al., 2002). Following Kohonen (2001), this is done in three steps:

Step 1. Determine two eigenvectors v_1 and v_2 with the largest eigenvalues from the covariance matrix of all data Ω.

Step 2. Let v_1 and v_2 span a two-dimensional linear subspace and fit a rectangular array along it, where the two dimensions are the eigenvectors and the center coincides with the mean of Ω. Hence, the direction of the long side is parallel to the longest eigenvector v_1 with a length of 80% of the length of v_1. The short side is parallel to v_2 with a length of 80% of the length of v_2.

Step 3. Identify the initial value of the reference vectors $m_i(0)$ with the array points, where the corners of the rectangle are $\pm\, 0.4v_1 \pm 0.4v_2$.

Following the initialization, the batch training algorithm operates through a specified number of iterations 1, 2, ..., t in two steps. In the first step, each input data vector x is assigned to the BMU m_c:

$$\left\| x - m_c(t) \right\| = \min_i \left\| x - m_i(t) \right\| \tag{1.43}$$

We employ a semi-supervised version of the SOM by also including class information when determining the BMU. In the second step, each reference vector m_i (where $i = 1,2,...,M$) is adjusted by using the batch update formula:

$$m_i(t+1) = \frac{\sum_{J=1}^{N} h_{ic(j)}(t)x_j}{\sum_{J=1}^{N} h_{ic(j)}(t)} \tag{1.44}$$

where
  index j indicates the input data vectors that belong to unit c
  N is the number of the data vectors

The neighborhood function $h_{ic(j)} \in (0,1]$ is defined as the following Gaussian function:

$$h_{ic(j)} = \exp\left(-\frac{\|r_c - r_i\|^2}{2\sigma^2(t)}\right) \tag{1.45}$$

where

$\|r_c - r_i\|^2$ is the squared Euclidean distance between the coordinates of the reference vectors m_c and m_i on the two-dimensional grid

the radius of the neighborhood $\sigma(t)$ is a monotonically decreasing function of time t

The radius of the neighborhood begins as half the diagonal of the grid size $(\sigma = (X^2 + Y^2)^{1/2}/2)$ and goes monotonically toward the specified tension value $\sigma(t) \in (0,2]$.

1.7.3 Clustering of the SOM

The use of a second-level clustering of the SOM enables one to choose a large number of SOM units for visualization, while keeping the number of clusters low. It has also been shown that the SOM can increase the clustering quality (Vesanto and Alhoniemi, 2000). The second-level clustering is herein performed by using an agglomerative hierarchical clustering. The following modified Ward's criterion is used as a basis for measuring the distance between two candidate clusters (Ward, 1963):

$$d_{kl} = \begin{cases} \dfrac{n_k n_l}{n_k + n_l} \cdot \|c_k - c_l\|^2 \\ \infty \end{cases} \tag{1.46}$$

where

k and l represent two clusters, n_k and n_l are the number of data points in these clusters k and l

$\|c_k - c_l\|^2$ is the squared Euclidean distance between the respective cluster centers of k and l

The Ward clustering is modified only for the purpose of merging clusters with other topologically neighboring clusters by defining the distance between nonadjacent clusters as infinitely large. The algorithm starts with each unit as its own cluster and merges units for all possible numbers of clusters using the minimum Ward distance $(1, 2, ..., M)$.

1.7.4 *Identifying Systemic Financial Crises*

As an example, let us look at the dataset, an updated version of that in Janus and Riera-Crichton (2013). It consists of a database of systemic events and a set of vulnerability and risk indicators. The quarterly dataset consists of 28 countries (10 advanced and 18 emerging economies) for the time period from January 1990 to February 2011. Hence, the data vector $x_j \in \Re$ is formed of the class variables $x_{cla} \in \Re$ and the indicator vector $x_{ind} \in \Re$ for each quarter and country in the sample.

The identification of systemic financial crises is performed with a financial stress index (FSI). The FSI is a country-specific composite index that covers segments of the money market, equity market, and foreign exchange market of the domestic financial market:

1. The spread of the 3-month interbank rate over the 3-month government bill rate (Ind_1)
2. Negative quarterly equity returns (Ind_2)
3. The realized volatility of the main equity index (as average daily absolute changes over a quarter) (Ind_3)
4. The realized volatility of the nominal effective exchange rate (Ind_4)
5. The realized volatility of the yield on the 3-month government bill (Ind_5)

Each indicator $j(Ind_j)$ of the FSI for country i at quarter t is transformed into an integer from 0 to 3 according to the quartile of the country-specific distribution, while the transformed variable is denoted as $q_{j,i,t}(Ind_{j,i,t})$. The FSI is computed for country i at time t as a simple average of the transformed variables as follows:

$$\text{FSI}_{i,t} = \frac{\sum_{j=1}^{5} q_{j,i,t}(Ind_{j,i,t})}{5} \tag{1.47}$$

In practice, we create a binary *crisis* variable, denoted as C0 that takes a value 1 in the quarter when the FSI is above the threshold of the 90th percentile of its country-specific distribution $\Omega_i^{90th}(\text{FSI}_{i,t})$ and 0 otherwise. That is,

$$C0_{i,t} = 1 \text{ if } \text{FSI}_{i,t} > \Omega_i^{90th}(\text{FSI}_{i,t})$$

$$C0_{i,t} = 0 \quad \text{otherwise}$$

To describe either the financial stability cycle or risk, one needs to create a set of other class variables besides the crisis variable, such as a *precrisis* class variable and *postcrisis* class variable and others. In fact, the WireVis system was created in cooperation with financial analysts at the Bank of America to address fundamental problems they had in understanding their vast flow of financial transactions.

Its initial purpose was to seek and discover fraud, especially wire transfer fraud. Although WireVis initially focused on wire fraud, its capabilities are general, and appropriately extended versions are now being considered for general financial analysis including risk and customer analyses, which represents a very difficult investigative analysis problem. William et al. (2009) regarded that some tools must be exploratory, because new modes of deception are tried all the time and must help human analysts see odd patterns over time in financial transactions.

Chapter 2

From Micheal to Heckscher–Ohlin: ODE for Capital Account Liberation

2.1 General Theory of Ordinary Differential Equations

2.1.1 *Basic Concepts of Ordinary Differential Equations*

Ordinary differential equations (ODE) serve as mathematical models for many exciting *real-world* problems not only in science and technology but also in diverse fields such as economics, psychology, defense, and demography. Rapid development in the theory of differential equations and in its applications to almost every branch of knowledge has resulted in a continued interest in its study by students in many disciplines. This has given ODEs a distinct place in mathematics curricula all over the world, and it is now being applied in almost every scientific field, especially in finance.

An ODE may be defined as an equation that involves a single unknown function of a single variable and some finite number of its derivatives.

In general, in many practice problems, we are asked to find out all of the functions that satisfy a certain condition that involves one or more derivatives of the unknown function. We can reformulate our definition of a differential equation as follows.

Definition 2.1 (Differential Equation). Let F be a function of $n + 2$ variables. Then the equation

$$F(x, y, y', \ldots, y^{(n)}) = 0$$

is called an ODE of order n for the unknown function y. The order of the equation is the order of the highest order derivative that appears in the equation.

Definition 2.2 (Solution). Let J be an interval and R the set of all real numbers. A function $y(x): J \to R$ is called a solution to the differential equation (2.1) (in J) if y is differentiable in J, the graph of y is a subset of D, and (2.1) holds, that is, if $(x, y(x)) \in D$ and

$$y'(x) = f(x, y(x)) \quad \text{for all } x \in J \tag{2.1}$$

The differential equation (2.1) has a simple geometric interpretation. If $y(x)$ is an integral curve of (2.1) that passes through a point (x_0, y_0) (i.e., $y(x_0) = y_0$, $y'(x_0) = f(x_0, y_0)$). This leads naturally to the notions of line element and direction field, which we will now define. We interpret a numerical triple of the form (x, y, p) geometrically in the following way.

Definition 2.3 (Line Element). Let (x, y) be a point in the xy-plane, and the third component p the slope of a line through the point (x, y) (α, with $\tan \alpha = p$, is the angle of inclination of the line). Such a triple (or its geometric equivalent) is called a line element.

Definition 2.4 (Direction Field). The collection of all line elements of the form $(x, y, f(x, y))$, that is, those with $p = f(x, y)$, is called a direction field.

The connection between direction fields and the differential equation (2.1) can be expressed in geometric terms as follows: A solution $y(x)$ of a differential equation *fits* its direction field, that is, the slope at each point on the solution curve agrees with the slope of the line element at that point. To put it another way, if $y(x)$ is a solution in J, then the set of line elements $(x, y(x), y'(x))$, with $x \in J$, is contained in the set of all line elements $(x, y, f(x, y))$, $(x, y) \in D$.

The strategy of sketching a few of the line elements in the direction field and then trying to draw curves that fit these line elements can be used to get a rough idea of the nature of the solutions to a differential equation. This procedure suggests quite naturally the view that for each point (ξ, η) in D there is exactly one solution curve $y(x)$ passing through that point.

Definition 2.5 (The Initial Value Problem). Let a function $f(x, y)$, defined on a set D in the (x, y)-plane, and a fixed point $(\xi, \eta) \in D$ be given. A function $y(x)$ is sought that is differentiable in an interval J with $\xi \in J$ such that

$$y'(x) = f(x, y(x)) \text{ in } J \tag{2.2}$$

$$y(\xi) = \eta \tag{2.3}$$

Equation 2.3 is called an initial condition. Naturally, in (2.2), it is assumed that the graph $y = \{(x, y(x)) : x \in J\} \subset D$ (otherwise the right-hand side of (2.2) would not even be defined). Then, we introduce a few kinds of equations.

2.1.1.1 Equation with Separated Variables: $y' = f(x)g(y)$

This class of equations can also be solved by using the method of separation of variables. We will describe this method first in heuristic terms. One goes from the given equation to the following equation $dy/g(y) = f(x)dx$, and then by integrating the equation

$$\int \frac{dy}{g(y)} = \int f(x)dx \tag{2.4}$$

a solution can be obtained by solving Equation 2.4 for y. In order to get the solution that passes through the point (ξ, η), it is necessary to choose the limits of integration such that Equation 2.4 is satisfied when $x = \xi$, $y = \eta$. This is accomplished by setting

$$\int_{\eta}^{y} \frac{ds}{g(s)} = \int_{\xi}^{x} f(t)dt \tag{2.5}$$

The following theorem gives conditions under which this procedure is permitted. It concerns with the initial value problem

$$y' = f(x)g(y), \quad y(\xi) = \eta \tag{2.6}$$

under the following general hypothesis:

(H): $f(x)$ is continuous in an interval J_x; $g(y)$ is continuous in J_y; and $\xi \in J_x$, $\eta \in J_y$.

Theorem 2.1 Let η be an interior point of J_y with $g(y) \neq 0$ and let (H) hold true. Then there exists a neighborhood of ξ in which the initial value problem (2.6) has a unique solution $y(x)$. The solution can be obtained from Equation 2.5 by solving for y.

2.1.1.2 Homogeneous Differential Equation: $y' = f(y/x)$

Using the substitution $u(x) = y(x)/x$ $(x \neq 0)$ and calculating the derivative, one obtains a differential equation for $u(x)$ with separated variables,

$$u' = \frac{f(u) - u}{x} \tag{2.7}$$

To this end, one can see immediately that every solution $u(x)$ of (2.7) leads to a solution $y(x) = x \cdot u(x)$ of the given differential equation.

2.1.1.3 Fractional Homogeneous Differential Equation: $y' = f\left(\dfrac{ax + by + c}{\alpha x + \beta y + \gamma}\right)$

In the case when the determinant $\begin{vmatrix} a & b \\ \alpha & \beta \end{vmatrix} = 0$, that is, where $a = \lambda\alpha$ and $b = \lambda\beta$, the equations can be reduced to one of the types we have already considered. If this determinant is not zero, then the linear system of equations

$$ax + by + c = 0$$
$$\alpha x + \beta y + \gamma = 0 \tag{2.8}$$

has a unique solution (x_0, y_0). If a new system of coordinates $(\bar{x}, \bar{y})$ is introduced by translating the origin to the point (x_0, y_0),

$$\bar{x} = x - x_0, \quad \bar{y} = y - y_0$$

Then, in the new coordinate system, we have the following differential equation

$$\frac{d\bar{y}}{d\bar{x}} = f\left(\frac{a + b\bar{y}/\bar{x}}{\alpha + \beta\bar{y}/\bar{x}}\right)$$

This equation is just the special case $c = \gamma = 0$ of the original equation. It is homogeneous and can be solved using the techniques mentioned in Section 2.1.1.2.

Example 2.1: Population Growth Model

Let $y(t)$ be the size of a population at time t. If the relative growth rate of the population per time unit is denoted by $c = c(t, y)$, then we have $y' = cy$. In any ecological system, the resources available to support a certain form of life are always limited; and this in turn places a limit on the size of the population that can be sustained in the system. The number N denotes the size of the largest population that can be supported by the system and is called the carrying capacity of the ecosystem. We consider a sequence of three single-population models that incorporate the following assumptions: The relative population growth rate depends only on y (that is to say, not explicitly on t) and goes to zero as the population approaches N. In particular, we assume that $c = c(y)$ is given by one of the following:

$$c(y) = \alpha(N - y)^k \quad k = 0, 1, 2$$

To illustrate these ideas, we will model the human population of the earth and choose the year 1969 as the starting point ($t = 0$, with t measured in years). Let y_0 denote the population of the earth in the year 1969 and c_0 the relative annual population growth rate for the year 1969. These are given by $y_0 = 3.55 \cdot 10^9$ and $c_0 = 0.02$. From the condition $c(0) = c_0$, it follows that $\alpha = c_0(N - y_0)^{-k}$.

If we measure $y(t)$ in multiples of y_0, that is, we set

$$y(t) = y_0 u(t), \quad N = \beta y_0$$

where β gives the carrying capacity in multiples of y_0, then we obtain the initial value problem

$$u' = c_0 \left(\frac{\beta - u}{\beta - 1} \right)^k \cdot u, \quad u(0) = 1 \ (k = 0,1,2) \tag{2.9}$$

If $k = 0$, Equation 2.9 reduces to the equation $u' = c_0 u$, which produces the well-known exponential growth function $u(t) = e^{c_0 t}$. For the other cases,

$$k = 1: \quad c_0 t = (\beta - 1)\int_1^u \frac{ds}{s(\beta - s)} = \frac{\beta - 1}{\beta} \log\left(\frac{(\beta - 1)u}{\beta - u} \right)$$

$$k = 2: \quad c_0 t = (\beta - 1)^2 \int_1^u \frac{ds}{s(\beta - s)^2} = \left(\frac{\beta - 1}{\beta} \right)^2 \left\{ \log\left(\frac{(\beta - 1)u}{\beta - u} \right) + \frac{\beta}{\beta - u} - \frac{\beta}{\beta - 1} \right\}$$

Solving these equations for u is easy when $k = 1$ but difficult when $k = 2$. However, for many questions, solving for u is not necessary. For instance, we can calculate the doubling time by putting $u = 2$. For the case of $k = 0$, the population of the year 1969 doubles in $50 \cdot \log 2 = 34.7$ years, and if $\beta = 5$ is used, it doubles in $50 \cdot (4/5) \cdot (8/3) = 39.2$ years in the case $k = 1$ and in 44.8 years for $k = 2$.

2.1.1.4 The Logistic Equation

The equation with $k = 1$ is called the logistic equation. It was initially proposed in as early as 1838 by the Belgian mathematician Pierre Francois Verhulst (1804–1849). We will consider this equation in more detail by using different notations:

$$u' = u(b - cu) \quad \text{for } b, c > 0 \tag{2.10}$$

Using the methods described in formula (2.4), one obtains the solutions

$$u_\gamma = \frac{b}{c} \cdot \frac{1}{1 + \gamma e^{-bt}} \quad \text{for } \gamma \neq 0 \tag{2.11}$$

as well as two stationary solutions $u \equiv 0$ and $u \equiv b/c$ (the reader should check this). These are all the solutions. On the one hand, every initial condition $u(t_0) = u_0$ can be satisfied by one of these solutions. On the other hand, exactly one solution goes through each point.

There is a simple result shown below.

Theorem 2.2 (a) Every solution u of (2.10) with $u(t_0) > 0$ remains positive for $t > t_0$ and tends to be b/c as $t \to \infty$. (b) $u''_\gamma = 0$ if and only if $u_\gamma = b/2c$.

In population models, u_γ with $\gamma > 0$ describes the growth of the population, and b/c is the carrying capacity β. We will now check how the world population $y(t) = y_0 u(t)$ has grown since 1969 ($t = 0$) according to this model. Recall that $c_0 = 0.02$. We have $u_\gamma(0) = 1$, and we obtain $\gamma = \beta - 1$ from (2.11). Under the assumption $\beta = 3$ ($b = 0.03$), one obtains a population of $y = 5.157$ billion for the year 1990 ($t = 21$) and with $\beta = 5$ ($b = 0.025$) $y = 5.273$ billion. The actual population size in 1990 was 5.321 billion. The assumption $\beta = 5$ gives a better approximation, though $\beta = 3$ corresponds to the carrying capacity $N = y_0 \beta \approx 10$ billion, which is sometimes used by demographers.

The point of inflection marks the turning point where the second derivative becomes negative, and hence the point beyond which the yearly population growth rate begins to decrease. It occurs in this model at $\beta/2$ according to (b). Applying this result to the world population under the assumption that $\beta = 3$ would mean that we have already passed this point ($N/2 = y_0 \beta/2 \approx 5$ billion). The situation fits $\beta = 5$ better. One should, however, not forget that we are dealing with the simplest growth model with bounded growth.

2.1.1.5 Homogeneous Equation: $Ly := y' + g(x)y = 0$

This is an equation with separated variables that can be solved by using the separated variables techniques discussed in Section 2.1.1.1. From formula (2.4) we obtain the family of solutions

$$y(x;c) = c \cdot e^{-G(x)} \quad \text{with } G(x) = \int_\xi^x g(t)dt \, (\xi \in J \text{ fixed}) \tag{2.12}$$

Notice that every solution is given by (2.14) and can also be verified directly: If ϕ is a solution of $L\phi = 0$ and $u(x) := e^{G(x)}\phi(x)$, then $u' = e^{G(x)}(g\phi + \phi') = 0$; that is, u is constant and, hence, ϕ has the form (2.12).

It exists in all of J, the unique solution satisfying the initial condition $y(\xi) = \eta$ is given by

$$y(x) = \eta \cdot e^{-G(x)} \quad \text{with } G(x) = \int_{\xi}^{x} g(t)dt \tag{2.13}$$

2.1.1.6 Nonhomogeneous Equation: Ly=h(x)

Solutions to the nonhomogeneous equation can be obtained with the help of an ansatz that goes back to Lagrange, the method of variation of constants. In this method, the constant C in the general solution $y(x; C) = Ce^{-G(x)}$ of the homogeneous equation is replaced by a function $C(x)$. The calculation of an appropriate choice of $C(x)$ gives a solution of the nonhomogeneous equation. Indeed, the ansatz

$$y(x) = C(x)e^{-G(x)} \quad \text{with } G(x) = \int_{\xi}^{x} g(t)dt$$

leads to

$$Ly \equiv y' + gy = (C' - gC + gC)e^{-G(x)} = C'e^{-G(x)}$$

Hence, $Ly = h$ holds if and only if

$$C(x) = \int_{\xi}^{x} h(t)e^{G(t)}dt + C_0 \tag{2.14}$$

Theorem 2.3 If the functions $g(x)$, $h(x)$ are continuous in J and $\xi \in J$, then the initial value problem

$$Ly = y' + g(x)y = h(x), \quad y(\xi) = \eta \tag{2.15}$$

has exactly one solution:

$$y(x) = \eta \cdot e^{-G(x)} + e^{-G(x)} \int_{\xi}^{x} h(t)e^{G(t)}dt \tag{2.16}$$

2.1.1.7 Bernoulli's Equation: $y' + g(x)y + h(x)y^\alpha = 0$, $\alpha \neq 1$

This differential equation, named after Jacob Bernoulli (1654–1705), can be transformed into a linear equation. To this end, let $z = y^{1-\alpha}$. Then the given equation becomes a linear differential equation:

$$z' + (1-\alpha)g(x)z + (1-\alpha)h(x) = 0 \qquad (2.17)$$

Equation 2.17 can then be solved by using Theorem 2.3. Conversely, if $z(x)$ is a positive solution of (2.12), then the function $y(x) = (z(x))^{1/(1-\alpha)}$ is a positive solution of Bernoulli's differential equation. For $\eta > 0$, the initial condition $y(\xi) = \eta$ transforms into $z(\xi) = \eta^{1-\alpha} > 0$. By Theorem 2.1, this condition uniquely defines a solution z of (2.12). Hence, each initial value problem for the Bernoulli equation with a positive initial value at the point ξ is uniquely solvable.

2.1.1.8 Riccati's Equation: $y' + g(x)y + h(x)y^2 = k(x)$

In this equation, which is named after the Italian mathematician Jacopo Francesco Riccati (1676–1754), the functions $g(x)$, $h(x)$, $k(x)$ are assumed to be continuous in an interval J. Except in special instances, the solutions cannot be given in any closed form. However, if one solution is known, then the remaining solutions can be explicitly calculated. To this end, let us consider the difference of two solutions y and ϕ, $u(x) = y(x) - \phi(x)$. It satisfies the linear differential equation

$$u' + gu + h(y^2 - \phi^2) = 0$$

Since $y^2 - \phi^2 = (y - \phi)(y + \phi) = u(u + 2\phi)$, we have

$$u' = [g(x) + 2\phi(x)h(x)]u + h(x)u^2 = 0$$

Thus, the difference satisfies a Bernoulli differential equation, which can be converted by using the techniques described in Section 2.1.1.3 into the following linear differential equation

$$z' - [g(x) + 2\phi(x)h(x)]z = h(x), \quad \text{where } z(x) = \frac{1}{u(x)} \qquad (2.18)$$

Equation 2.18 can then be solved by using Theorem 2.3.

To summarize, if a solution $\phi(x)$ of the Riccati's equation is known, then all of the other solutions can be obtained in the form

$$y(x) = \phi(x) + \frac{1}{z(x)} \tag{2.19}$$

where $z(x)$ is an arbitrary solution of the linear equation (2.18).

2.1.2 Systems with Constant Coefficients

In this section, we suppose that $A = (a_{ij})$ in the following homogeneous linear system

$$y' = Ay \tag{2.20}$$

is a constant complex matrix. Solutions can be obtained by using the ansatz

$$y(t) = c \cdot e^{\lambda t} = \begin{pmatrix} c_1 e^{\lambda t} \\ \vdots \\ c_n e^{\lambda t} \end{pmatrix} \tag{2.21}$$

where λ and c_i are complex constants. Substituting $y = c \cdot e^{\lambda t}$ into (2.20) leads to $y' = \lambda c e^{\lambda t} = A c e^{\lambda t}$; that is, $y(t)$ is a solution of (2.20) if and only if

$$Ac = \lambda c \tag{2.22}$$

Definition 2.6 (Eigenvalues and Eigenvectors). A vector c ($c \neq 0$) that satisfies Equation 2.22 is called an eigenvector of the matrix A; the number λ is called the eigenvalue of A corresponding to c.

We recall a couple of facts from linear algebra. Equation 2.22 or what amounts to the same thing

$$(A - \lambda I)c = 0 \tag{2.23}$$

is a linear homogeneous system of equations for c. This system has a nontrivial solution if and only if

$$\det(A - \lambda I) = \begin{vmatrix} a_{11} - \lambda & a_{12} & \cdots & a_{1n} \\ a_{21} & a_{22} - \lambda & \cdots & a_{2n} \\ \vdots & \vdots & \ddots & \vdots \\ a_{n1} & a_{n2} & \cdots & a_{nn} - \lambda \end{vmatrix} = 0 \tag{2.24}$$

In other words, the eigenvalues of A are the zeros of the polynomial $P_n(\lambda) = \det(A - \lambda I)$, called the characteristic polynomial. This polynomial is of degree n.

Thus, it has n (real or complex) zeros, where each zero is counted according to its multiplicity. An eigenvector $c \neq 0$ corresponding to a zero λ (an eigenvector $\neq 0$ by definition) is obtained by solving the system (2.23). It is determined only up to a multiplicative constant. The set $\sigma(A)$ of eigenvalues is called the spectrum of A.

Theorem 2.4 (Complex Case). The function (λ, c, A *complex*, $c \neq 0$) $y(t) = c \cdot e^{\lambda t}$ is a solution of Equation 2.20 if and only if λ is an eigenvalue of the matrix A and c is a corresponding eigenvector.

The solutions $y_i(t) = e^{\lambda_i t} c_i$ ($i = 1, \ldots, p$) are linearly independent if and only if the vectors c_i are linearly independent. In particular, they are linearly independent if all eigenvalues $\lambda_1, \ldots, \lambda_p$ are distinct. Thus, if A has n linearly independent eigenvectors, then the system obtained in this manner is a fundamental system of solutions.

Theorem 2.5 (Real Case). If $\lambda = \mu + iv$ ($v \neq 0$) is a complex eigenvalue of a real matrix A and $c = a + ib$ is a corresponding eigenvector, then the complex solution $y = c e^{\lambda t}$ produces two real solutions:

$$u(t) = \operatorname{Re} y = e^{\mu t} \{a \cos vt - b \sin vt\}$$

$$v(t) = \operatorname{Im} y = e^{\mu t} \{a \sin vt + b \cos vt\}$$

Suppose that there are $2p$ distinct, nonreal eigenvalues $\lambda_1, \ldots, \lambda_p$; $\lambda_{p+1} = \overline{\lambda}_1, \ldots, \lambda_{2p} = \overline{\lambda}_p$ and q distinct real eigenvalues $\lambda_i (i = 2p + 1, \ldots, 2p + q)$. If for the $2p$ distinct, nonreal eigenvalues, one constructs $2p$ real solutions

$$u_i = \operatorname{Re} c_i e^{\lambda_i t}, \quad v_i = \operatorname{Im} c_i e^{\lambda_i t} \quad (i = 1, \ldots, p)$$

in the manner described earlier and q real solutions y_i corresponding to the q distinct real eigenvalues by using (2.22), then the resulting $2p + q$ solutions are linearly independent. A corresponding result also holds if some of the λ_i are equal, that is, if there are eigenvalues of multiplicity greater than 1, if the $2p + q$ real solutions obtained after splitting into real and imaginary parts. In particular, if A has n linearly independent eigenvectors, then one obtains a real fundamental system.

The independence of these real solutions follows from the fact that the original solutions $y_i = c_i e^{\lambda_i t}$ ($i = 1, \ldots, 2p + q$) are linearly independent by Theorem 2.4 and can be represented as linear combinations of the real solutions.

Real systems for $n = 2$. We consider the following real system for $y = (x, y)^T$

$$\begin{pmatrix} x \\ y \end{pmatrix}' = A \begin{pmatrix} x \\ y \end{pmatrix}, \quad A = \begin{pmatrix} a_{11} & a_{12} \\ a_{21} & a_{22} \end{pmatrix}$$

under the assumption that $D = \det A \neq 0$. This implies that $\lambda = 0$ is not an eigenvalue. The corresponding characteristic polynomial

$$P(\lambda) = \det(A - \lambda I) = \lambda^2 - S\lambda + D \quad \text{with } S = trA = a_{11} + a_{22}$$

has characteristic roots

$$\lambda = \frac{1}{2}(S - \sqrt{S^2 - 4D}), \quad \mu = \frac{1}{2}(S + \sqrt{S^2 - 4D})$$

2.1.2.1 Real Normal Forms

Our first goal is to show that every real system can be reduced by means of a real affine transformation to one of the following normal forms:

$$R(\lambda, \mu) = \begin{pmatrix} \lambda & 0 \\ 0 & \mu \end{pmatrix}, \quad R_a(\lambda) = \begin{pmatrix} \lambda & 1 \\ 0 & \lambda \end{pmatrix}, \quad \kappa(\alpha, \omega) = \begin{pmatrix} \alpha & \omega \\ -\omega & \alpha \end{pmatrix}$$

where λ, μ, α, ω are real numbers with $\mu \neq 0$ and $\omega > 0$. If $S^2 > 4D$, we have the real case (R) or (R_a), which occurs depending on whether $\lambda = \mu$ has two linearly independent eigenvectors. We construct now the affine transformation C.

Case (R): There are two (real) eigenvectors c, d with $Ac = \lambda c$, $Ad = \mu d$. If $C = (c, d)$, then $C^{-1}AC = R(\lambda, \mu)$.

Case (R_a): We have $\lambda = \mu$ and only one eigenvector c. However, as is shown in linear algebra, there is a vector d linearly independent of c such that $(A - \lambda I)d = c$. The matrix $C = (c, d)$ again satisfies $C^{-1}AC = R_a(\lambda)$.

Case (κ): $\mu = \bar{\lambda}$; hence $Ac = \lambda c$ and $A\bar{c} = \bar{\lambda}\bar{c}$. The matrix $(c, \bar{c})$ transforms the system to the normal form $B = diag(\lambda, \bar{\lambda})$. However, we want to find a real normal form. This can be obtained as follows.

Let $c = a + ib$, $\lambda = \alpha + i\omega$ with $\omega > 0$. Separating the equation $Ac = \lambda c$ into real and imaginary parts leads to

$$\left.\begin{array}{l} Aa = \alpha a - \omega b \\ Ab = \alpha b + \omega a \end{array}\right\} \Leftrightarrow A(a,b) = (Aa, Ab) = (a,b)\begin{pmatrix} \alpha & \omega \\ -\omega & \alpha \end{pmatrix}$$

Since $c, \bar{c}$ are linearly independent and can be represented in terms of a, b, it follows that a and b are also linearly independent; that is, the matrix $C = (a, b)$ is regular and transforms the system to the real form $\kappa(\alpha, \omega)$.

We investigate now each of these cases and construct a phase portrait of the differential equation, from which the global behavior of the solutions can be seen.

1. $A = R(\lambda, \mu)$ with $\lambda \le \mu < 0$. The solutions of the system $x' = \lambda x, y' = \mu y$ are given by $(x(t), y(t)) = (ae^{\lambda t}, be^{\mu t})(a, b$ real), their trajectories by

$$\left(\frac{x}{a}\right)^{\mu} = \left(\frac{y}{b}\right)^{\lambda} \quad (a, b \ne 0 \text{ with } x/a, y/b > 0)$$

The special cases $a = 0$ or $b = 0$ are simple. All solutions tend to 0 as $t \to \infty$. In the case $\lambda = \mu$, the trajectories are half-lines, and in the general case, corresponding power curves. The origin is called a (stable) node.

2. $A = R_a(\lambda)$ with $\lambda < 0$. From $x' = \lambda x + y, y' = y$, one obtains

$$x(t) = ae^{\lambda t} + bte^{\lambda t}, \quad y(t) = be^{\lambda t}$$

For $a = 0$ (this means that $(x(0), y(0)) = (0, b)$), we have $x = ty$ and $\lambda t = \log(y/b)$. Thus, the trajectories are given by

$$\lambda x = y \log\left(\frac{y}{b}\right) \quad \text{for } b \ne 0 \left(\text{with } \frac{y}{b} > 0\right)$$

The positive and negative x-axes are also trajectories. Here, too, all solutions tend to the origin as $t \to \infty$, which is again called a (stable) node.

3. $A = R(\lambda, \mu)$ with $\lambda < 0 < \mu$. The solutions and their trajectories are determined formally as in (a). However, the phase portrait has a completely different appearance. There are only two trajectories that point toward the origin ($b = 0$). All of the other solutions (with $(a, b) \ne 0$) tend to infinity; $x^2(t) + y^2(t) \to \infty$ as $t \to \infty$. The origin is called a saddle point.

4. $A = \kappa(\alpha, \omega)$ with $\alpha \le 0$. It is easy to check that

$$(x_1, y_1) = e^{\alpha t}(\cos \omega t, -\sin \omega t), \quad \text{and} \quad (x_2, y_2) = e^{\alpha t}(\sin \omega t, \cos \omega t)$$

are two solutions of (2.20), from which one can construct a fundamental system $X(t)$ with $X(0) = I$. Using complex notation, in which complex numbers are identified with pairs of real numbers, these solutions are represented by $z_1(t) = e^{\alpha t}e^{-i\omega t}$ and $z_2(t) = iz_1(t)$. In this notation, the form of the trajectories can be read off.

If $\alpha = 0$, the trajectory is a circle around the origin; it is traced out in the negative sense with circular frequency ω. If $\alpha < 0$, then an additional factor $e^{\alpha t}$ is included and the trajectories are spirals that approach the origin. In the case $\alpha = 0$, the origin is called a center and in the case $\alpha < 0$ a (stable) vortex.

5. Switching from t to $-t$. If (x, y) is a solution of (2.20), then the pair of functions $\bar{x} = x(-t), \bar{y} = y(-t)$ is a solution of a related equation, in which A is replaced by $-A$. The functions (x, y) and $(\bar{x}, \bar{y})$ have the same trajectories, only the direction of the arrows is reversed. This takes care of all possible cases.

6. *Summary.* Table 2.1 summarizes the properties of the origin in various cases.

Table 2.1 Classification of the Critical Point of Linear Systems

Eigenvalues	Type of Critical Point
$\lambda_1 < \lambda_2 < 0$	Asymptotically table node
$\lambda_1 = \lambda_2 < 0$	Stable node or focus
$\lambda_1 < 0 < \lambda_2$	Unstable saddle point
$\lambda_1 = \lambda_2 > 0$	Unstable node or focus
$\lambda_1 > \lambda_2 > 0$	Unstable node
$\lambda_1 < \lambda_2 = a \pm bi\ (a < 0)$	Stable focus
$\lambda_1 < \lambda_2 = a \pm bi\ (a > 0)$	Unstable focus
$\lambda_1, \lambda_2 = \pm bi$	Unstable or stable, center or focus

Note: The entries in brackets will be explained in Section 2.1.2.2.

2.1.2.2 Stability

We consider the following homogeneous linear system

$$y' = A(t)y \quad \text{in} \quad J = [a, +\infty) \tag{2.25}$$

and assume that $A(t)$ is continuous in J. The zero solution $y(t) \equiv 0$ is called stable if all solutions are bounded in J, asymptotically stable if every solution tends to 0 as $t \to 0$, and unstable if there exists a solution that is unbounded in J.

If the zero solution is stable and if $X(t)$ denotes the fundamental system with $X(a) = I$, then there exists $\kappa > 0$ such that $|X(t)| \le \kappa$ for $t \ge a$. If y is the solution with the initial value $y(a) = c$, then $|y(t)| = |X(t)c| \le \kappa|c|$. Thus, stability means, roughly speaking, that solutions with small initial values remain small for all future time. The differential equation $y' = A(t)y$ is also sometimes called stable, asymptotically stable, or unstable, when the zero solution is stable, asymptotically stable, or unstable, respectively.

If the matrix A is constant, then Table 2.2 gives us a complete description of the stability behavior of the zero solution.

Table 2.2 Stability of the Zero Solution

Asymptotically stable	If Re $\lambda < 0$ for all $\lambda \in \sigma(A)$
Stable	If Re ≤ 0 for $\lambda \in \sigma(A)$ and if $m'(\lambda) = m(\lambda)$ for all eigenvalues λ with Re $\lambda = 0$
Unstable	In all other cases, that is, if there exists a $\lambda \in \sigma(A)$ with Re $\lambda > 0$ or Re $\lambda = 0$ and $m'(\lambda) < m(\lambda)$

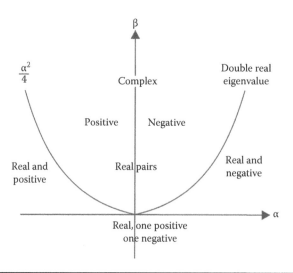

Figure 2.1   Stability of the zero solution.

In the first two cases, there exists a fundamental matrix $Y(t)$, which tends to 0 as $t \to 0$ or remains bounded, respectively. The same holds true then for an arbitrary solution, since every solution can be represented in the form $y(t) = Y(t)Y^{-1}(a)y(a)$. In the third case, there exists an unbounded solution $y = ce^{\lambda t}$ with Re $\lambda > 0$ or $y = p(t)e^{i\omega t}$ with real ω and degree $p \geq 1$ (Figure 2.1).

2.2 Dynamic Path of Nonperforming Loans: First-Order ODE

2.2.1 Introduction

At the moment, the trend of economic globalization and financial market integration is strong and unstoppable, capital flows play an increasingly important role in the international economy. Therefore, the importance of realizing open capital accounts, which is to achieve freedom of capital movement in the international economy, cannot be ignored. Firstly, capital account liberalization can increase the efficiency of resource allocation and strengthen domestic capital accumulation, which can in turn increase economic growth.

From a national perspective, in developed countries, marginal product of capital is low, while it is the high marginal product of capital of developing countries that encourages the capital in developed countries to flow into developing countries. These flows can improve both the output of the capital and the capacity of resource utilization, leading to consequent GDP growth. At the same time, foreign capital investment will increase the domestic capital stock and accelerate the domestic capital accumulation. This is a root cause for developing countries to open their capital accounts. Secondly, capital account liberalization can increase the number of investment options for the domestic residents. They can invest in overseas capital markets, and diverse risk and increase earnings. Finally, capital account openness enhances restraint on domestic monetary policy and fiscal policy so that they can no longer make intermediation of funds on their own way of inflation. However, because fiscal and monetary policies are still immature to a certain extent, that leads to financial market volatility. When such fluctuations in the environment of financial transmissions are built over a long time, they will evolve into turbulences in financial markets and the number of nonperforming bank loans will rise, and turbulences in financial markets provide a good breeding ground for financial crises to break out. When induced by individual sporadic events, the bank closures may occur anywhere, thus triggering the breakout of financial crises. Therefore, it is necessary to study the factors of the nonperforming bank loans and the dynamics of how these loans could potentially accumulate in-depth when a potential financial crisis is still in its bud.

Capital account liberalization, increasing instability in the financial system, is likely to increase the ratio of nonperforming loans. Capital account liberalization, in a variety of expectations, the flow of short-term capital into the bank, can quickly lead to a comfortable liquidity of banks. When supervision is insufficient, banks will be driven by the interests of compulsively generated loans, and even to the poor efficiency, bad credit business loans, and to high-risk industries. Such situation can easily result in a decline in the quality of credit assets so that asset bubbles are inflated. Once in the scenario of a reversed capital movement, from the initial inflow to outflow, the asset bubbles will burst, producing a large number of nonperforming loans and a security crisis of the banking system. Thus, in the process of capital account liberalization, studying the factors that affect the nonperforming bank loans is also very important.

2.2.2 Hypothesis

In this section, L represents the total number of loans, N that of nonperforming loans. On one hand, the total balance of L loans, making the consumption C required by the production of relevant goods and services, is made possible by increased consumption utility. On the other hand, nonperforming loans (with N) that are representative of a nation is rather a *financial pollution*. It produces a negative effect. Assume that the role and the form of the available functions are expressed as follows:

$$U = C(L) + P(N).$$

Namely, the total utility function of society is:

$$U = C^{\eta} - vN^{m} = uL^{\theta} - vN^{m}$$

where, $0<\theta<1$, $0<\eta<1$, $v>0$. Total number L of loans will produce a positive effect uL^θ, and the nonperforming loans have a negative effect $-vN^m$.

Because $u>0$, $L>0$, $(\partial U/\partial L)=u\theta L^{\theta-1}>0$, $(\partial^2 U/\partial L^2)=u\theta(\theta-1)L^{\theta-1}<0$. It means that the marginal utility of the total lending L is greater than 0, the marginal utility of total number of loans L is diminishing, that is, the utility U of the society is a concave function of total number L of loans.

Because $m>1$, $N>0$, $(\partial U/\partial N)=-vmN^{m-1}<0$, $(\partial^2 U/\partial N^2)=-vm(m-1)N^{m-2}<0$. That is, the marginal utility of the nonperforming loans N is less than 0, the marginal utility of the nonperforming loans is diminishing. That is, the utility U of the society is a concave function of the nonperforming loans N. For simplicity, this presentation takes the following two assumptions.

H_1 (positive factor): M_p indicates that the positive factors, which affect the nonperforming loans, mainly include two important factors: The first one is a good system of property rights. By strengthening the bank's efforts of internal management, the property rights system, which improves the level of nonperforming loans, can affect the rate of change of size, and by strengthening the bank's efforts of internal management, the rate of change over time $\dot{N}$ of the number of nonperforming loans N is reduced.

The second assumption is about the nonperforming loans. In particular, $(\partial \dot{N}/\partial N)<0$. It means that as the number N of nonperforming loans increases, the rate $\dot{N}$ of change over time decreases.

H_2 (negative factor): M_n indicates that the negative factors that affect the nonperforming loans include mainly the following five key factors.

The first is the extent D of opening up, satisfying $(\partial \dot{N}/\partial D)>0$. Its means: The rate of change over time of nonperforming loans N is an increasing function of D. The greater D is, the bigger $\dot{N}$ is. When the extent D of opening up is greater, the higher possibility the *financial virus* from abroad has access to the country's financial system; the more possible domestic enterprises become vulnerable to shocks and to face greater foreign competitions; and the more uncertain economic factors are, the more money enterprises may lose. Hence, $\dot{N}$ is greater. The second is about the total number of loans L. Here, $\dot{N}$ is an increasing function of L. The greater L is, the greater $\dot{N}$ is. The third is about $i-r$, which is the difference of the nominal interest rate and the actual earnings of business. In particular, it means that $\dot{N}$ is an increasing function of $i-r$. The greater $i-r$ is, the higher value $\dot{N}$ takes. The nominal interest rate represents the cost of financing. So, the higher the cost of financing is, or the lower the real return of corporates is, the greater the difference $i-r$ becomes. That is, the greater the scale of loss is, the higher value $\dot{N}$ takes. The fourth one is about $|\Delta G|$, which is on the absolute value of the economic growth rate. It means that our discussion is divided into two possible situations:

1. Suppose that economic growth rate satisfies $\Delta G \gg 0$. That means that ΔG increases substantially. In an overheating economy, people are afraid of losing opportunities to make money so that they launch various projects using

bank loans even when the prospect of profitability is not good. That results in elevated values for $\dot{N}$

2. Suppose that the economic growth rate satisfies $\Delta G \ll 0$. This means that ΔG decreases substantially. When the economy is depressed, the supply exceeds the demand, prices fall, the sales of products stalls, and corporate profits decline. All of these problems lead to increasing values of $\dot{N}$.

The fifth is the animal spirit of entrepreneurs S, satisfying $(\partial \dot{N}/\partial S) > 0$. It means that $\dot{N}$ is an increasing function of S. The greater S is, the higher value $\dot{N}$ takes, because each increase in the animal spirit of entrepreneurs tends to lead to increasing amount of entrepreneurs' nonrational behaviors. With increasing volatility of the economic cycle (entrepreneurs' nonrational behaviors), or excessive investment, or over tightening investment, the growth rate of nonperforming loans increases.

This section ignores the change in exchange rates as caused by changes in the number of nonperforming loans, as this factor has both positive and negative influences.

On one hand, if $\Delta E > 0$, then $(\partial \dot{N}/\partial \Delta E) > 0$. That means that when the local currency devalues, the local currency-denominated foreign debts increase, which adds to the debt burden of enterprises so that the possibility for the enterprise to lose increases. Therefore, $\dot{N}$ is greater. On the other hand, if $\Delta E < 0$, then $(\partial \dot{N}/\partial \Delta E) > 0$. That means that the currency appreciates; through adjusting the ratio of foreign and domestic debts, the purpose of reducing the debt burden of corporations can be materialized so that the number of nonperforming bank loans could possibly be reduced.

2.2.3 The Model

Consider the model proposed by Zeng (2009),

$$\dot{N} = -\alpha L - \beta A - \delta N + h|\Delta G| + \gamma(i - r) + fD + aS + b\Delta E \qquad (2.26)$$

where

$\dot{N}$ is nonperforming loans
δ is the elasticity coefficient of nonperforming loans
L is the total loan balance
α is the elasticity coefficient of the total loan balance
A is the effort of strengthening the bank's internal management
β is the elasticity coefficient of the effort of strengthening the bank's internal management
$|\Delta G|$ is the absolute value of the change in economy growth
h is the elasticity coefficient of the absolute value of the change in economy growth
$i - r$ is difference of the nominal interest rate and actual earnings of business

γ is the flexibility coefficient of the difference of the nominal interest rate and actual earnings of business

D is the extent of opening-up

f is the flexibility coefficient of the opening degree. S stands for the animal spirit of entrepreneurs

a is the flexibility coefficient of the animal spirit of entrepreneurs

ΔE is the change in exchange rate

b is the flexibility coefficient of the change in exchange rate

We introduce a concept called the red line of nonperforming loans. In particular, the so-called red line stands for the maximum amount of nonperforming loans banks are allowed to provide as imposed by the government in order to maintain financial stability. Let us rewrite (2.15) as follows:

$$\dot{N} = -\delta N(\bar{N} - N) + f(M_p, M_n) \tag{2.27}$$

where M_p stands for the positive factors that affect the nonperforming loans. It includes the totality of nonperforming loans itself, the banking system of property rights, etc. And, M_n represents the negative factors that impact nonperforming loans. It includes openness, the absolute value of the magnitude of change in economic growth, the animal spirit of entrepreneurs, and so on.

2.2.4 Analysis

Case 1: $f(M_p, M_n) = -\alpha M_p(t)$

In this case, Equation 2.27 can be turned into the following form

$$\frac{dN}{N(\bar{N} - N)dt} = -\delta - \alpha M_p(t) \tag{2.28}$$

From the given initial condition $N|_{t=t_0} = N_0$, by integrating Equation 2.28, we obtain

$$\ln\frac{N}{\bar{N} - N} - \ln\frac{N_0}{\bar{N} - N_0} = -\bar{N}\delta(t - t_0) - \alpha\bar{N}\int_{t_0}^{t} M_p(t)dt$$

Hence, $N(t)$ can be solved as follows:

$$N(t) = \frac{\bar{N}\exp(-\bar{N}\delta(t - t_0))}{\exp(-\bar{N}\delta(t - t_0)) + ((\bar{N}/N_0) - 1)\exp\left(\alpha\bar{N}\int_{t_0}^{t} M_p(t)dt\right)} \tag{2.29}$$

Theorem 2.6

(i) If $\delta > 0$, then there exists limit $\lim_{t \to \infty} N(t) = 0$, and

(ii) If $\delta = 0$, then there are two situations:

    (1) If $\int_{t_0}^{t} M_p(t)dt$ divergences, then $\lim_{t \to \infty} N(t) = 0$, and

    (2) If $\int_{t_0}^{t} M_p(t)dt$ converges, then

$$\tilde{N} = \lim_{t \to \infty} N(t) = \frac{\bar{N}}{1 + ((\bar{N}/N_0) - 1)\exp\left(\alpha\bar{N}\int_{t_0}^{t} M_p(t)dt\right)}$$

It is ready to see that $0 < \tilde{N} < \bar{N}$.

Case 2: $f(M_p, M_n) = \beta M_p(t)$,

    In this case, Equation 2.27 can be turned into the following form

$$\frac{dN}{N(\bar{N} - N)dt} = -\delta + \beta M_p(t) \tag{2.30}$$

For the given initial condition $N|_{t=t_0} = N_0$, we can integrate Equation 2.30 and get its solution as follows:

$$N(t) = \frac{N_0\bar{N}\exp\left\{-\bar{N}\delta(t - t_0) + \beta\bar{N}\int_{t_0}^{t} M_n(t)dt\right\}}{(\bar{N} - N_0) + N_0\exp\left\{-\bar{N}\delta(t - t_0) + \beta\bar{N}\int_{t_0}^{t} M_n(t)dt\right\}} \tag{2.31}$$

Theorem 2.7

(i) If $\int_{t_0}^{t} M_n(t)dt$ converge, then there exists limit $\lim_{t \to \infty} N(t) = 0$, and

(ii) If $\int_{t_0}^{t} M_n(t)dt$ diverge, then there exists limit $\lim_{t \to \infty} N(t) = \bar{N}$.

2.2.5 Conclusions

From a global point of view, capital account liberalization will direct savings toward more efficient investment projects on a global scale so that it can improve the efficiency of global capital allocation. It also plays a promotion role in the financial system mainly in the following two ways. Firstly, capital account liberalization can help rapidly to expand the size of the financial system of the host country, thereby contributing to economic and financial processes. Secondly, capital account liberalization helps to reduce and even eliminate the enormous costs of capital controls in the host country. In addition, capital account liberalization can promote industrial upgrades, technological advances, and improve the overall strength of the host country's industry.

According to the aforementioned discussion and different combinations of situations, there are four possibilities:

Case 1: The integral of M_p diverges while the integral of M_n converges. So, the overall effect of the function $f(M_p, M_n)$ depends on the positive factor M_p. Therefore, after sufficient long period of continued effort, M_p will play a major role, banks' nonperforming loans can be effectively controlled.

Case 2: The integral of M_p converges while the integral of M_n diverges. The overall effect of the function $f(M_p, M_n)$ depends on the negative factor M_n. Therefore, after a sufficient long period of time, banks' nonperforming loans encroaches the red line, which indicates that a financial crisis could erupt at any time moment.

Case 3: Both of the integrals of M_p and M_n are convergent. Then the overall effect of the function $f(M_p, M_n)$ is similar to what is described in Theorem 2.7 (2) (ii). That is, after sufficient long period of evolution, the number of banks' nonperforming loans can be stabilized at a certain level below the red line. The particular level of stabilization depends on the intensities of the positive and negative factors.

Case 4: Both of the integrals of M_p and M_n are divergent. The overall effect of the function $f(M_p, M_n)$ will appear similar to the findings and conclusions of the two scenarios in a complex situation, which is needed to be investigated further.

To sum up, when positive factors are dominant, it is beneficial for banks to reduce their nonperforming loans. That is, in terms of the factors that dominate the strengthening of a bank's internal management and the reform of the property rights system, the number of nonperforming loans itself, namely, A, N plays a dominant role. It can reduce the number of the bank's bad loans; the bank's control of nonperforming loans is between zero and the red line of the bank's nonperforming loans. The upper limit of the number of nonperforming loans will not exceed the red line. However, if negative aspects dominate banks, that is, a number of macroeconomic environments deteriorate (such as the openness D, change in the absolute magnitude of economic growth rate $|\Delta G|$, animal spirit of entrepreneurs S, the change ΔE of the exchange rate will increase the number of nonperforming loans), the level of deterioration exceeds the bank's internal management and the reform effort of the property rights system, its positive effect will be weakened, the situation of nonperforming loans becomes difficult to improve, although the bank's internal management is further strengthened. Thus, the number of the bank's nonperforming loans at the macro environment is determined by the microeconomic environment and the internal control mechanism. If good microeconomic factors dominate, the number of non-performing loans will decrease. But, if a vicious macroeconomic environment dominates, the number of nonperforming loans will increase.

Therefore, under the condition of capital account liberalization, to reduce the number of a bank's nonperforming loans, first, one needs to strengthen the bank's internal management and property rights system. Secondly, it needs to strengthen the exchange rate, improve the forecast of interest rates, and accelerate the introduction of advanced interest rate, exchange rate risk measurement models, and internal control mechanism. Thirdly, it needs to design a series of financial derivative instruments as soon as possible so that banks can choose their particular ways to mitigate exchange rate risk.

2.3 Stock Market's Liquidity Risk: Second-Order ODE

2.3.1 Introduction

On August 21, 1998, Long Term Capital Management (LTCM) announced that "the company is on the verge of bankruptcy," eventually resulting in a liquidity crisis in the global financial market. Afterward, regulators and the financial industry reflected on this crisis, found that one of the causes of the crisis is that the high profit LTCM generated in the early years was achieved by taking on a higher level of liquidity risk. As one of the largest hedge fund companies at that time, LTCM used a lot of leverage in order to maximize its profit. A vicious cycle was generated by its aggressive measure. Firstly, when the margin is required to be increased with additional capital, they had to liquidate their positions as soon as possible with losses. However, as the liquidity is insufficient in the traded assets and the markets, its trading behavior affected the assets' prices, led to negative changes in the assets' values, and further produced additional requirements for margin. That worsened the liquidity risks of the assets and the liabilities the company faced.

Based on how different investors influence different regions of the market, the liquidity risk can be divided into endogenous and exogenous liquidity risks; the comparison between them is listed as Table 2.3.

It can be seen from Table 2.3 that the investors who hold assets of higher liquidity will face lower endogenous liquidity risk and those who hold a bigger position size will face higher endogenous liquidity risk. The main object of this section is the endogenous liquidity risk of the investors in the securities market.

For a long period of time, an increasing number of scholars have noticed the existence of liquidity risk. It is believed that because the market does not

Table 2.3   Comparison of Endogenous and Exogenous Liquidity Risk

Risk Category	Exogenous Liquidity Risk	Endogenous Liquidity Risk
Influence factor	All the factors that affect the overall market liquidity	Investors' individual factors
Region of influence	All the investors throughout the market	Individual investors in the market
Avoidance measure	Configurations found in different markets (stock market, bond market, money market, etc.)	Hold higher liquidity assets; possess smaller position sizes; formulate appropriate trading strategies

represent perfect flows of capital, information, etc., some assumptions of any theoretical model do not hold true in practice. By considering the management of the liquidity of a company assets, Holmstrom and Tirole (1996) proposed the CAPM based on liquidity. Cetin et al. (2003) established the arbitrage pricing theory by considering the liquidity risk. Brennan and Subrehmanyam (1996) pointed out that the stock returns are contained in the compensation of liquidity risk. The empirical test of Wu et al. (2003) on the Chinese stock market also confirmed the *illiquid compensation* hypothesis. Cetin et al. (2002) put forward the option pricing theory based on their considering the liquidity risk. Wu et al. (2004) proposed a hybrid pricing model based on the industry market and the capital market, where the gains in the stock price are decomposed into the income from the industry market and the revenue from the liquidity of the capital market, the result that combines the liquidity factor into CAPM turns out to be very good.

In this section, factors of the liquidity risk are inserted into the model in order to further improve the model's ability to explain the reality.

2.3.2 The Model

Consider that an investor holds a stock position X in the initial zero moment. Due to the impact of external shocks, he holds a stock position Y after T trading days, the stock realized a gain of $X-Y$ during the holding period. At time moment t, $t \in [0, T]$, assume that the investor holds a stock position $x(t)$. For the investor with a mean-variance utility, the functional of his loss utility is

$$W[x(t)] = E[EC] + \frac{\lambda}{2}V[EC] = \frac{1}{2}\gamma(X^2 - Y^2) + \int_0^T \left[\beta v^2(t) + \frac{\lambda}{2}\sigma^2 x^2(t) \right] dt$$

where the investor's optimal strategy of variance $x^*(t)$ satisfies

$$x^*(t) = \arg\min_{x(t)} W[x(t)]$$

Let

$$J[x(t)] = \int_0^T \left[\beta v^2(t) + \frac{\lambda}{2}\sigma^2 x^2(t) \right] dt, \quad F[x(t)] = \beta v^2(t) + \frac{\lambda}{2}\sigma^2 x^2(t)$$

In order to obtain the minimum of $J[x(t)]$, $F[x(t)]$ should satisfy the following Euler equation:

$$\frac{\partial F}{\partial x} - \frac{d}{dt}\left(\frac{\partial F}{\partial \dot{x}} \right) = 0$$

From the fact that $v(t) = -(dx(t)/dt)$, it follows that x satisfies the homogeneous second-order differential equation with constant coefficients

$$\ddot{x} - \frac{\lambda \sigma^2}{2\beta} x = 0 \quad \text{and} \quad x(0) = X, x(T) = Y \tag{2.32}$$

where
  λ is the investor's risk aversion coefficient (when $\lambda > 0$, the investor is risk aversion and when $\lambda < 0$, the investor is risk preference)
  σ is the volatility
  β is the instant shock relative to the scale factor of the transaction velocity (any instant shock is a linear function of the trading velocity $\beta v(t)$)

1. When $\lambda > 0$, denote $\rho = \sqrt{\lambda \sigma^2 / 2\beta}$. Then the general solution of Equation 2.32 is

$$x(t) = \left(\frac{Y}{\sinh(\rho T)} - \frac{X}{\tanh(\rho T)} \right) \sinh(\rho t) + X \cosh(\rho t)$$

2. When $\lambda = 0$, the general solution of second-order differential equation (2.32) is

$$x(t) = (Y - X)\frac{t}{T} + X$$

3. When $\lambda < 0$, the general solution of second-order differential equation (2.32) is

$$x(t) = \left(\frac{Y}{\sin \rho T} - \frac{X}{\tan \rho T} \right) \sin \rho t + X \cos \rho t$$

2.3.3 Exogenous Shocks

Case 1: Let us consider that changes in the index of the stock market during the holding period have an impact on the investor's realized strategy. The model then is modified to

$$\ddot{x} - \rho^2 x = b e^{-\gamma t} \quad (\gamma > 0) \tag{2.33}$$

where $b > 0$ is the amplitude that the index exceeds the investor's expectation during the holding period, γ is the attenuation rate of the index's impulse impact.

$$\text{Denote } \rho_+ = \sqrt{\frac{\lambda}{2\beta}} \sigma \quad (\lambda > 0), \quad \rho_- = \sqrt{\frac{-\lambda}{2\beta}} \sigma \quad (\lambda < 0)$$

Theorem 2.8

(i) If $\lambda = 0$, then the general solution of Equation 2.33 is

$$x(t) = X + \frac{(Y - X)t}{T} + \frac{b}{\gamma^2}(e^{-\gamma t} - 1) + \frac{bt(1 - e^{-\gamma T})}{\gamma^2 T} \tag{2.34}$$

(ii) If $\lambda < 0$, then the general solution of Equation 2.33 is

$$x(t) = \left(X - \frac{b}{\gamma^2 - \rho_-^2} \right) \cos \rho_- t + \left(Y - \frac{b e^{-\gamma t}}{\gamma^2 - \rho_-^2} \right) \frac{\sin \rho_- t}{\sin \rho_- T}$$

$$- \left(X - \frac{b}{\gamma^2 - \rho_-^2} \right) \frac{\sin \rho_- t}{\tan \rho_- T} + \frac{b e^{-\gamma t}}{\gamma^2 - \rho_-^2} \tag{2.35}$$

(iii) If $\lambda < 0$, the general solution of Equation 2.33 has two different forms.
(1) When $\gamma \neq \rho$, the general solution of Equation 2.33 is

$$x(t) = \left(X - \frac{b}{\gamma^2 - \rho_+^2} \right) \cosh(\rho_+ t) - \left(X - \frac{b}{\gamma^2 - \rho_+^2} \right) \frac{\sinh(\rho_+ t)}{\tanh(\rho_+ T)}$$

$$+ \left(Y - \frac{b}{\gamma^2 - \rho_+^2} e^{-\gamma t} \right) \frac{\sinh(\rho_+ t)}{\sinh(\rho_+ T)} + \frac{b}{\gamma^2 - \rho_+^2} e^{-\gamma t} \tag{2.36}$$

(2) When $\gamma = \rho$, the general solution of Equation 2.33 is

$$x(t) = X \cosh(\rho_+ t) - X \frac{\sinh(\rho_+ t)}{\tanh(\rho_+ T)} + \left(Y - \frac{bTe^{-\gamma T}}{\gamma^2 T - \rho_+^2 T - 2\gamma} \right)$$

$$\times \frac{\sinh(\rho_+ t)}{\sinh(\rho_+ T)} + \frac{bTe^{-\gamma T}}{\gamma^2 T - \rho_+^2 T - 2\gamma} e^{-\gamma t} \tag{2.37}$$

Case 2: Let us consider that the changes in the index of the stock market during the holding period have an impact on periodic oscillation of the investor's realized strategy. Then, the model is modified to

$$\ddot{x} - \frac{\lambda\sigma^2}{2\beta} x = be^{-\gamma t} \cos\omega t \quad b > 0, \gamma > 0, \omega > 0 \tag{2.38}$$

where

$b > 0$ is the amplitude that the index exceeds the investor's expectation during the holding period

γ is the attenuation rate of the index's impulse impact

ω is the index's fluctuation cycle

$$\text{Denote } \rho_+ = \sqrt{\frac{\lambda}{2\beta}}\sigma \quad (\lambda > 0), \quad \rho_- = \sqrt{\frac{-\lambda}{2\beta}}\sigma \quad (\lambda < 0)$$

Theorem 2.9

(i) If $\lambda > 0$, this case indicates that the investor is risk averse so that the general solution of Equation 2.38 is

$$x(t) = (X - b_+ \cos\varphi_+)\cosh(\rho_+ t) + (b_+ \cos\varphi_+ - X)\frac{\sinh(\rho_+ t)}{\tanh(\rho_+ T)}$$

$$+[Y - b_+ e^{-\gamma T}\cos(\omega T + \varphi_+)]\frac{\sinh(\rho_+ t)}{\sinh(\rho_+ T)} + b_+ e^{-\gamma t}\cos(\omega t + \varphi_+) \tag{2.39}$$

where

$$b_+ = \frac{b}{\sqrt{\left(\gamma^2 - \omega^2 - \rho_+^2\right)^2 + 4\omega^2\gamma^2}}, \quad \tan\varphi_+ = \frac{2\omega\gamma}{\gamma^2 - \omega^2 - \rho_+^2}$$

(ii) If $\lambda < 0$, this case indicates that the investor is risk appetite so that the general solution of (2.38) is

$$x(t) = (X - b_- \cos\varphi_-)\cos(\rho_- t) + \left(\frac{Y - b_- e^{-\gamma t}\cos(\omega T + \varphi_-)}{\sin(\rho_- T)} - \frac{X - b_- \cos\varphi_-}{\tan(\rho_- T)} \right)$$

$$\times \sin(\rho_- t) + b_- e^{-\gamma t}\cos(\omega t + \varphi_-) \tag{2.40}$$

where

$$b_- = \frac{b}{\sqrt{\left(\gamma^2 - \omega^2 - \rho_-^2\right)^2 + 4\omega^2 \gamma^2}}, \quad \tan\varphi_- = \frac{\gamma^2 - \omega^2 + \rho_-^2}{2\omega\gamma}$$

(iii) If $\lambda = 0$, the general solution of Equation 2.38 is

$$x(t) = (Y - b_0 e^{-\gamma t}\cos(\omega t + \varphi_0) + X + b_0 \cos\varphi_0)\frac{t}{T} + X - b_0 \cos\varphi_0$$

$$+ b_0 e^{-\gamma t}\cos(\omega t + \varphi_0) \tag{2.41}$$

where

$$b_0 = \frac{b}{\sqrt{(\gamma^2 - \omega^2)^2 + 4\omega^2\gamma^2}}, \quad \tan\varphi_0 = \frac{2\omega\gamma}{\gamma^2 - \omega^2}$$

2.3.4 Numerical Example

On July 3, 2013, the State Council of China's national cabinet approved the establishment of the Shanghai Free Trade Area (FTA), where several laws regulating foreign investment and trade would not be applied. The Shanghai FTA, a brainchild of Premier Li Keqiang, will eventually become a place where foreign trades, investments and capital can freely flow in and out. Clearly, Li wants the FTA to be an *experimental field* for the country's economic future, just as Shenzhen was in the 1980s. On September 29, 2013, the Shanghai Free Trade Zone was formally established. Firstly, the construction of the zone, in fact, is China's initiative to choose a new open pilot. Its core is to use the zone to force speeding up of the economic reform in China. Therefore, the construction of the FTA will lead and promote the introduction of finance, taxation, trade, government management, and a series of reform measures. These reform measures will have a huge demonstration effect for the rest of the nation. Secondly, the FTA is an important engine to help upgrade China's economy; it will help boost foreign trade and stabilize economic development, creating a favorable environment for China's economic transformation and upgrading. And by establishing the FTA in line with the development trend of the international trade, the FTA has tremendous long-term economic growth potential. The listed companies within the jurisdiction will enjoy substantial positive impacts on operation of the trade zone. The *New York Times* evaluated the zone as follows: "Shanghai FTA provides the signal of financial reform, a bridgehead of relaxation of capital controls, and the convertibility of Yuan." The establishment of the Shanghai FTA is also a major piece of good news for the local industry, the stocks local to Shanghai, such as Waigaoqiao (600648) and Lujiazui (600663), had enjoyed strong momentum before the FTA was established with successive daily limits pulling the share prices to new highs.

The market reacted positively to the news, as shown with such Shanghai FTA-concept stocks as Waigaoqiao (600648), Lujiazui (600663), and Pudong Jinqiao (600639). After July 4, 2013, riding on the news of the favorable policies recently introduced by the government, all of the stocks related to Shanghai FTA continued to rise. Particularly for Pudong Jinqiao, there appeared unprecedented 15 consecutive daily limits. However, within the half year following that drastic rise, the share price was halved. Considering the fact that the irrational capital of short-term speculation is often affected by beneficial polices, we extend the model from Zhong (2004) to model (2.33) by using the parameter b to describe the strength of influence of the good news on this plant, while r is used to describe the duration of the influence of the good news. These two parameters are estimated by using the method of regression through fitting the closing prices of the stock.

Assume that an investor had 10 million shares of the FTA concept stock over a period of time since the good news initially appeared, and is ready to reduce the holding to 2 million within 10 trading days. That is, we have $X = 10$ (million) and $Y = 2$ (million). We can see from Figure 2.2 that when the volatility indicates little risk and the attenuation rate is high, the best strategy for the investor to employ is to reduce the number of shares he holds to 5 million in the first trading day, and then in the following 9 trading days he reduces on the average to the target holding number of 2 million shares (see Figure 2.2a). When the volatility indicates low risk and the attenuation rate is small, the best strategy for the investor to use is to reduce the shares he holds to 4 million shares in the first two trading days, and in the following 8 trading days, he reduces on the average to the target holding number of 2 million shares (see Figure 2.2b and c). When the volatility indicates high risk and the attenuation rate is small, the best strategy for the investor is to reduce the number of shares he holds to 1.5 million on the eighth trading day after a fall has appeared, and in the following 2 trading days, he increases on the average to his target holding number of 2 million shares (see Figure 2.2d).

As one can see from Figure 2.3, when the attenuation rate is high, the best strategy for the investor to consider is to reduce the number of shares he holds to 5 million in the first two trading day, and in the following 8 trading days, he reduces on the average to the target holding number to 2 million shares (see Figure 2.3a and b). When the attenuation rate is small, the best strategy for the investor is to reduce the number of shares he holds to 2 million shares on the tenth trading day after a fall has appeared (see Figure 2.3c and d).

As one can see from Figure 2.4, when the volatility indicates low risk and the attenuation rate is high, the best strategy for the investor is to reduce the number of

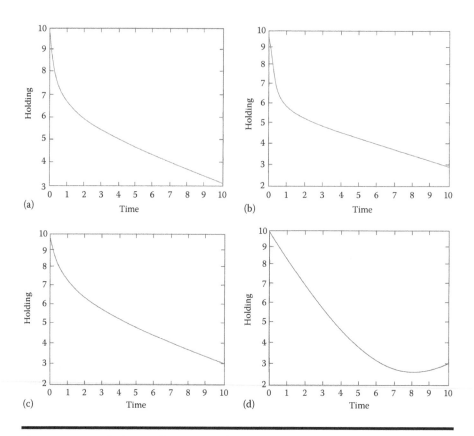

Figure 2.2 Solutions of **(2.32)**, where the investor reduces his holding **(a)** to 5 million shares on first day then to 2 million in the next 9 days; **(b and c)** to 4 million shares on first two days then to 2 million shares in the next 8 days; **(d)** to 1.5 million on the eighth day after a fall appeared, and then increases to his target holding of 2 million shares in the next 2 days.

shares he holds to 3 million in the first trading day, and in the following 9 trading days, he reduces on the average to the target holding number to 2 million shares (see Figure 2.4a). When the volatility indicates low risk and the attenuation rate is small too, the best strategy for the investor to use is to reduce the number of shares he holds to 6 million in the first two trading day, and in the following 8 trading days, he reduces on the average to the target holding number to 2 million shares (see Figure 2.4b and c). When the volatility indicates high risk and the attenuation rate is small, the best strategy for the investor is to reduce the number of shares he holds to 4 million in the first 2 trading days, in the following 5 trading days, he basically retains that position while cautiously increasing the number of shares he holds, and in the last 3 or 4 trading days, he slowly reduces to the target holding number to 2 million shares (see Figure 2.4d).

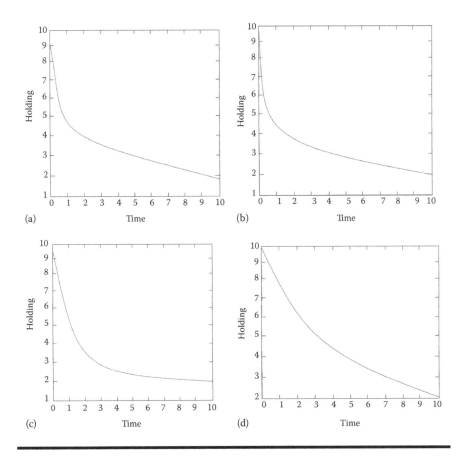

Figure 2.3 **Solutions of (2.33), where the investor reduces his holding (a and b) to 5 million in the first two days, and to 2 million shares in the next 8 days (c); to 2 million shares on the tenth day after a fall appeared (d).**

2.4 Stability of Michael Model under Capital Control: Two-Dimensional Systems (I)

2.4.1 Introduction

The research of stabilization under the condition of fixed exchange rate rule and delayed fiscal adjustment has attracted much attention. In this section, we use the qualitative theory of ODEs to discuss the equilibrium point on the two-dimensional consumption space of traded goods and real money balances by analyzing the existence condition and the stabilization of the equilibrium point.

Usually, an anti-inflation program is the implementation of an exchange rate rule without an immediate reduction in the fiscal deficit. That is why such program would render such a rule sustainable in the long run. While much of the literature

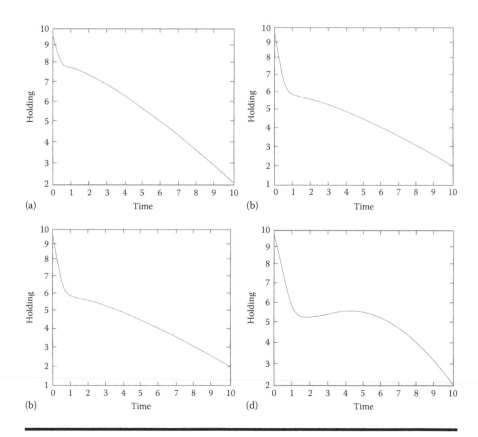

Figure 2.4 Solutions of (2.38), where the investor reduces his holding (a) to 3 million shares on first day, then to 2 million in the next 9 days; (b and c) to 6 million shares on first two days, then to 2 million in the following 8 days; (d) to 4 million in first two days, retains that position while cautiously increasing his holding in the next 5 days, and then slowly reduces to 2 million shares in last three or four days.

has focused on the effects of an anticipated abandonment of the exchange rate rule, some recent work has also considered the effects of anticipated deficit reduction policies that allow the exchange rate rule to survive. Ellis and Auernheimer (1996) discussed the stabilization under capital controls and compared the details of the results with those obtained under perfect capital mobility. This work seems to be a natural line of research, because it completes the analysis of the response to a successful stabilization program in which the components of the program proceed in stages, and addresses a far more common case.

In this section, x_T is used to represent the fixed flow of traded good income, s the flow of lump sum government transfers, T the flow of lump sum taxes, and ε the relative price of the traded good in terms of the nontraded good. This relative price is defined as the real exchange rate.

2.4.2 The Model

The model presented in this section is one developed for a small open economy with both traded and nontraded goods. The general price level is given by the price index:

$$P = E^{\sigma} P_H^{(1-\sigma)} \tag{2.42}$$

where

E stands for the nominal exchange rate

P_H is the price of the nontraded good

σ is the weight given to the traded good in the price index

From (2.41), the inflation rate π is

$$\pi = \sigma \pi_T + (1 - \sigma) \pi_H \tag{2.43}$$

where

π_H represents the rate of increase in the price of the nontraded good

π_T represents both the rate of price increase of the traded good and the rate of devaluation

The price of the nontraded good is assumed to be perfectly flexible. So the market clearing condition

$$x_H = c_H + g_H \tag{2.44}$$

holds true at all times. The output of the nontraded good is fixed and denoted as x_H; c_H stands for the private nontraded good consumption, and g_H is the government consumption of the nontraded good.

Individuals are identical, infinitely lived, and have perfect foresight. The representative individual maximizes the present value of his utility, which depends on the consumption of the two commodities and the real money stock, and it is assumed to be separable in consumption and money. The utility functional then takes the form

$$\int_0^{\infty} [u(c_T, c_H) + v(m)] e^{-\delta t} \, dt \tag{2.45}$$

where

δ is the constant rate of time preference

c_T is the consumption of the traded good

m is the balance of real money, defined as nominal money balance deflated by the price index (2.42)

It is assumed that both $u(\cdot)$ and $v(\cdot)$ are strictly concave and satisfy Inada type conditions.

The individual's flow budget constraint, expressed in terms of the traded good, is

$$x_T + \frac{(x_H + s)}{\varepsilon} = c_T + \frac{(c_H + T)}{\varepsilon} + \varepsilon^{(\sigma-1)} m\pi + \varepsilon^{(\sigma-1)} \dot{m} \qquad (2.46)$$

where

x_T stands for the fixed flow of traded good income

s is the flow of lump sum government transfers

T is the flow of lump sum taxes

ε is the relative price of the traded good in terms of the nontraded good

This relative price is defined as the real exchange rate.

Maximization of (2.45) subject to (2.46) and the initial level of m leads to the following conditions:

$$\varepsilon = \frac{u_T(c_T, c_H)}{u_H(c_T, c_H)} \qquad (2.47)$$

$$\dot{c}_T = \left(\frac{-u_T}{u_{TT}}\right)\left(u_T^{-\sigma} u_H^{(\sigma-1)} v_m - (\delta + \pi_T)\right) \qquad (2.48)$$

where Equation 2.46 determines the real exchange rate, and Equation 2.47 governs the private consumption of the traded good, and from this equation, the steady-state demand for real money balances is implicitly defined by

$$v_m(m^*) = (\delta + \pi_T) u_T \left(c_T^*, c_H^*\right)^{\sigma} u_H \left(c_T^*, c_H^*\right)^{(1-\sigma)} \qquad (2.49)$$

where * represents steady-state values.

The government budget constraint, expressed in terms of the traded good, is given by

$$g_T + \frac{g_H}{\varepsilon} + \dot{A} + \frac{s}{\varepsilon} = Ar + \varepsilon^{(\sigma-1)} m\pi + \varepsilon^{(\sigma-1)} \dot{m} + \frac{T}{\varepsilon} \qquad (2.50)$$

where

g_T stands for the government consumption of the traded good

A is the stock of net foreign assets

r is the exogenous real interest rate on foreign assets

The flow of government transfers is equal to the component of the inflation tax resulting from nontraded good price inflation:

$$s = (1 - \sigma)\pi_H \varepsilon^\sigma m \tag{2.51}$$

Equation 2.43, the fixed exchange rate rule $\pi_T = 0$, and (2.44) imply that (2.46), (2.47), and (2.48) can be rewritten as follows:

$$\varepsilon = \frac{u_T(c_T, x_H - g_H)}{u_H(c_T, x_H - g_H)} \tag{2.52}$$

$$\dot{m} = (x_T - c_T)\varepsilon^{(1-\sigma)} + (g_H - T)\varepsilon^{-\sigma} \tag{2.53}$$

and

$$\dot{c}_T = \left(\frac{-u_T}{u_{TT}}\right)\left(u_T^{-\sigma}u_H^{(\sigma-1)}v_m - \delta\right) \tag{2.54}$$

2.4.3 Stability Analysis

Next, let us discuss the existence of the equilibrium point (c_T^*, m^*) of (2.53) and (2.54). If the equilibrium point (c_T^*, m^*) exists, it satisfies the following set of equations:

$$v_m \varepsilon^{1-\sigma} + \delta u_T = 0 \tag{2.55}$$

$$(x_T - c_T)\varepsilon + g_H = T \tag{2.56}$$

This set of equations can be rewritten as

$$v_m = -\delta u_{TT} \varepsilon^{\sigma-1} \tag{2.57}$$

$$\varepsilon = \frac{T - g_H}{x_T - c_T} \tag{2.58}$$

In this section, we consider a simple situation where $\varepsilon = ax + b$. Suppose that $u(c_T, c_H) = c_T^\alpha c_H^\beta$ and $v_m = m^{\sigma-1}$. Then we get a quadratic equation with one unknown x below:

$$ax^2 + (b - ax_T)x - bx_T + (T - g_H) = 0 \qquad (2.59)$$

Let the discriminant be written $\Delta = (b + ax_T)^2 - 4a(T - g_H)$. We discuss the existence of the equilibrium point according to the signal of $T - g_H > 0$. In order to discuss the stabilization of the equilibrium point, let us consider the Taylor expansion around the steady state. By retaining only the first item, we get

$$\dot{c}_T = \frac{v_m \left[(\sigma - 1)U_T^{-\sigma}U_{TT}U^{\sigma-1}{}_H + (1 - \sigma)U_T^{1-\sigma}U_H^{\sigma-2}U_{HT} \right] + \delta U_{TT}}{U_{TT}} \left(c_T - c_T^* \right)$$

$$- \frac{U_T^{1-\sigma}U_H^{\sigma-1}v_{mm}}{U_{TT}} (m - m^*)$$

$$\dot{m} = \varepsilon^{-\sigma} \left[(1 - \sigma)(x_T - c_T)\varepsilon_T - \frac{\sigma(g_H - T)\varepsilon_T}{\varepsilon} - \varepsilon \right] \left(c_T - c_T^* \right)$$

with the coefficient matrix:

$$A = \begin{bmatrix} \dfrac{v_m \left[(\sigma - 1)U_T^{-\sigma}U_{TT}U_H^{\sigma-1} + (1 - \sigma)U_T^{1-\sigma}U_H^{\sigma-2}U_{HT} \right] + \delta U_{TT}}{U_{TT}} & \dfrac{U_T^{1-\sigma}U_H^{\sigma-1}v_{mm}}{U_{TT}} \\[4mm] \varepsilon^{-\sigma} \left[(1 - \sigma)(x_T - c_T)\varepsilon_T - \dfrac{\sigma(g_H - T)\varepsilon_T}{\varepsilon} - \varepsilon \right] & 0 \end{bmatrix}$$

In a linear double variable case, a necessary and sufficient condition for stabilization is given as follows:

$$tr(A) = a_{11} + a_{22} = \frac{v_m \left[(\sigma - 1)U_T^{-\sigma}U_{TT}U_H^{\sigma-1} + (1 - \sigma)U_T^{1-\sigma}U_H^{\sigma-2}U_{HT} \right] + \delta U_{TT}}{U_{TT}} < 0$$

$$|A| = a_{11}a_{22} - a_{12}a_{21} = -\frac{U_T^{1-\sigma}U_H^{\sigma-1}v_{mm}}{U_{TT}} \varepsilon^{-\sigma}$$

$$\times \left[(1 - \sigma)(x_T - c_T)\varepsilon_T - \frac{\sigma(g_H - T)\varepsilon_T}{\varepsilon} - \varepsilon \right] > 0$$

Denote $c_T = x$ and $\bar{a} = -(\delta u_{TT})^{1/(\sigma-1)}$ $(\sigma \neq 1)$. As may be seen based on the knowledge introduced in Section 2.1, we obtain the following results.

Theorem 2.10 Suppose $T-g_H>0$ and $a>0$. Then the following hold true.

(i) If $\Delta<0$, then there does not exist any equilibrium point.

(ii) If $\Delta=0$, then there exists a unique equilibrium point:

$$(c_T^*,m^*)=\left(\frac{ax_T-b}{2a},\frac{ax_T+b}{2\bar{a}}\right)$$

(iii) If $\Delta>0$ and $b<\dfrac{T-g_H}{x_T}$, then there exist two equilibrium points:

$$E_1(c_T^*,m^*)=\left(\frac{ax_T-b+\sqrt{\Delta}}{2a},\frac{ax_T+b+\sqrt{\Delta}}{2\bar{a}}\right),$$

$$E_2(c_T^*,m^*)=\left(\frac{ax_T-b-\sqrt{\Delta}}{2a},\frac{ax_T+b-\sqrt{\Delta}}{2\bar{a}}\right)$$

(iv) If $\Delta>0$ and $b>\dfrac{T-g_H}{x_T}$, then there exists unique equilibrium point:

$$(c_T^*,m^*)=\left(\frac{ax_T-b+\sqrt{\Delta}}{2a},\frac{ax_T+b+\sqrt{\Delta}}{2\bar{a}}\right)$$

Theorem 2.11 Suppose $T-g_H>0$ and $a<0$. Then the following hold true.

(i) If $b<\dfrac{T-g_H}{x_T}$, then there does not exist any equilibrium point.

(ii) If $b>\dfrac{T-g_H}{x_T}$, then there exists a unique equilibrium point:

$$(c_T^*,m^*)=\left(\frac{ax_T-b+\sqrt{\Delta}}{2a},\frac{ax_T+b+\sqrt{\Delta}}{2\bar{a}}\right)$$

Theorem 2.12 Suppose $T-g_H<0$ and $a>0$. Then there exists a unique equilibrium point:

$$(c_T^*,m^*)=\left(\frac{ax_T-b+\sqrt{\Delta}}{2a},\frac{ax_T+b+\sqrt{\Delta}}{2\bar{a}}\right)$$

Theorem 2.13 Suppose $T - g_H < 0$ and $a < 0$. Then the following hold true.

(i) If $\Delta < 0$, then there does not exist any equilibrium point.

(ii) If $\Delta = 0$, then there exists a unique equilibrium point:

$$(c_T^*, m^*) = \left(\frac{ax_T - b}{2a}, \frac{ax_T + b}{2\bar{a}} \right)$$

(iii) If $\Delta > 0$, then there exist two equilibrium points:

$$E_1(c_T^*, m^*) = \left(\frac{ax_T - b + \sqrt{\Delta}}{2a}, \frac{ax_T + b + \sqrt{\Delta}}{2\bar{a}} \right)$$

$$E_2(c_T^*, m^*) = \left(\frac{ax_T - b - \sqrt{\Delta}}{2a}, \frac{ax_T + b - \sqrt{\Delta}}{2\bar{a}} \right)$$

Now, let us discuss the stability of the two equilibrium points in Theorem 2.10 (iii). Firstly, let us calculate three partial derivatives at the equilibrium point.

$$\left(\frac{\partial Q}{\partial x} \right)_{E_1} = -(ax_T - b + \sqrt{\Delta}) + (ax_T - b) = -\sqrt{\Delta} < 0$$

$$\left(\frac{\partial P}{\partial y} \right)_{E_1} = \left(-\frac{1}{u_{TT}} \right)(\sigma - 1) \left(\frac{ax_T + b + \sqrt{\Delta}}{2\bar{a}} \right)^{\sigma - 2} \frac{ax_T + b + \sqrt{\Delta}}{2}$$

$$= \delta(\sigma - 1) \left(\frac{ax_T + b + \sqrt{\Delta}}{2} \right)^{\sigma - 1} \bar{a}$$

$$\left(\frac{\partial Q}{\partial x} \right)_{E_1} = \left(-\frac{1}{u_{TT}} \right) \left[a \left(\frac{ax_T + b + \sqrt{\Delta}}{2\bar{a}} \right)^{\sigma - 1} - a\delta\sigma \left(\frac{ax_T + b + \sqrt{\Delta}}{2} \right)^{\sigma - 1} \right]$$

$$= -a\delta \left(\frac{ax_T + b + \sqrt{\Delta}}{2} \right)^{\sigma - 1} \left(1 + \frac{\sigma}{u_{TT}} \right)$$

Theorem 2.14 Suppose that $T - g_H > 0$, $a > 0$, and $\Delta > 0$. Then the stability of the two equilibrium points is just opposite.

Proof. Because $\left(\frac{\partial Q}{\partial x} \right)_{E_1} \left(\frac{\partial P}{\partial y} \right)_{E_1} \cdot \left(\frac{\partial Q}{\partial x} \right)_{E_2} \left(\frac{\partial P}{\partial y} \right)_{E_2} < 0$, the conclusion is true.

Substitute the previous three partial derivatives into its eigenvalue equation,

$$
\begin{vmatrix}
\dfrac{\partial P}{\partial x} - \lambda & \dfrac{\partial P}{\partial y} \\[2mm]
\dfrac{\partial Q}{\partial x} & \dfrac{\partial Q}{\partial y} - \lambda
\end{vmatrix} = 0
$$

It is ready to obtain the following results according to the theory of ODEs.

Theorem 2.15 Suppose that $T-g_H > 0$, $a > 0$, and $\Delta > 0$. Then the following hold true:

(i) If $u_{TT} + \sigma > 0$ and $(\sigma - 1)\bar{a} < 0$, then equilibrium point E_1 is stable.
(ii) If $u_{TT} + \sigma < 0$ and $(\sigma - 1)\bar{a} < 0$, then equilibrium point E_1 is unstable.
(iii) If $(\sigma - 1)\bar{a} > 0$, then equilibrium point E_1 is a saddle point (unstable).

Theorem 2.16 Suppose that $T-g_H > 0$, $a > 0$, and $\Delta > 0$. Then the following hold true:

(i) If $u_{TT} + \sigma > 0$ and $(\sigma - 1)\bar{a} < 0$, then the equilibrium point E_2 is stable.
(ii) If $u_{TT} + \sigma < 0$ and $(\sigma - 1)\bar{a} < 0$, then the equilibrium point E_2 is unstable.
(iii) If $(\sigma - 1)\bar{a} < 0$, then the equilibrium point E_2 is a saddle point (unstable).

Proof. Because

$$
\left(\frac{\partial Q}{\partial x} \right)_{E_2} = -(ax_T - b - \sqrt{\Delta}) + (ax_T + b) = \sqrt{\Delta} > 0
$$

$$
\left(\frac{\partial P}{\partial y} \right)_{E_2} = \delta(\sigma - 1)\left(\frac{ax_T + b - \sqrt{\Delta}}{2} \right)^{\sigma - 1} \bar{a}
$$

and

$$
\left(\frac{\partial P}{\partial y} \right)_{E_2} = -a\delta\left(\frac{ax_T + b - \sqrt{\Delta}}{2} \right)^{\sigma - 1}\left(1 + \frac{\sigma}{u_{TT}} \right)
$$

by substituting the three partial derivatives into its eigenvalue equation, one can derive the results in Theorem 2.16 readily.

Theorem 2.17 Suppose that $T-g_H>0$, $a>0$, $\Delta>0$, and $b>\dfrac{T-g_H}{x_T}$. Then the following hold true:

(i) If $u_{TT}+\sigma>0$ and $(\sigma-1)\bar{a}<0$, then the equilibrium point E is stable.
(ii) If $u_{TT}+\sigma<0$ and $(\sigma-1)\bar{a}<0$, then the equilibrium point E is unstable.
(iii) If $(\sigma-1)\bar{a}>0$, then the equilibrium point E is a saddle point (unstable).

Theorem 2.18 Suppose that $T-g_H>0$, $a<0$, $\Delta>0$, and $b>\dfrac{T-g_H}{x_T}$. Then the following hold true:

(i) If $u_{TT}+\sigma>0$ and $(\sigma-1)\bar{a}<0$, then the equilibrium point E is unstable.
(ii) If $u_{TT}+\sigma<0$ and $(\sigma-1)\bar{a}<0$, then the equilibrium point E is stable.
(iii) If $(\sigma-1)\bar{a}>0$, then the equilibrium point E is a saddle point (unstable).

Theorem 2.19 Suppose that $T-g_H<0$, $a>0$, and $\Delta>0$. Then the following hold true:

(i) If $u_{TT}+\sigma>0$ and $(\sigma-1)\bar{a}<0$, then the equilibrium point E is unstable.
(ii) If $u_{TT}+\sigma<0$ and $(\sigma-1)\bar{a}<0$, then the equilibrium point E is stable.
(iii) If $u_{TT}+\sigma>0$, then the equilibrium point E is a saddle point (unstable).

Theorem 2.20 Suppose that $T-g_H<0$, $a<0$, and $\Delta=0$. Then the following hold true:

(i) If $u_{TT}+\sigma>0$, then the equilibrium point E is unstable.
(ii) If $u_{TT}+\sigma<0$, then the equilibrium point E is stable.

Theorem 2.21 Suppose that $T-g_H<0$, $a<0$, and $\Delta>0$. Then the following hold true:

(i) If $u_{TT}+\sigma>0$ and $(\sigma-1)\bar{a}<0$, then the equilibrium point E_1 is unstable.
(ii) If $u_{TT}+\sigma<0$ and $(\sigma-1)\bar{a}<0$, then the equilibrium point E_1 is stable.
(iii) If $(\sigma-1)\bar{a}>0$, then the equilibrium point E_1 is a saddle point (unstable).

Theorem 2.22 Suppose that $T-g_H<0$, $a<0$, and $\Delta>0$. Then the following hold true:

(i) If $u_{TT}+\sigma>0$ and $(\sigma-1)\bar{a}<0$, then the equilibrium point E_2 is stable.
(ii) If $u_{TT}+\sigma<0$ and $(\sigma-1)\bar{a}<0$, then the equilibrium point E_2 is unstable.
(iii) If $(\sigma-1)\bar{a}>0$, then the equilibrium point E_2 is a saddle point (unstable).

When the real exchange rate fluctuates in a linear fashion, if the weight of the traded good in the utility function is larger than two and the economic development of the whole world is in a relatively low level, the commodities traded on international commodity markets are the necessities. At this time, if the developing countries imported a large number of the necessities, while developed countries controlled their imports of goods, then the stable state the economy reached represents a stable node. Under this situation, the developed countries are the main exporting countries, while the developing countries are the main importing countries, and the economy is in a low-level equilibrium stable state. This equilibrium point is a stable node and is also asymptotically stable. The economy slowly returns to a stable state after a period of time departing from the stability. If the developed countries imported a large number of the necessities, while developing countries controlled their imports of goods, then the stable state the economy reached is a stable vortex. Under this situation, the developing countries are the main exporting countries, while the developed countries are the main importing countries, and the economy is in a low-level equilibrium stable state. This equilibrium point is stable vortex and is also asymptotically stable. The economy spontaneously moves closer to the equilibrium point and eventually reaches a steady state.

If the weight of the traded good in the utility function is two, then the entire world is under the same fluctuation of the real exchange rate. If the weight of the traded good in the utility function is less than one, and the economic development of the whole world is in a relatively high level, then the commodities traded on the international commodity market are the main goods, even including luxury goods. At this time, if the developing countries control their imports of the traded goods, while developed countries expand the quantity of the traded goods that they import, then the stable state the economy reaches represents a stable node. Under this situation, the developing countries are the main exporting countries, while the developed countries are the main importing countries, and the economy is in a high-level equilibrium stable state. This equilibrium point is a stable node and is also asymptotically stable. The economy slowly returns to a stable state after a period of time departing from the stability. If the developing countries imported a large number of traded goods, while developed countries controlled their imports of goods, then the stable state the economy reaches is a stable vortex. Under this situation, the developing countries are the main importing countries, while the developed countries are the main exporting countries, and the economy is in a high-level equilibrium stable state. This equilibrium point is a stable vortex and is also asymptotically stable. The economy spontaneously moves closer to the equilibrium point and reaches a steady state.

The economy will reach an ideal state. Under this situation, all countries have neither trade surplus nor trade deficit; the quantity of each country's exports is equal to that of its imports. This equilibrium point represents a center and is stable. This is an ideal result.

At the same time, if the weight of the traded good in the utility function is between one and two, that is to say, during the time when the economic growth is elevated from a low level to a higher level, there exists an unstable area, where the economy cannot reach an equilibrium state. Along with the development, the economy gradually goes out of the uneven areas, and reaches a new equilibrium and stable state. This also explains the existence of periodic economy crises.

2.4.4 Conclusions

In Ellis and Auernheimer (1996), the exchange rate changes decreasingly with the consumption of the traded good, based on the assumption that all the traded goods are normal. In comparison, we relaxed this assumption in this section. That is, by allowing the foreign exchange rates to change bidirectionally, the traded goods can also include abnormal goods. On this basis, we derived a variety of equilibrium points and the corresponding stability conditions for the economy, and we discussed the stability type of each equilibrium point as well. According to the previous analysis, no matter whether it is about developed or developing countries, both can reach their respective equilibria at the stages of stable states. The stability of such equilibrium is only related to the weight of the traded good in the utility function and the consumption of the traded goods. The factor that influences the consumption of the traded goods is manifested as the fixed flow of the traded-good income. In specific circumstances, the economy can reach equilibrium and stabilization by controlling the consumption of the traded goods and the fixed flow of the traded-good income. At the same time, it is also found that during the time when the economic growth develops from a low level upward, there exists an unstable area. That is to say, there will be a period of economic imbalances in the process of economic development, which might lead to economic crises. At this point, through appropriate adjustments, the economy can reach a new equilibrium and stability at a higher level. Reversely, if the adjustments are not introduced appropriately, the economy will fall back to the earlier state of stability. This represents both an opportunity and a challenge for developing countries.

2.5 Exchange Rate Fluctuations under Capital Control: Two-Dimensional Systems (II)

2.5.1 Introduction

In this section, we look at the stabilization under the situation of fixed exchange rate rule and delayed fiscal adjustment. By using the qualitative theory of ODEs, we discuss the consumption of tradable goods and the exchange rate on the two-dimensional plane where equilibrium point(s) are studied, and we further analyze the impact of exchange rate volatility on trade.

We consider the effects of a stabilization program that consists of an exchange rate rule and a subsequent fiscal deficit reduction that allows the exchange rate rule to be maintained. To this end, we assume that capital is completely fixed, and for simplicity, that the *exchange rate rule* takes the particular form of a fined exchange rate. Ellis and Auernheimer (1996) established an interesting result that in the case of a tax-based stabilization capital mobility allows Ricardian equivalence to hold true. He suggested that the timing at which the fiscal reform (which is essentially a change from borrowing to neutral tax financing) takes place does not influence the adjustment. The result for the private sector is essentially the same as if the reform was implemented contemporaneously with the adoption of the exchange rate anchor. He found that without capital mobility this *neutrality* of the timing of a tax increase cannot be obtained, since Ricardian equivalence does not hold true. He also found that the extent to which the timing of fiscal reforms relying on expenditure reductions influences the adjustment does not depend on whether there is capital mobility. However, Ellis and Auernheimer (1996) did not fully discuss the existence and stabilization of the equilibrium point of the tradable goods and real money balances. In this section, we use ODEs and solve this problem thoroughly.

2.5.2 Stability Analysis

In this section, we discuss a more general situation than that discussed in Section 2.4. For the sake of a more accurate simulation of the exchange rate fluctuation, we consider $u(c_T, c_H) = c_T^\alpha c_H^\beta$, $v_m = m^{\sigma-1}$, and $\varepsilon = ax^{n-1} + bx + c$ $(c_T = x)$. In this situation, we discuss the stability of the equilibrium point (c_T^*, m^*) according to Equations 2.53 and 2.54. The x-coordinate of the equilibrium point of Equations 2.53 and 2.54 satisfies the following algebraic equation:

$$ax^n - ax_T x^{n-1} + bx^2 + (c - bx_T)x + T - g_H - cx_T = 0 \qquad (2.60)$$

In the light of intermediate value theorem, we obtain following results about the existence of the equilibrium point of Equations 2.53 and 2.54.

Lemma 2.1 If $cx_T > 0$ and $0 < T - g_H < cx_T$, then the algebraic equation (2.60) has at least one real root in the interval $[0, x_T]$.

Lemma 2.2 If $cx_T < 0$, $T - g_H < 0$, and $T - g_H > cx_T$, then the algebraic equation (2.60) has at least one real root in the interval $[0, x_T]$.

Lemma 2.3 If $a > 0$ and $\max\{T - g_H, T - g_H - cx_T\} < 0$, then the algebraic equation (2.60) has at least one real root in the interval $[x_T, +\infty)$.

Lemma 2.4 If $a < 0$ and $\min\{T - g_H, T - g_H - cx_T\} > 0$, then the algebraic equation (2.60) has at least one real root in the interval $[x_T, +\infty)$.

Lemma 2.5 (Hurwitz Theorem). A necessary and sufficient condition for all the roots of following nth order algebraic equation with real coefficients

$$\lambda^n + P_1\lambda^{n-1} + \cdots + P_{n-1}\lambda + P_n = 0$$

to have negative real parts is that the following Hurwitz determinant

$$\Delta_1 = \frac{b}{a} - x_T, \Delta_2 = \begin{vmatrix} P_1 & P_3 \\ P_0 & P_2 \end{vmatrix}, \Delta_3 = \begin{vmatrix} P_1 & P_3 & P_5 \\ P_0 & P_2 & P_4 \\ 0 & P_1 & P_3 \end{vmatrix},\ldots,$$

$$\Delta_n = \begin{vmatrix} P_1 & P_3 & \cdots & P_{2n-1} \\ P_0 & P_2 & \cdots & P_{2n-2} \\ \cdots & \cdots & \cdots & \cdots \\ 0 & 0 & \cdots & P_n \end{vmatrix} = P_n\Delta_{n-1}$$

is not less than zero.

Now, let us look at several special cases.

Case 1: $n = 3$. In this case, the algebraic equation (2.60) can be rewritten as

$$x^3 + \left(\frac{b}{a} - x_T\right)x^2 + \left(\frac{c - bx_T}{a}\right)x + \frac{T - g_H - cx_T}{a} = 0 \qquad (2.61)$$

After simple calculations, the Hurwitz determinant becomes

$$\Delta_1 = \frac{b}{a} - x_T, \quad \Delta_2 = \frac{1}{a^2}[(b - ax_T)(c - bx_T) - a(T - g_H - cx_T)]$$

$$\Delta_3 = \frac{1}{a^3}(T - g_H - cx_T)[(b - ax_T)(c - bx_T) - a(T - g_H - cx_T)]$$

According to the previous lemmas and Hurwitz theorem, it is ready to prove the following results.

Theorem 2.23

(i) If $\dfrac{b}{a} < x_T < \dfrac{a}{c}(T - g_H)$, $b^3 - 4abc + 4a^2(T-g_H) < 0$, and $a > 0$, $b > 0$, $c < 0$, or

(ii) If $\dfrac{b}{a} < x_T < \dfrac{a}{c}(T - g_H)$, $b^3 - 4abc + 4a^2(T-g_H) > 0$, and $a < 0$, $b > 0$, $c < 0$,

then the equilibrium point (c_T^*, m^*) is stable.

Theorem 2.24

(i) When $a < 0$ and $b < 0$, if $(T - g_H) < \min\{0, cx_T\}$, $ax_T < b$, and $c^2 < b(T - g_H)$, or

(ii) When $a > 0$ and $b < 0$, if $\dfrac{b}{a} < x_T < \dfrac{1}{c}(T - g_H)$ and $(ax_T^2 + c)^2 < 4x_T(T - g_H) > 0$,

then the equilibrium point (c_T^*, m^*) is stable.

Case 2: $n = 4$. In this case, the algebraic equation (2.59) can be rewritten as

$$x^4 - x_T x^3 + \frac{b}{a}x^2 + \left(\frac{c - bx_T}{a}\right)x + \frac{T - g_H - cx_T}{a} = 0 \qquad (2.62)$$

In this case, the Hurwitz determinant becomes

$$\Delta_1 = -x_T < 0, \quad \Delta_2 = -\frac{c}{a} < 0, \quad \Delta_3 = -\frac{c^2 - \left(x_T^2 + b\right)x_T c + x_T^2(T - g_H)}{a^2} < 0$$

$$\Delta_4 = -\frac{T - g_H - cx_T}{a}\Delta_3 < 0$$

Because the above determinants are not less than zero, according to the previous lemmas and Hurwitz theorem, it is ready to prove the following results.

Theorem 2.25 If (i) $x_T > 0$, (ii) $c > 0$, (iii) $T - g_H > \max\left\{cx_T, \dfrac{1}{4}(x_T^2 + b)^2\right\}$, then the equilibrium point (c_T^*, m^*) is stable.

Case 3: $n = 5$. In this case, the algebraic equation (2.59) can be rewritten as

$$x^5 - x_T x^4 + \frac{b}{a}x^2 + \left(\frac{c - bx_T}{a}\right)x + \frac{T - g_H - cx_T}{a} = 0 \qquad (2.63)$$

After some simple calculations, the Hurwitz determinant becomes

$$\Delta_1 = -x_T < 0, \quad \Delta_2 = -\frac{b}{a} < 0$$

$$\Delta_3 = \frac{x_T}{a}[x_T(c - bx_T) - (T - g_H - cx_T)] - \left(\frac{b}{a}\right)^2 < 0$$

$$\Delta_4 = -\frac{T - g_H - cx_T}{a}\left(-\frac{b}{a}\right) + \frac{c - bx_T}{a}\Delta_3 < 0$$

Because the above determinants are not less than zero, according to the previous lemmas and Hurwitz theorem, it is ready to prove the following results.

Theorem 2.26 If (i) $x_T > 0$; (ii) $\dfrac{b}{a} > 0$;

(iii) $\dfrac{x_T}{a}[x_T(c - bx_T) - (T - g_H - cx_T)] - \left(\dfrac{b}{a}\right)^2 < 0$;

(iv) $\dfrac{c - bx_T}{a} \Delta_3 < \dfrac{b(T - g_H - cx_T)}{a^2}$,

then the equilibrium point (c_T^*, m^*) is stable.

Case 4: $n = 6$. In this case, the algebraic equation (2.60) can be rewritten as

$$x^6 - x_T x^5 + \frac{b}{a}x^2 + \left(\frac{c - bx_T}{a}\right)x + \frac{T - g_H - cx_T}{a} = 0 \qquad (2.64)$$

So, the Hurwitz determinant becomes

$$\Delta_1 = -x_T < 0, \quad \Delta_2 = 0, \quad \Delta_3 = -\frac{c}{a}x_T < 0, \quad \Delta_4 = -\left(\frac{c}{a}\right)^2 < 0$$

$$\Delta_5 = -\frac{1}{a^2}\left[(T - g_H - cx_T)^2 x_T^3 + \frac{c^2}{a}(c - bx_T)\right] < 0, \quad \Delta_6 = \frac{T - g_H - cx_T}{a}\Delta_5 < 0$$

Because they are not less than zero, according to the previous lemmas and Hurwitz theorem, it is ready to prove the following results.

Theorem 2.27 If (i) $x_T > 0$; (ii) $\dfrac{c}{a} > 0$;

(iii) $\dfrac{b(T - g_H - cx_T)}{a} > 0$;

(iv) $(T - g_H - cx_T)^2 x_T^3 + \dfrac{c^2}{a}(c - bx_T) > 0$,

then the equilibrium point (c_T^*, m^*) is semistable.

Case 5: $n = 7$. In this case, the algebraic equation (2.60) can be rewritten as

$$x^7 - x_T x^6 + \frac{b}{a}x^2 + \left(\frac{c - bx_T}{a}\right)x + \frac{T - g_H - cx_T}{a} = 0 \qquad (2.65)$$

In this case, the Hurwitz determinant becomes

$$\Delta_1 = -x_T < 0, \quad \Delta_2 = 0, \quad \Delta_3 = -\frac{b}{a} < 0, \quad \Delta_4 = -\left(\frac{b}{a}\right)^2 < 0$$

$$\Delta_5 = -\frac{b^3}{a^3} - \frac{x_T(c - bx_T)}{a^2}\left[-(c - bx_T)x_T^2 + \left(\frac{b}{a} - x_T\right)(T - g_H - cx_T)\right] < 0$$

$$\Delta_6 = -\frac{c - bx_T}{a}\Delta_5 + \frac{T - g_H - bx_T}{a}D < 0$$

$$\Delta_7 = \frac{T - g_H - cx_T}{a}\Delta_6 < 0$$

where

$$D = \frac{(c - bx_T)(T - g_H - cx_T)}{a^2}(1 + x_T) + \frac{(T - g_H - cx_T)^2}{a^2} - \frac{b(b - cx_T)^2}{a_2}x_T^2$$

Because the above determinants are not less than zero, according to the previous lemmas and Hurwitz theorem, it is ready to prove the following results.

Theorem 2.28 If (i) $x_T > 0$; (ii) $\frac{b}{a} > 0$;

(iii) $\frac{c - bx_T}{a} > 0$;

(iv) $\frac{T - g_H - cx_T}{a^2} > 0$;

(v) $abx_T^3 - bcx_T + (b - ax_T)(T - g_H) > 0$;

(vi) $D < 0$,

then the equilibrium point (c_T^*, m^*) is semistable.

2.5.3 Conclusions

Ellis and Auernheimer (1996) showed that the movement of the exchange rate on paper is a decreasing function of the tradable goods. This conclusion is established on the basis that the tradable goods are normal. In comparison, our study on the stability of Michael's model under capital control relaxes this assumption so that the exchange rate can move in either direction, and the tradable goods do not have to be normal goods. On this basis, we have derived a variety of economic and corresponding steady-state stability conditions, and discussed the stability of each type of equilibrium.

According to the analysis, it is concluded that both the developed and developing countries can reach their respective equilibria in a periodic steady state. The stability of the equilibria is only affected by the weight of the consumption and the consumption of tradable goods in the utility function, whereas the consumption of tradable goods affects the performance of tradable goods income flow. In specific cases, by controlling the consumption of tradable goods, the revenue inflows of tradable goods can help to reach economic equilibria and the stability of the equilibrium state. Meanwhile, it is also found that a steady state at a low level and a steady state at a high level exist around an unstable region. That is, in the process of economic development, there will be periods of economic imbalances, or known as economic crises. At such times, if equilibrium is reached through appropriate adjustments, then the economy will be situated in a high level of stability; on the contrary, if an adjustment is done improperly, the economy will return to a lower level of stability. For developing countries, such periods of economic imbalances represent both an opportunity and a challenge.

Based on the previous analysis, we further assumed that the exchange rate can fluctuate freely, from which we obtained the conclusion that under the condition of capital controls, fluctuations in the exchange rate will lead to government deficit and trade deficit in the current account.

2.6 Dynamic Optimization of Competitive Agents: Three-Dimensional Systems

2.6.1 Introduction

Both existence and stability of equilibria have found many successful applications in economics or finance. For instance, Aseda et al. (1995) constructed an analytically tractable endogenous growth model of money and banking, where money provides *liquidity services* to facilitate transactions and banks convert nonreserve deposits into productive capital. He examined both the long- and short-run effects of changes in the money growth rate or the reserve requirement ratio. In response to a change in the required reserve ratio, the inflation rate and the growth rates of capital, real balances, and consumption need not adjust monotonically along the transition path. While the balanced growth equilibrium may be either a saddle or a source locally, the global dynamical system exhibits flip bifurcation. In the light of the quantitative theory of ODEs, Greiner and Semmler (1996) embedded the Taylor interest rate rule in a simple macroeconomic model with Calvo contracts. He derived conditions under which the adjustment of the economy is characterized by a unique saddle-path and showed that the conditions required for this to be the case are more stringent when the authorities adopt the Taylor rule. In both cases, the possible failure of the saddle-path condition arises when there are debt-deflation effects in the IS curve. If interest rates are set according to the Taylor rule, then debt-deflation is always enough to cause the failure of the saddle-path condition. However, when interest rates are determined by the LM curve, then it is possible that the real balance effect from the LM curve may offset the debt-deflation effect and produce a saddle-path.

Using a *sticky-price* monetary model, Wirl (1996) focused on an empirical model of the rational expectations version of the asset-market approach to exchange rate determination, and formulated the model by a difference-equations system. He obtained explicit solutions as functions of forcing variables extending to past dates. This *backward-looking* characteristic of the solution is in stark contrast to conventional *forward-looking* models, and alleviates empirical investigation because of the need for past data only. This discrete dynamic model is superior to the corresponding continuous model, because its solutions neither exhibit the empirically unfounded *overshooting* behavior, nor have saddle-point instability. Rather, the exchange rate is shown to follow an oscillatory path with asymptotic stability, and this seems to replicate the actual movements closely approximated by a random walk process. Wirl (1997) studied capital growth rate and found that its dynamics has a saddle-point trajectory that converges to a unique steady state. Along the transition path, the growth rate exhibits exponentially decreasing oscillations.

Bellman and Cooke (1963) and Boldrin and Rustichini (1994) showed us an excellent example in the study of three-species communities. The authors analyzed such dynamical behaviors as boundedness, existence of periodic orbits, persistence, as well as stability. The long-term coexistence of the three interacting species is addressed, and the stability analysis of the model shows that the biologically most relevant equilibrium point is globally asymptotically stable whenever it satisfies a certain criterion. Practical implications are explored and related to real-life populations.

In this section, on the basic work of Wirl (1997), we look for conditions about stability or instability of responding equations in the three-dimensional space in the light of the theory of ODEs.

2.6.2 *The Model*

Wirl (1997) considered a competitive economy where the representative agent solves the following dynamic optimization problem:

$$\max_{\{u(t)\}} \int_0^\infty \exp(-rt)v(x(t),u(t),y(t))dt \tag{2.66}$$

$$\dot{x}(t) = f(x(t),u(t),y(t)), \quad x(0) = x_0. \tag{2.67}$$

The agent considers the evolution of $y(t)$ as exogenous data, because y is negligibly affected by the representative agent's actions due to the supposition of a competitive market.

We select the associated Hamiltonian as follows:

$$H = v(x, u, y) + zf(x, u, y)$$

Suppose that the matrix H satisfies strictly the concave conditions, that is, $H_{uu} < 0$, $H_{xx} \leq 0$, as well as $H_{uu}H_{xx} - H_{ux}^2 \leq 0$. Therefore, the following conditions are sufficient for an optimal interior decision, denoted by u^*:

$$H_u = 0 \Rightarrow u^* = U(x, z; y)$$

$$U_x = -\frac{H_{ux}}{H_{uu}}$$

$$U_y = -\frac{H_{xy}}{H_{uu}}$$

$$U_z = -\frac{f_u}{H_{uu}}$$

It is straightforward for us to obtain the optimality condition about the representative agent's optimization problem, which can be amended for the evolution of the externality result in the following three ODEs:

$$\dot{x} = f(x, U(x, \lambda, y), y)$$

$$\lim_{t \to \infty} \exp(-rt)x(t)\lambda(t) = 0$$

$$\dot{y} = g(x, U(x, \lambda, y), y), \quad y(0) = y_0$$

According to the theory of ODEs, the stability of the dynamic system is determined by the characteristic values of determinant of the matrix J.

$$J = \frac{1}{H_{uu}} \begin{pmatrix} f_x H_{uu} - f_u H_{ux} & -f_u^2 & f_y H_{uu} - f_u H_{uy} \\ H_{ax}^2 - H_{xx}H_{uu} & (r - f_x)H_{uu} + f_u H_{ux} & H_{ux}H_{uy} - H_{xy} \\ g_x H_{uu} - g_u H_{ux} & -f_u g_u & g_y H_{uu} - g_u H_{uy} \end{pmatrix}$$

2.6.3 Analysis

We rewrite the previous Wirl model as follows:

$$
\begin{cases}
\dot{x} = a + mx + ny + \sqrt{p - \dfrac{z}{k}} \\[2ex]
\dot{y} = qx - \delta y \\[1.5ex]
\dot{z} = (r - m)z - 1
\end{cases}
\tag{2.68}
$$

First of all, let us discuss its equilibrium points. To this end, we have to solve the following equations:

$$
\begin{cases}
a + mx + ny + \sqrt{p - \dfrac{z}{k}} = 0 \\[2ex]
qx - \delta y = 0 \\[1.5ex]
(r - m)z - 1 = 0
\end{cases}
$$

If $m \ne r$, we obtain the unique equilibrium point as follows:

$$
x^* = \frac{\delta}{m\delta + ny} \left(\sqrt{p - \frac{1}{k(r - m)}} - a \right)
$$

$$
y^* = \frac{q}{m\delta + ny} \left(\sqrt{p - \frac{1}{k(r - m)}} - a \right)
$$

$$
z^* = \frac{1}{r - m}
$$

Then, let us consider the stability of the linear Wirl model:

$$
\begin{cases}
\dot{x} = \bar{a} + mx + ny + lz \\[1.5ex]
\dot{y} = qx - \delta y \\[1.5ex]
\dot{z} = (r - m)z - 1
\end{cases}
\tag{2.69}
$$

where

$$l = \frac{1}{2k}\left(p - \frac{1}{k(r-m)}\right)^{-1/2} \quad \text{and} \quad \bar{a} = a + mx^* + ny^* + lz^*$$

The corresponding coefficient matrix can be obtained as follows:

$$J = \begin{pmatrix} m & n & l \\ q & -\delta & 0 \\ 0 & 0 & r-m \end{pmatrix} \tag{2.70}$$

From expression (2.69), it follows that $l = 0$ approximately as $p \gg 1/(r-m)$. At this situation, the stability of Equation 2.68 in the three-dimensional space is equivalent to that of Equation 2.69 in the two-dimensional space as follows:

$$\begin{cases} \dot{x} = \bar{a} + mx + ny \\ \dot{y} = qx - \delta y \end{cases} \tag{2.71}$$

from which the equilibrium point (x^*, y^*) is obtained as follows:

$$x^* = -\frac{a\delta}{m\delta + nq}, \quad y^* = -\frac{aq}{m\delta + nq}$$

The characteristic equation of the linear part of Equation 2.71 can be written as

$$\lambda^2 - (m-\delta)\lambda - (m\delta + nq) = 0 \tag{2.72}$$

In the light of the theory of ODEs, one can obtain the following five results.

Theorem 2.29 If $p \gg \dfrac{1}{r-m}$, then the states of the equilibrium point (x^*, y^*) can be classified as follows:

(i) If $\Delta > 0$, then when $\delta > m$, the equilibrium point is an unstable node; when $\delta < m$, the equilibrium point is a stable node.

(ii) If $\Delta < 0$, then when $\delta > m$, the equilibrium point is an unstable focus; when $\delta < m$, the equilibrium point is stable focus.

(iii) If $\Delta = 0$, then when $\delta > m$, the equilibrium point is unstable node or focus; when $\delta < m$, the equilibrium point is either a stable node or focus.

(iv) If $\delta = m$, then when $m\delta + nq < 0$, the equilibrium point is a center.

(v) If $m\delta + nq > 0$, then the equilibrium point is a saddle point.

(vi) If $\delta + m = n = 0$, then when $\delta > 0$, the equilibrium point is an unstable degenerating node; when $\delta < 0$, the equilibrium point is a stable degenerating node.

Now, let us consider the characteristic equation of matrix J:

$$\begin{vmatrix} \lambda - m & -n & -l \\ -q & \lambda + \delta & 0 \\ 0 & 0 & \lambda + m - r \end{vmatrix} = 0 \tag{2.73}$$

According to the signs of roots of the characteristic equation (2.73), we have following results.

Theorem 2.30 If $r < m < \delta$ and $m\delta + nq < 0$, then the equilibrium point is local stable.

Proof. In fact, the characteristic equation (2.73) is cubic. However, it can be decomposed as

$$[\lambda - (r - m)][\lambda^2 + (\delta - m)\lambda - (m\delta + nq)] = 0 \tag{2.74}$$

Therefore, it is ready to look for three roots of characteristic equation (2.73).

$$\lambda_1 = r - m, \quad \lambda_2 = \frac{m - \delta}{2} + \frac{1}{2}\sqrt{\Delta}, \quad \lambda_3 = \frac{m - \delta}{2} - \frac{1}{2}\sqrt{\Delta}$$

where

$$\Delta = (\delta + m)^2 + 4nq$$

It is obvious to see that when the conditions of Theorem 2.29 are satisfied, the signs of the three roots are negative, hence the conclusion of Theorem 2.30 is true.

Theorem 2.31 If $r > m$, then the equilibrium point is unstable.

Proof. If $r > m$, then $\lambda_1 = r - m > 0$. It means that one of the roots of the characteristic equation (2.73) is a positive number. Therefore, it is impossible that the signs of the three roots are all negative; hence, conclusion of Theorem 2.31 is true.

In the same way, we also have:

Theorem 2.32 If $r < m$ and $m\delta + nq > 0$, then the equilibrium point is unstable.

Theorem 2.33 If $r < m$ and $\delta < m$, then the equilibrium point is unstable.

2.6.4 Conclusions

In terms of methodology, this section only looked at the special case of $l = 0$. In this situation, the states of the equilibrium points of the third order linear system degenerate into those of a second-order linear system. Therefore, by making use of the analysis method of plane linear systems, we can derive all of the conclusions on the stability and instability of the third-order linear system (2.69).

As for the general case when $l \neq 0$, the corresponding discussions can be carried out by employing the method introduced in the first section of this chapter. All the relevant details are left to the reader as an exercise.

2.7 Dynamic Heckscher–Ohlin Model: Four-Dimensional Systems

2.7.1 Introduction

The long-term rapid economic growth seen in some East Asian economies has been a hot research topic in the field of macroeconomics in recent years. A large body of the published work has been devoted to the explanation of this extraordinary growth phenomenon from different perspectives and angles. Of the literature, one of the leading contentions is the claim that (physical/human) capital accumulation has been the (main) source of the growth miracle. While the observed high savings rates made large-scale investments possible, people may wonder why personal savings rates in these economies have been so high. Are the high savings rates contributing to the catching-up and overtaking processes these economies experienced? In this section, we try to resolve these questions by using a four-dimensional system.

Since the high savings rates observed in these economies are too striking to be explained by the conventional consumption-smoothing hypothesis alone, we include agents' status-seeking behavior as an additional argument. That is, if social status is represented by wealth-holding, then agents' saving today not only benefits their future consumption (the conventional *indirect effect*) but also benefits the agents' current preference by having a higher social status (a new *direct effect*). The fact that people relate their social status to felicity has long been realized. Empirical studies also reveal that a positive connection exists between individuals' social status and happiness. The reason that individuals' status-seeking behavior affects economic performance lies in its distortion effect on resource allocation. For example, in the case where wealth represents status, status-seeking behavior makes agents shift resources from goods consumption to capital accumulation, which may benefit or damage an economy's long-run growth depending on the model setting.

On the other hand, these East Asian economies are also characterized by large volumes of international trade. Therefore, we employ an international trade model with two countries trading the produced goods. To be specific, we adopt a standard two-country, two-factor dynamic Heckscher–Ohlin (H–O) model. The usage of this framework can be justified by recalling the world economic reality these East Asian economies faced several decades ago. Though even developing economies today have access to the international credit market, this was rarely true in the past. Our following discussion reveals that the H–O model is a powerful framework, not only in theory but also in practice.

The intuition behind the uniqueness result can be understood as follows. In the standard models, wealth accumulation affects individuals' preference through its indirect effect, benefiting future consumption only. Under the H–O structure, in which two countries own the same technology and preference functions, the wealth effect on preference will be the same. There is no incentive for the two countries to have different capital accumulation processes. Hence, the two countries will keep their initial wealth distribution and the initial poor/rich country will be poor/rich forever. In this section, however, in addition to the indirect effect, a direct effect of capital accumulation on individuals' preference in the form of status-seeking is also considered. Different from the standard models, a poor country with lower status, through status-seeking, can catch up with the rich by accumulating more capital.

A unique steady state is important due to the need for policy evaluation. The effects of policy changes can be evaluated by conducting comparative statics. But one can do so only when the existence of a unique steady state is ensured. For example, to evaluate the effects of trade policies, we can calculate the derivatives of the steady state with respect to, say, the tariff rate. However, this approach is meaningful only when the existence of a unique steady-state equilibrium is ensured. Otherwise, if there is more than one steady state, any parameter change may shift the economy from one steady state to another unrelated one, which implies meaningless policy predictions.

In the process of globalization, the extent of this status-seeking behavior has unavoidably expanded across borders. Therefore, in this section, we assume that households relate their wealth to a world average level, which is a combination of domestic and foreign average wealth holdings.

2.7.2 The Model

Hu and Shimomura (2007) considered a two-country (home and foreign) world, where two goods, a pure consumption good (good 1) and a pure investment good (good 2), are produced. The production of these two goods requires labor and capital as inputs. Denote the endowments of labor, initial physical capital of the home country respectively as l and $k(0)$, and those of the foreign country respectively as the ones with asterisks, l^* and $k^*(0)$. Assuming away population growth, then l and l^* are constant. Since physical capital can be accumulated, k and k^* are time-varying with the initial endowment $k(0)$ and $k^*(0)$. In the following, for simplicity, the time subscript is omitted. Suppose that there is a continuum of households dwelling in each country. If the number of households inside a country is normalized to one, then we can use the same symbol to represent either the aggregate or an individual or even the average level of the variable. Following the conventional Heckscher–Ohlin structure, we assume the two countries own the same technology and have the same preference.

The optimization problem of the representative household in the home country is, for given average world wealth level K, to maximize its aggregate utility

$$\max_c \int_0^\infty \left[u(c) + v\left(\frac{k}{K}\right) \right] e^{-pt} \, dt$$

subject to the budget constraint

$$\dot{k} = G(k, p; l) - pc - \delta k \tag{2.75}$$

As distinct from these previous works, we assume households relate their wealth to the average level of the world wealth. The average wealth level in the home country is defined as $K = \alpha k + (1 - \alpha)k^*$. Here, k and k^* are the average capital levels in the home and foreign countries, respectively. The parameters $\alpha(\in [0, 1])$ and $(1 - \alpha)$ represent the weights of domestic households put on home and foreign wealth levels, respectively. That is, a higher α means that domestic households are more concerned about their own country's wealth-holding. In the polar case of $\alpha = 1$, domestic households do not care about the wealth level in the foreign country. Similarly, the average wealth level in the foreign country can be defined as $K^* = \alpha^* k^* + (1 - \alpha^*)k$ with $\alpha^* \in ([0, 1])$ being the weight the foreign households put on their own domestic households' average capital.

Here $u(c)$ is the usual utility term representing the preference from the consumption c of goods, while $v(s)$, $s = k/K$, is a status term, which appears in the utility function, because the agent cares about its status in society. Notice again that the index of status takes a relative wealth form. Both u and v have the usual monotonous, concave properties, $u' > 0$, $u'' < 0$, and $v' > 0$, $v'' < 0$. And $p > 0$ is the rate of time preference. The function $G(k, p; l)$ represents the total production of the country given the labor endowment l and the initial capital level $k(0)$, p is the price of the consumption good in units of the investment good. In the case of incomplete specialization, we have $G(k, p; l) = r(p)k + w(p)l$, where $r(p)$, $w(p)$ are respectively the factor prices expressed with commodity price p. And $\delta > 0$ is a constant depreciation rate of physical capital.

Combining the market-clearing condition with the equilibrium conditions obtained for households and firms, we can obtain a closed-form, four-dimensional dynamic system:

$$\dot{k} = G(k, p; l) - \delta k - p\phi(p\lambda)$$

$$\dot{k}^* = G(k^*, p; l^*) - \delta k^* - p\phi(p\lambda^*)$$

$$\dot{\lambda} = [\rho + \delta - G_k(k, p; l)]\lambda - \frac{v'(s)}{K}$$

$$\dot{\lambda}^* = [\rho + \delta - G_{k^*}(k^*, p; l^*)]\lambda^* - \frac{v'(s^*)}{K^*}$$

2.7.3 Local Stability Analysis

It is obvious that, in the present general dynamic equilibrium model, there is only one steady state in which the production in the two countries is incompletely specialized. That is, regardless of whether a country is initially poor or rich, in the long run, all countries have the same steady state. To finish the characterization of the equilibrium dynamics, we will examine whether or not the convergence path to the steady state exists. In the neighborhood of the steady state (k^e, $k^{*\,e}$, λ^e, $\lambda^{*\,e}$), as obtained earlier, let us consider the linearization equation of the previous dynamic system around the steady state,

$$(\dot{k}, \dot{k}^*, \dot{\lambda}, \dot{\lambda}^*)^T = J(k - k^e, k^* - k^{*e}, \lambda - \lambda^e, \lambda^* - \lambda^{*e})^T$$

where

$$J = \begin{pmatrix} r - \delta & 0 & -p^2\phi' & 0 \\ 0 & r - \delta & 0 & -p^2\phi' \\ a & b & \rho + \delta - r & 0 \\ b & a & 0 & \rho + \delta - r \end{pmatrix}$$

with

$$a = k^{-2}(1-\alpha)v'(1)\left[\frac{\alpha}{1-\alpha} + \mu(1)\right], \quad b = k^{-2}(1-\alpha)v'(1)[1-\mu(1)]$$

$$m = 2r''(p)k + 2w''(p)l - 2\lambda\phi' > 0, \quad \text{where } m > 0$$

because $\phi' = (\mu'')^{-1} < 0$ and $r''(p)k + w''(p)l > 0$.

The characteristic equation of J is

$$[x^2 - \rho x + \Delta_1]\left[x^2 - \rho x + \frac{\Delta_2}{m}\right] = 0 \qquad (2.76)$$

where

$$\Delta_1 = (r - \delta)(\rho + \delta - r) + k^{-2}p^{-2}\phi'v'(1)[\alpha(1 - \alpha)^{-1} + \mu(1)]$$

$$\Delta_2 = m[(r - \delta)(\rho + \delta - r) + k^{-2}p^2\phi'v'(1)] + 2\lambda\phi'[r(\varepsilon_r - 1) + \delta][r(\varepsilon_r - 1) + \rho + \delta]$$

Notice the condition $r(\varepsilon_r - 1) + \rho + \delta < 0$. This condition can be satisfied, for example, when $\rho > 0$ is not too large.

Theorem 2.34 If $r(\varepsilon_r - 1) + \rho + \delta < 0$ and $\eta \le \min\{2(1-\alpha)\mu(1) + (2\alpha-1), 1\}$, then we have $\Delta_1 < 0$ and $\Delta_2 < 0$.

Theorem 2.35 If $r(\varepsilon_r - 1) + \rho + \delta < 0$ and $\eta \le \min\{2(1-\alpha)\mu(1) + (2\alpha-1), 1\}$, then the unique steady state of the present general dynamic model is locally saddle-point stable.

2.7.4 Conclusions

Our equilibrium dynamics analysis illustrates the function of agents' status-seeking behavior in singling out a unique steady state in a two-country dynamic general equilibrium model. The framework can be used to explain economic realities. For example, to understand why Japanese save so much compared with Americans, their different attitudes toward status-seeking may provide one explanation. On the other hand, how this difference in extent of status-seeking affects the trade pattern between the two countries is another interesting issue.

This framework can also be used to perform analysis on policy effects. Firstly, it is an interesting issue to see how one country's fiscal policy affects the performance of another country. We notice that most work in the fiscal policy literature until now has limited the discussion to the closed economy framework, although a few work by some scholars have made efforts in this direction. Secondly, it would be interesting to compare the policy effects between open and closed economy frameworks. Once difference occurs, with an increasingly integrated world economy, our understanding of policy effects until now may have to be revised. Thirdly, the present structure can also be used to investigate the policy effects with one country's domestic (fiscal) policy and international trade policy considered at the same time.

2.8 Instability: Risk of Capital Flow

2.8.1 Introduction

While capital flows tend to be a catalyst for economic development, the economic effects of capital flows is still a controversial topic. Scholars from different countries have built empirical analysis models to explain the effects of capital account openness on economic growth with the following ideas put forward: the promotion theory, the unrelated theory, and the uncertainty theory.

The promotion theory is represented by Quinn, where there exists a positive correlation between changes in indicators of capital account openness and economic growth (Quinn, 1997). He builds empirical analysis models to derive the conclusion, and the result is statistically significant. Whether in developed or emerging market countries, capital account openness and economic growth are positively correlated (Quinn and Toyoda, 2008). In order to test whether capital account openness will lead to higher growth, they researched capital account and financial account openness measures of 94 countries since 1950. Time series, cross-sectional least squares method, and sector GMM estimation were used to study the growth rate from 1955 to 2004. They find interesting evidence that only FDI has a positive effect on growth and that FPI has an unfavorable, if not negative, effect on growth (Shen et al., 2010). These authors re-examined the relationship between international capital flows by employing panel data for 80 countries that cover the time period of 1976–2007. They provided suggestions that countries should strive to develop sound financial markets, including the trading activities of capital markets, to enhance the effects of international capital flows on economic growth.

The relationship between capital flows and economic growth is not significant in unrelated theory. Capital account openness and its nexus with economic growth of Indian economy have been dealt with by Chandra (2011). In his study, the cointegration subtests are able to find evidence in favor of long-run relationship between net capital flows and nonagricultural GDP in the case of Indian economy. The Granger causality test suggests that coefficients are not statistically significant in the regression equations. Based on these tests, neither can they make any claims about the predictability of growth from capital inflows nor can they infer that capital inflows have been due to the growth factors.

International capital flows may not improve economic growth in uncertainty theory. Eswar et al. (2007) documented the recent phenomenon of *uphill* flows of capital from nonindustrial to industrial countries. They found that there is a positive correlation between current account balances and growth among nonindustrial countries, implying that a reduced reliance on foreign capital is associated with higher growth. In no case do they find any evidence that an increase in foreign capital inflows directly boosts growth. They provide some evidence that even successful developing countries have limited absorptive capacity for foreign resources.

They suggest that greater caution toward foreign capital inflows might be warranted if domestic financial sector is underdeveloped.

The reasons for inconsistency among previous studies include the omission of destabilizing risks of capital flows and the failure to account for the impact of crises on growth. Even if some studies do pay attention to risks, their interest is largely in the risk faced by external investors. Not many studies seriously address the risk of volatile capital flows faced by financially weak countries. So many countries still stick with capital controls to some extent and choose a limited opening based on their financial features simply because so many financial crises have occurred under widened market openness. Differing choices lead to different economic performance in terms of growth and instability. Our emphasis is put on the importance of sounder financial sectors for healthier market openness that can lead to higher growth and lower instability.

2.8.2 Instability σ

2.8.2.1 σ–g Relations

This section focuses on the risk of instability in capital flows. We all know that the risk of instability is a function of financial openness and weakness, that is, $\sigma = \sigma(a, b)$, where $\partial\sigma/\partial(a, b) > 0$ and σ measures instability. Let a be an index where a higher a represents a more widely opened capital market. And a lower b represents a sounder financial sector. Countries with wider capital openness and a weaker financial sector represent a greater level of instability.

We take g as economic growth for $g = g(a, b)$, which initially rises with more capital flows under wider openness, then reaches a maximum at an intermediate level of capital flow, and finally declines as external capital continues to enter the country under far wider openness. Developing countries cannot quickly strengthen their financial sectors, since institutional factors are hard to change overnight. Hence, it is reasonable to treat b as an underlying parameter while taking a as a choice variable. Based on these assumptions, we can infer the relationship between g and σ and record it as $g = g_{b_0}(\sigma)$.

Using Taylor formula to expand $\sigma = \sigma(a, b)$ and $g = g(a, b)$ to the second order at $a = 0$ and $b = b_0$ gives

$$\sigma(a, b_0) = \sigma_0 + m_1 a + n_1 a^2 \tag{2.77}$$

$$g(a, b_0) = g_0 + m_2 a + n_2 a^2 \tag{2.78}$$

where

$$m_1 = \sigma'(a, b_0)\big|_{a=0}, \quad n_1 = \frac{\sigma''(a, b_0)}{2!}\bigg|_{a=0}$$

and

$$m_2 = g'(a, b_0)|_{a=0}, \quad n_2 = \frac{g''(a, b_0)}{2!}\bigg|_{a=0}$$

Eliminating the coefficient of a^2 gives

$$a = \frac{n_2 \sigma - n_1 g - n_2 \sigma_0 + n_1 g_0}{m_1 n_2 - m_2 n_1}$$

Putting a into (2.77) gives

$$p_1 g = n_2 (n_2 \sigma - n_1 g)^2 + p_2 (n_2 \sigma - n_1 g) + p_3$$

where

$$p_1 = (m_1 n_2 - m_2 n_1)^2, \quad p_2 = m_1 m_2 n_2 - m_2^2 n_1 - 2n_2^2 \sigma_0 + 2n_1 n_2 g_0$$

and

$$p_3 = (m_1 n_2 - m_2 n_1)^2 g_0 - m_2 (m_1 n_2 - m_2 n_1)(n_2 \sigma_0 - n_1 g_0) + n_2 (n_2 \sigma_0 - n_1 g_0)^2$$

Reordering the equation to obtain

$$Ag^2 + Bg + C = 0 \tag{2.79}$$

where

$$A = n_1^2 n_2, \quad B = -\left(2n_1 n_2^2 \sigma + p_2 n_1 + p_1\right), \text{ and } C = n_2^3 \sigma^2 + p_2 n_2 \sigma + p_3$$

By looking at the discriminant ($\Delta = B^2 - 4AC$), we have the following three results.

Theorem 2.36 If $\Delta < 0$, then Equation 2.79 has no solution, implying there is no real functional relationship in $g = g_{b_0}(\sigma)$.

This result implies that if an open economy becomes more unstable when attacked, then its growth rate will slow down a lot, the greater negative relationship between growth and instability exists, the greater impact instability makes on the growth.

Theorem 2.37 If $\Delta = 0$, Equation 2.79 has a unique solution:

$$g = g_{b_0}(\sigma) = \frac{n_2}{n_1} \sigma + \frac{p_2 n_1 + p_1}{2n_1^2 n_2}$$

This theorem shows that g is a linear function of σ and there exists a negative correlation between g and σ for its coefficient $\frac{n_2}{n_1} < 0$. This result implies that economic instability may happen if an open economy permits volatile capital flows without reversing its financial vulnerability. But how instability affects the growth is complex.

Theorem 2.38 If $\Delta > 0$, Equation 2.78 has two different solutions:

$$g = g_1(\sigma) = \frac{n_2}{n_1}\sigma + \frac{p_2 n_1 + p_1}{2n_1^2 n_2} + \delta_1 \sqrt{\sigma + \delta_2}$$

and

$$g = g_2(\sigma) = \frac{n_2}{n_1}\sigma + \frac{p_2 n_1 + p_1}{2n_1^2 n_2} - \delta_1 \sqrt{\sigma + \delta_2}$$

where

$$\delta_1 = \frac{1}{n_1}\sqrt{\frac{p_1}{n_1}}$$

and

$$\delta_2 = \frac{(p_2 n_1 + p_1)^2 - 4n_1^2 n_2 p_3}{4n_1 p_2 n_2^2}$$

We can roughly draw pictures of $g = g_1(\sigma)$ and $g = g_2(\sigma)$. The curve $g = g_1(\sigma)$ characterizes the risky opportunities of growth for a country, which can be open more widely to capital flows in the hope of seeking faster growth but has to face greater instability due to its financial vulnerability to potential adverse shocks. However, excessive openness will lead to instability and lower growth. From the concave curve of $g = g_2(\sigma)$, we know that if economic instability is high, then continued economic recession will be an inevitable price for some underdeveloped countries to pay, if these countries permit disoriented openness without first strengthening financial sector substantially.

Throughout the development of the world economy, most countries are still in line with the relationships presented in Figure 2.5. This figure means that economy grows with capital flows; but if a country does not strengthen the construction of its financial market, wide openness will make the economy more unstable. Such dramatic fluctuations in economic growth will lead to recession.

2.8.2.2 Trade-Off Optimization

The curve of $g = g_b(\sigma)$ characterizes the risky opportunities of growth for a country. The welfare function $u = u_c(\sigma, g)$ satisfies the properties $MU_\sigma < 0$ and $MU_g > 0$,

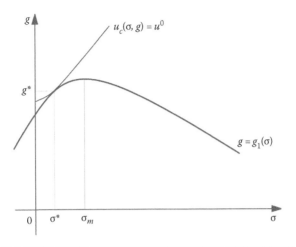

Figure 2.5 Curve $g = g_1(\sigma)$.

which imply that lower stability and faster growth lead to greater utility. The curvature parameter $c > 0$ measures the degree of risk aversion, where a smaller c indicates lower risk aversion. The indifference curve *ICs* is obtained by setting the welfare function equal to a constant. Each country has a different attitude toward risk (with a distinct c), and displays a different shape of *ICs*, where a smaller c exhibits flatter *ICs*.

For each country, only one of its *ICs* is tangent to its opportunity curve in Figure 2.5. The tangency point gives its optimal trade-off between instability and growth. However, none of its *ICs* are tangent to its opportunity curve in Figure 2.6, similarly to a linear function.

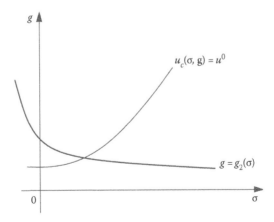

Figure 2.6 Curve $g = g_2(\sigma)$.

Next, we look for the trade-off optimization of Figure 2.5. Using Taylor formula to expand $\sqrt{\sigma + \delta_2}$ to $\sqrt{\sigma + \delta_2} = 1 + (1/2)(\sigma + \delta_2 - 1) - (1/8)(\sigma + \delta_2 - 1)^2 + p_4$ of $g = g_1(\sigma)$ gives

$$g = g_1(\sigma) = \frac{n_2}{n_1}\sigma + \frac{p_2 n_1 + p_1}{2 n_1^2 n_2} + \delta_1 \sqrt{\sigma + \delta_2} = -\alpha(\sigma - \sigma_m)^2 + \beta \qquad (2.80)$$

where

$$\alpha = -\frac{1}{8}\delta_1, \quad \sigma_m = \frac{4 n_2}{\delta_1 n_1} + 3 - \delta_2, \quad \beta = \frac{p_2 n_1 + p_1}{2 n_1^2 n_2} + \frac{3}{2}\delta_1 + \delta_1 p_4 + \frac{2 n_2^2}{\delta_1 n_1} + \frac{n_2}{n_1}(3 - \delta_2)$$

α is an intercept coefficient

$\sigma_m > 0$ is the location parameter

$\sigma_m \pm \sqrt{\beta/\alpha}$ are the two horizontal intercepts

$g(\sigma_m) = \beta$ is the maximal growth

Assume that the welfare function takes the form of $u_c(\sigma, g) = g \exp(-c\sigma)$, where the curvature parameter $c > 0$ measures the degree of risk aversion. By assuming that the welfare function $u_c(\sigma, g)$ is equal to u^0, we can infer *IC* as an exponential function having the required property of convexity:

$$g(\sigma) = u^0 e^{c\sigma} \qquad (2.81)$$

Letting (2.80) be equal to (2.81) gives

$$-\alpha(\sigma - \sigma_m)^2 + \beta = u^0 e^{c\sigma} \qquad (2.82)$$

Differentiating (2.80) and (2.81) with respect to σ and setting them equal to each other yields the tangency condition

$$-2\alpha(\sigma - \sigma_m) = u^0 c e^{c\sigma} \qquad (2.83)$$

Plugging (2.82) into (2.83) gives

$$-2\alpha(\sigma - \sigma_m) = c[-\alpha(\sigma - \sigma_m)^2 + \beta] \qquad (2.84)$$

From (2.84), we obtain $\sigma^* = \sigma_m + (1 - \lambda)/c$. Then plugging it into (2.80) gives $g^* = 2\alpha(\lambda - 1)/c^2$, where $\lambda = \sqrt{1 + c^2 \beta/\alpha} > 1$.

The optimal trade-off between instability and growth (σ^*, g^*) is therefore derived from risky opportunities and the welfare function.

2.8.2.3 b-Effect and c-Effect

We get optimal trade-off between instability and growth (σ^*, g^*) from Figure 2.5, what conclusions can we draw from the trade-off optimization?

Theorem 2.39 Assume that the welfare function takes the form of $u_c(\sigma, g) = g\exp(-c\sigma)$, and the risky opportunities of growth take the form of $g = g_{b_0}(\sigma) = -\alpha(\sigma - \sigma_m)^2 + \beta$. Then, $dg^*/d\sigma^*|_c > 0$.

Proof. The optimal trade-off between instability and growth (σ^*, g^*) is given by $g^* = 2\alpha(\lambda - 1)/c^2$ and $\sigma^* = \sigma_m + (1 - \lambda)/c$. Differentiating them with respect to c gives $\partial\sigma^*/\partial c = (1 - \lambda)/(\lambda c^2) < 0$, $\partial g^*/\partial c = -2\alpha(\lambda - 1)^2/(c^3\lambda) < 0$, then one obtains $dg^*/d\sigma^*|_c = (\partial g^*/\partial c)/(\partial\sigma^*/\partial c) > 0$.

Theorem 2.39 is called the *c*-effect. Assume that country A and country A' share the same opportunity curve but differ in policy preferences in Figure 2.7. The utility parameter is smaller for A' than for A: $c' < c$. That means that country A' has lower risk aversion. The optimal trade-off of country A' is between higher instability and faster growth made possible by wider openness a'^* due to less risk averse preferences, as compared with a^* for country A (i.e., $a > 1/2$). The *c*-effect tells us that while less prudence causes wider openness (α), greater instability appears for higher growth if the finance sector is weak.

Theorem 2.40 Given the same condition as Theorem 2.39, the following condition holds true: $dg^*/d\sigma^*|_b < 0$.

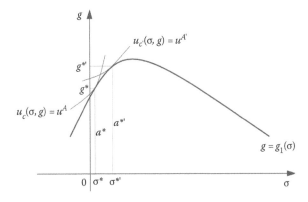

Figure 2.7   The c-effect.

Proof. One obtains $\alpha'(b) > 0$, $\sigma'_m(b) > 0$, and $\beta'(b) < 0$ from the image of the quadratic function $g(\sigma) = -\alpha(\sigma - \sigma_m)^2 + \beta$. Differentiating $\sigma^* = \sigma_m + (1-\lambda)/c$ and $g^* = 2\alpha(\lambda - 1)/c^2$, respectively, with respect to α, β, σ_m gives

$$\frac{\partial \sigma^*}{\partial \alpha} > 0, \frac{\partial \sigma^*}{\partial \alpha} > 0, \frac{\partial g^*}{\partial \alpha} > 0, \frac{\partial \sigma^*}{\partial \beta} < 0, \frac{\partial g^*}{\partial \beta} > 0, \frac{\partial \sigma^*}{\partial \sigma_m} > 0, \text{ and } \frac{\partial g^*}{\partial \sigma_m} = 0$$

By taking the total derivative of $g^* = g[\alpha(b), \beta(b), \sigma_m(b)] = g(b)$, one can derive that $dg^*/db < 0$, if $|\beta'(b)|/\alpha'(b) > (1-\lambda)^2/c^2$. So, it follows that

$$\frac{dg^*}{db} = \frac{\partial g^*}{\partial \alpha} \alpha'(b) + \frac{\partial g^*}{\partial \beta} \beta'(b) + \frac{\partial g^*}{\partial \sigma_m} \sigma'_m(b) = \frac{(1-\lambda)^2}{c^2 \lambda} \alpha'(b) + \frac{1}{\lambda} \beta'(b)$$

from which $d\sigma^*/db > 0$ is established. Therefore, we have the conclusion that $dg^*/d\sigma^*|_b = (dg^*/db)/(d\sigma^*/db) < 0$.

Theorem 2.40 is called the *b*-effect. Because stronger financial conditions promote growth $(dg^*/db < 0)$, if there is a stronger impact on β than on α while instability is reduced unconditionally $(d\sigma^*/db > 0)$, there then exists a negative relationship between growth and instability $(dg^*/d\sigma^*|_b < 0)$. The *b*-effect implies that a desirable combination of lower level of instability and higher level of growth can be accomplished but only via strengthened financial conditions.

2.8.3 Empirical Examples

This section confronts our theory with real-world observations about the effects of capital flows on economic growth and instability. We collect the data of the BRICS countries (including China, Brazil, Russia, India, and South Africa) that cover the time period of 2000–2010. We divide the standard deviation of GDP growth by the mean rate of this growth, and take the result as our index for instability.

This check for the empirical validity of our theory is confined by a data display of σ–g relations, Figure 2.8.

The σ–g space is divided into four quadrants by the world average indexes of growth $W_g = 3.90$ and instability $W_\sigma = 1.56$. Three countries, China, India, and Brazil, fall into Quadrant I with the best performance: low instability and high growth. China's performance is most decisive with its growth reaching 16.81% and instability staying low at 0.4. This success arises from the fact that its capital account has been opened gradually; and in the process of opening, the GDP maintains a rapid growth with price level kept stable, its international competitiveness continues to strengthen, foreign exchange reserves become more abundant, and its financial markets have been strengthened gradually. India's performance is also

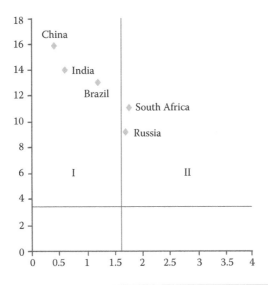

Figure 2.8 **Economic growth and instability of BRICS countries. (From The Wind Database.)**

impressive. Gradual capital account opening provides the financial security needed for the country's economic development and stimulates the development potential of its economy. During the process of capital account opening, the government played a very mature and rational role, and has followed the predetermined steps and not easily to make any promise. Generally speaking, capital account opening in Brazil is successful. After the opening, the GDP grows stably, the problem of hyperinflation has been resolved, exports grow rapidly, its international competitiveness has been greatly strengthened, its international reserves have become more abundant, and its financial markets have also been greatly developed. Even so, its exchange rate and short-term foreign capital flows are still unstable, and the price level is still not quite stable.

South Africa and Russia are situated in Quadrant II characterized by both high growth and high instability. South Africa's foreign currency reserves are insufficient, making it hard to provide strong support for a convertible currency. Its exchange rate is too volatile, which inhibits the inflows of foreign capital. In the inflows of foreign capital has mainly been the foreign portfolio investment, which is vulnerable to the attack of short-term capital flows. On the other hand, Russia's openness is successful in general. But it took a radical way, which has caused economic turmoils and a huge loss of social welfare, making Russian economy lose a *decade*.

BRICS countries all fall into Quadrant I and Quadrant II, which implies that they share the same great economic growth but differ in instability (Figure 2.8). China, Brazil, and India have all opened gradually, while their financial markets are all strengthened along the way. But Russia and South Africa took radical approaches, although their rates of economic growth have been greatly improved, but their weak financial sectors have been left with great instability. We find that a desirable combination of lower instability and higher growth can be accomplished but only via strengthened financial conditions.

2.8.4 Conclusion

In this section, we investigated the optimal trade-offs between economic growth and instability by opening up economies based on their policy preferences about risky growth opportunities offered by international capital flows. Policy choices of many countries make these countries care about their potential economic instability due to their financial vulnerabilities when they seek high growth potentials via risky capital flows.

This section reveals the c-effect of capital flows, which tells us that when an economy without a sound financial sector is widely open under policies of lower risk aversion, large instability must be endured in exchange for the potential of high rates of economic growth. This section also establishes the b-effect, implying that high instability is an inevitable price paid for rapid economic growth if the host countries permit wide capital market opening without strengthening their financial sectors first.

Chapter 3

From European to Asian Option: PDE for Capital Account Liberation

3.1 General Method of Parabolic Partial Differential Equations of Second Order

In this section, we briefly introduce the general method for solving a sort of second order parabolic differential equations, which has found many applications in the study of various kinds of options.

Assume that W_t is a standard Brown motion, r is a nonrisk interest rate, q is the profit of the underlying asset, and σ is the volatility of the underlying asset. Suppose that the underlying asset price S_t satisfies the following stochastic differential equation:

$$dS_t = (r - q)S_t dt + \sigma S_t dW_t$$

Then the Asian call option of the arithmetic average, whose strike price is K and date of expiration is T, satisfies

$$C_T = (J_T - K)^+, \quad \text{where } J_T = \frac{1}{T} \int_0^T S_\xi d\xi$$

According to the principle of reproducibility, we derive a second-order parabolic partial differential equation (PDE) and an initial condition as follows:

$$\begin{cases} \dfrac{\partial C}{\partial t} + \dfrac{S-J}{t}\dfrac{\partial C}{\partial J} + \dfrac{\sigma^2}{2}S^2\dfrac{\partial^2 C}{\partial S^2} + (r-q)S\dfrac{\partial C}{\partial S} - rS = 0 \text{ in } \Sigma \\ C(S,J,T) = (J-K)^{+} \\ \Sigma = \left\{ (S,J,t) \middle| 0 < S < +\infty, 0 < J < +\infty, 0 \le t < T \right\} \end{cases}$$

(3.1)

To simplify Equation 3.1, let us take the following independent variable substitution:

$$\begin{cases} x = \dfrac{TK - tJ}{S}e^{-(r-q)r} - \dfrac{1}{r-q}\left(1 - e^{-(r-q)r}\right) \\ \tau = T - t \\ C(S,J,t) = \dfrac{S}{T}e^{-q\tau}V(x,\tau) \end{cases}$$

(3.2)

By plugging (3.2) into Equation 3.1, we obtain

$$\begin{cases} \dfrac{\partial V}{\partial \tau} - \dfrac{\sigma^2}{2}\left[x + Q(\tau)\right]^2\dfrac{\partial^2 V}{\partial x^2} = 0, & -\infty < x < \infty, \tau > 0 \\ V(x,0) = (-x)^{+}, & -\infty < x < \infty \end{cases}$$

(3.3)

where $Q(\tau) = \left(1/(r-q) \right)\left[1 - e^{-(r-q)\tau} \right]$. According to Fichera theory of PDEs, we know that $V(x,\tau) = -x$, if $x < -Q(\tau)$. Therefore, it is sufficient for us to look for an approximation solution of $V(x,\tau)$ when $x > -Q(\tau)$.

Without loss of generality, we assume $r > q$ and denote $\alpha = 1/(r-q)$. Therefore, $V(x,\tau)$ is the solution of the following problem:

$$\begin{cases} \dfrac{\partial V}{\partial \tau} - \dfrac{\sigma^2}{2}\left[x + Q(\tau)\right]^2\dfrac{\partial^2 V}{\partial x^2} = 0, & -\alpha < x < +\infty, 0 < \tau \le T \\ V(-\alpha,\tau) = \alpha, & 0 < \tau \le T \\ V(x,0) = (-x)^{+}, & -\alpha < x < +\infty \end{cases}$$

(3.4)

The comparison principle in PDEs tells us the following fact: For problem (3.4), its solution $V(x,\tau) \to 0$, as $x \to +\infty$. Therefore, we assume that $V = V_0 + V_1$, where V_0 is an approximation solution of V, and V_1 an error item of V_0, where V_0 satisfies the following:

$$
\begin{cases}
\dfrac{\partial V_0}{\partial \tau} - \dfrac{\sigma^2}{2} Q^2(\tau) \dfrac{\partial^2 V_0}{\partial x^2} = 0, & -\alpha < x < +\infty, 0 < \tau \le T \\
V_0(-\alpha, \tau) = \alpha, & 0 < \tau \le T \\
V_0(x, 0) = (-x)^+, & -\alpha < x < +\infty
\end{cases}
\tag{3.5}
$$

Now, we take the following function substitution:

$$
y = x + \alpha, \eta = \frac{\sigma^2 \alpha^3}{4}\left[-3 + 2(r-q)\tau + 4e^{-(r-q)\tau} - e^{-2(r-q)\tau}\right]
$$

$$
g(y, \eta) = V_0(x, \tau) - \alpha
\tag{3.6}
$$

By applying an odd expansion of the function g, we can get the following problem in the standard form:

$$
\begin{cases}
\dfrac{\partial g}{\partial \eta} - \dfrac{\partial^2 g}{\partial y^2} = 0, & -\infty < y < +\infty, \eta > 0 \\
g(y, 0) = \varphi(y), & -\infty < y < +\infty
\end{cases}
\tag{3.7}
$$

where

$$
\varphi(y) = \begin{cases}
-\alpha, & y > \alpha \\
-y, & -\alpha \le y \le \alpha \\
\alpha, & y < -\alpha
\end{cases}
$$

It is a straightforward task to solve Equation 3.7 by using the Poisson formula. That is, as $y \ge 0$, we have

$$
g(x, \eta) = \frac{1}{2\sqrt{\pi\eta}} \int_{-\infty}^{+\infty} \varphi(x)e^{-\left((y-x)^2/4\eta\right)}\, du
$$

$$
= \alpha\left[N\left(-\frac{\alpha+y}{\sqrt{2\eta}}\right) + N\left(\frac{\alpha-y}{\sqrt{2\eta}}\right) - 1\right] + \sqrt{\frac{\eta}{\pi}}\left[e^{-\left((y-\alpha)^2/4\eta\right)} - e^{-\left((y+\alpha)^2/4\eta\right)}\right]
$$

$$
- y\left[N\left(\frac{\alpha-y}{\sqrt{2\eta}}\right) - N\left(-\frac{\alpha+y}{\sqrt{2\eta}}\right)\right]
$$

where $N(x) = \dfrac{1}{\sqrt{2\pi}} \displaystyle\int_{-\infty}^{x} e^{-\left(u^2/2\right)} du$. By substituting g into the representation of $V_0(x,\tau)$, we obtain

$$V_0\left(x,\tau\right) = -xN\left(-\frac{x}{\sqrt{2\tau}}\right) + \sqrt{\frac{\tau}{\pi}} e^{-\left(x^2/4\eta\right)} + \left(x+2\alpha\right)N\left(-\frac{x+2\alpha}{\sqrt{2\tau}}\right)$$

$$-\sqrt{\frac{\tau}{\pi}} e^{-\left(\left(x+2\alpha\right)^2/4\eta\right)} \tag{3.8}$$

where η is represented by (3.6). Notice that some related partial derivatives are given as follows:

$$\frac{\partial V_0}{\partial x} = N\left(-\frac{x+2\alpha}{\sqrt{2\eta}}\right) - N\left(-\frac{x}{\sqrt{2\eta}}\right)$$

$$\frac{\partial V_0}{\partial \eta} = \frac{\partial^2 V_0}{\partial x^2} = \frac{1}{2\sqrt{\pi\eta}}\left[e^{-\left(x^2/4\eta\right)} - e^{-\left(\left(x+2\alpha\right)^2/4\eta\right)}\right]$$

$$= \frac{\sigma^2}{2} Q^2\left(\tau\right) \frac{\partial V_0}{\partial \tau} \tag{3.9}$$

Finally, we establish the fact that the error item V_1 satisfies the following initial-value problem:

$$\begin{cases} \dfrac{\partial V_1}{\partial \tau} - \dfrac{\sigma^2}{2}\left[x+Q(\tau)\right]^2 \dfrac{\partial^2 V_1}{\partial x^2} = \dfrac{\sigma^2 x}{4\sqrt{\pi\eta}}\left[x+2Q(\tau)\right]\left[e^{-\frac{x^2}{4\eta}} - e^{-\frac{\left(x+2\alpha\right)^2}{4\eta}}\right], \\ \qquad\qquad -\alpha < x < +\infty,\, 0 < \tau \le T \\ V_1\left(-\alpha,\tau\right) = 0, \qquad -\alpha < x < +\infty \\ V_1\left(x,0\right) = 0, \qquad\quad 0 < \tau \le T \end{cases} \tag{3.10}$$

3.2 Pricing of Carbon Emission Cost: Linear Parabolic PDEs (I)

In this section, a linearization method is applied to the STIRPAT model of China's carbon emission projections, where we use the option pricing formula to elicit the analytic solution of the potential cost of a nation's carbon emissions. Based on the analytic solution, we analyze the short-term and long-term marginal cost of carbon emissions, how the marginal cost changes over the level of carbon emission, and how the carbon emission cost changes over GDP. In the end, we establish the conclusion that carbon emission reduction should be considered in a long period of time.

3.2.1 Introduction

At present, mankind is confronted with three major challenges: climate change, environmental pollution, and resource constraints. The world economy will face a transition to a state of low carbon emission. The industrial competitiveness of the *Post-Copenhagen Era* will reflect low carbon emission and relevant economic competition in all respects. In this context, it is important to summarize the low carbon economy development model with Chinese characteristics and explore a new *Chinese Road* that encourages the development of a low carbon economy.

Currently, the STIRPAT model has been widely used, its core consists of multiple regressions. Based on the STIRPAT model and the panel data methodology, Li and Li (2010) calculated the carbon dioxide emission in 30 Chinese provinces, which can be divided into low, medium, and high emission regions. Their empirical results showed that the differences in the carbon dioxide emissions among the regions are clearly observable and have shown a trend of expansion. In addition, the level of China's carbon dioxide emission showed a clear phenomenon of path dependence, indicating that the existing carbon reduction policy is difficult to change. Zhu et al. (2010) extended the STIRPAT model, where the impacts from population, consumption, and technology on carbon emission are analyzed econometrically by using the ridge regression method. They showed that the population size, population, urbanization, and the level of people's consumption have significant impact on China's carbon emission, and the impact of changing demographics on carbon emissions has been higher than the influence of changes in population size.

From the point of view of analysis methods and based on the Environment Kuznets Curve (EKC) method, Lin and Jiang (2009) predicted the amount of CO_2 emission. Their projections showed that the year 2020 should be the knee point. There is also a large amount of literature that studied the impact of China's carbon emission factors based on logarithmic mean Di's index (LMDI) of Kaya identities such as Wang et al. (2005). The main conclusion of these works is that the impact of carbon emissions in China is a major factor of the descending order of economic development among other factors such as the level of energy production, the structure of energies, and the industrial structure. By using the contribution of China's per capita carbon emission factors in the time period of 1990–2007, Lin and Jiang (2009) found that the per capita GDP, energy intensity, and energy structure all have significant impacts on China's per capita carbon emission. The per capita GDP and energy intensity have the greatest impact on the per capita CO_2 emission, while the impact of the energy structure is relatively small.

Carbon emissions accounting is the basis of calculation for carbon emission reductions and carbon trading. Current related research on carbon emissions accounting include: accounting standards, accounting methods, and applications (Alessandro et al., 2005).

The research methodology on carbon emissions mainly includes the IPCC inventory method, International Council for Local Environmental Initiatives (ICLEI) methods, measurement method, material balance method, factor decomposition method, life cycle analysis, input–output analysis, and others. The IPCC inventory method is provided by the *Guidelines for National Greenhouse Gas Inventories* for estimating national greenhouse gas anthropogenic emissions and removals. It provides a variety of default emission factors (the default values) for reference (IPCC, 2006). Countries can also base on their respective domestic research or their respective uses of different models to determine emission factors of their country-specific emission sources. The method is simple and practical. The ICLEI methodology is ICLEI drown on the urban CO_2 emission experience from the climate change plan (CCP) action plan, developed as five milestone framework inventory accounting software to reduce emissions and greenhouse gas—Clean Air and Climate Protection (CACP). It mainly analyzes four department emissions: energy use, traffic condition, solid waste, and other emissions. The measurement method calculates gas emissions by collecting exhaust gases wands and measures their velocities, flow rates, concentrations, etc. The material balance is a scientific and effective method of computation. It helps with the conduction of quantitative analysis of the production of materials used in the process by combining emissions, production technology and management, resources (raw materials, water, energy) of the industrial sources, and systematically and comprehensively studies the production and emission in production processes (Ministry of Environmental Protection, 2001). The factor decomposition method is a more common tool for exploring the impact of carbon emissions. It is capable of studying how to reduce CO_2 emissions, conducting quantitative analysis of various factors that impact CO_2 emissions, and revealing relationships between CO_2 emissions and various contributing factors. Scholars from different nations have done a lot of research and built a variety of models. The common used models are the IPAT model (Ehrlich and Holden, 1971), the STIRPAT model (York et al., 2003), the LMDI decomposition method (Chunbo and David, 2008), and so on. The life cycle analysis/assessment (LCA) requires a detailed study of a life cycle's energy demand, raw materials used, and waste emissions into the environment resulting from various activities. It includes raw material recycling, mining, transportation, manufacturing/processing, distribution, use/reuse/maintenance, and the waste disposal. Since the study originated in resource development activities, one needs to track all processes of energy, raw materials used in the processes, and the formation of a full energy chain; it comprehensively conducts quantitative and qualitative analyses of gas emissions of each link in the chain. When using this method to the study of emissions of greenhouse gases of each active process, the chain of activities becomes the classification unit of the study object.

The input–output model generally consists of a set of linear equations that describe the complex linkages between economic sectors; through appropriate extensions, it includes the relationship between economic behavior and environment. This model has been used for environmental policy analysis, while its main use has been the analysis of CO_2 emissions inventory, implicit carbon emission of foreign trades, and industrial effects of reduction policies. However, the coefficients of the model are fixed; it is difficult to replace the relevant elements described in the model and to include changes in technology and behavior. This subsection makes use of linearization of the variable coefficients of the STIRPAT model of PDEs initially established for predicting carbon emissions, obtains an option pricing formula based on the STIRPAT model, analyzes the interpretation role of various factors on carbon emissions, and proves what should be the concerning factor in carbon emission reductions.

3.2.2 The Model

Ehrlich and Holden (1971) considered the impact of population pressure on the environment by using the IPAT equation. Later, scholars established the current widely used STIRPAT model on the basis of the IPAT equation by avoiding the nonlinear relationship between the independent variables and the dependent variable that cannot be measured by the IPAT model. The estimation equation of the STIRPAT model is

$$I = cP^{\beta_1} A^{\beta_2} T^{\beta_3} e^u \tag{3.11}$$

where
  I denotes the effect of carbon dioxide emission
  P denotes the population
  A denotes the wealth
  T denotes the technology

By taking logarithm to both sides of the previous equation, we get the following multivariate regression model on $\ln I$:

$$\ln I = \tilde{c} + \beta_1 \ln P + \beta_2 \ln A + \beta_3 \ln T + u \tag{3.12}$$

Emilio and Francesco included technical factors T in the error term. By letting A represent the total GDP, Y the real per capita GDP, then we have $A = PY$, and Equation 3.12 can be rewritten as

$$\ln I = \tilde{c} + (\beta_1 + \beta_2) \ln P + \beta_3 \ln Y + u \tag{3.13}$$

If we denote $b = \beta_1 + \beta_2$ and $c = \beta_3$, then the differential equation form of Equation 3.13 is

$$\frac{\dot{I}}{I} = b\frac{\dot{P}}{P} + c\frac{\dot{Y}}{Y} \tag{3.14}$$

where $\dot{I}, \dot{P}, \dot{Y}$ represent the derivatives of I, P, Y, respectively, with respect to t.

The logistics population growth model has been applied in many areas of research. If $P(P = P(t))$ indicates the population, P_0 denotes a base population, a constant ρ denotes the population growth rate, and assume that the environment can support a maximum population size of P_m, then the actual growth rate of the population size ρ will increase as the population P declines, and satisfies the following equation:

$$\rho(P) = \rho\left(1 - \frac{P}{P_m}\right) \tag{3.15}$$

Therefore, we can obtain the following differential equation and initial condition:

$$\begin{cases} \dfrac{dP}{dt} = \rho P\left(1 - \dfrac{P}{P_m}\right) \\ P(0) = P_0 \end{cases} \tag{3.16}$$

The solution of Equation 3.16 can be solved for as follows:

$$P(t) = \frac{P_m}{1 + \left(\left(P_m/P_0\right) - 1\right)e^{-\rho t}} \tag{3.17}$$

$$\frac{dP}{P} = \frac{\rho\left(\left(P_m/P_0\right) - 1\right)e^{-\rho t}}{1 + \left(\left(P_m/P_0\right) - 1\right)e^{-\rho t}} \tag{3.18}$$

Assume that the growth law of China's GDP Y satisfies the geometric Brownian motion, namely,

$$\frac{dY}{Y} = \mu dt + \sigma dW_t \tag{3.19}$$

where W_t is a standard Brownian motion, $E(dW_t) = 0$, $Var(W_t) = dt$, then from (3.14) and (3.18) we have

$$\frac{dI}{I} = f(t)dt + \hat{\sigma}\, dW_t \tag{3.20}$$

where

$$f(t) = \frac{c\mu + (b\rho + c\mu)\big(\big(P_m/P_0\big)-1\big)e^{-\rho t}}{1+\big(\big(P_m/P_0\big)-1\big)e^{-\rho t}}, \hat{\sigma} = c\sigma$$

If the carbon emission of a country must be reduced to a level below the specified threshold at the specified maturity date, and at that time the country's carbon emissions are not reduced to the target level, the part of excess must be purchased at price a dollars/t from the international market. This constraint can be abstracted as the final condition of an European call option whose trading object is the carbon dioxide emission I. The final value of the option is expenditures for the purchase of carbon credits at the maturity date T. If potential expenditure of the country is $C(I,t)$, then the potential expenditure at maturity date is $C = a(I_t - \bar{I})^+$. It is assumed that the risk-free rate is a constant r. So at time t, the conditional expectation of the potential expense is given as follows:

$$C(I,t) = E\left(a\big(I_T - \bar{I}\big)^+ e^{-r(T-t)} \mid I_{t_0} = I_0\right) \tag{3.21}$$

Because I is not financial asset itself, we cannot make these expenditures as risk-free expenditure through hedging. As is known from the Feynman–Kac formula that $C(I,t)$ is suitable for nonhomogeneous backward parabolic Cauchy problem, by using the formula we produce

$$\begin{cases} \dfrac{\partial C}{\partial t} + \dfrac{c\mu + (b\rho + c\mu)\big(\big(P_m/P_0\big)-1\big)e^{-\rho t}}{1+\big(\big(P_m/P_0\big)-1\big)e^{-\rho t}} I \dfrac{\partial C}{\partial I} + \dfrac{I^2}{2}c^2\sigma^2 \dfrac{\partial^2 C}{\partial I^2} - rC = 0 \\[2mm] C(I,T) = a(I_T - \bar{I})^+ \end{cases} \tag{3.22}$$

3.2.3 The Calculation

Denote

$$\tilde{r}(t) = \frac{c\mu + (b\rho + c\mu)\big(\big(P_m/P_0\big)-1\big)e^{-\rho t}}{1+\big(\big(P_m/P_0\big)-1\big)e^{-\rho t}}$$

After straightforward calculations, Equation 3.22 becomes

$$\frac{\partial C}{\partial t} + \frac{(c\sigma)^2}{2} I^2 \frac{\partial^2 C}{\partial I^2} + \tilde{r}(t) I \frac{\partial C}{\partial I} - rC = 0 \tag{3.23}$$

According to the method introduced in the previous section, we take the following independent variable transformation for (3.23): $x = \ln I$, $\tau = T - t$. Then Equation 3.23 can be changed into the following form:

$$-\frac{\partial C}{\partial \tau} + \frac{1}{2}(c\sigma)^2 \left(\frac{\partial^2 C}{\partial x^2} - \frac{\partial C}{\partial x} \right) + \tilde{r}(\tau) \frac{\partial C}{\partial x} - rC = 0$$

So, (3.22) can be rewritten as follows:

$$\begin{cases} \dfrac{\partial C}{\partial \tau} - \dfrac{1}{2}(c\sigma)^2 \dfrac{\partial^2 C}{\partial x^2} - \left[\tilde{r}(\tau) - \dfrac{1}{2}(c\sigma)^2 \right] \dfrac{\partial C}{\partial x} + rC = 0 \\ C\big|_{\tau=0} = a(e^x - \bar{I})^+ \end{cases} \tag{3.24}$$

By taking the function transformation $C = u e^{\alpha \tau + \beta x}$ for the Cauchy problem (3.24), we have

$$\frac{\partial u}{\partial \tau} - \frac{1}{2}(c\sigma)^2 \frac{\partial^2 u}{\partial x^2} - \left[\beta(c\sigma)^2 + \tilde{r}(\tau) - \frac{1}{2}(c\sigma)^2 \right] \frac{\partial u}{\partial x}$$

$$+ \left[r - \beta(\tilde{r}(\tau) - \frac{1}{2}(c\sigma)^2) - \frac{1}{2}(c\sigma)^2 \beta^2 + \alpha \right] u = 0 \tag{3.25}$$

In this section, we just consider the following two cases:

1. *Case 1*: When $t \to \infty$, $\lim\limits_{t \to \infty} \tilde{r}(t) = c\mu$.

The corresponding practical significance is that when time is sufficiently large and close to the specified due date, the country must consider reducing its carbon emission to a level below the predetermined threshold; otherwise, the part of excess must be purchased at price a dollars/t from the international market. Let us now first consider this situation.

In Equation 3.25, let

$$
\begin{cases}
\beta = \dfrac{1}{2} - \dfrac{c\mu}{(c\sigma)^2} = \dfrac{1}{2} - \dfrac{\mu}{c\sigma^2} \\[3mm]
\alpha = \beta\left[c\mu - \dfrac{1}{2}(c\sigma)^2 \right] + \dfrac{1}{2}(c\sigma)^2 \beta^2 - r
\end{cases}
$$

Then, we have

$$
\begin{cases}
\alpha = \dfrac{1}{2}c\mu - \dfrac{1}{8}(c\sigma)^2 - \dfrac{1}{2}\dfrac{\mu^2}{\sigma^2} - r \\[3mm]
\beta = \dfrac{1}{2} - \dfrac{\mu}{c\sigma^2}
\end{cases}
\tag{3.26}
$$

By combining this equation with (3.25) and its boundary conditions, we get the following standard Cauchy problem:

$$
\begin{cases}
\dfrac{\partial u}{\partial \tau} = \dfrac{1}{2}(c\sigma)^2 \dfrac{\partial^2 u}{\partial x^2} \\[3mm]
u\big|_{\tau=0} = ae^{-\beta x}(e^x - \bar{I})^+
\end{cases}
\tag{3.27}
$$

By using the Poisson formula, we obtain the following solutions:

$$
u(x,\tau) = \int_{-\infty}^{+\infty} \frac{1}{c\sigma\sqrt{2\pi\tau}} e^{-\left((x-\xi)^2/2c^2\sigma^2\tau\right)} \left[e^{-\beta\xi}(e^\xi - \bar{I})^+ \right] d\xi
$$

$$
= \int_{\ln\bar{I}}^{+\infty} \frac{1}{c\sigma\sqrt{2\pi\tau}} e^{-\left((x-\xi)^2/2c^2\sigma^2\tau\right)} \left[e^{(1-\beta)\xi} - \bar{I}e^{-\beta\xi} \right] d\xi
\tag{3.28}
$$

By noting that $\beta = (1/2) - (\mu/c\sigma^2)$, we obtain

$$
c(x,\tau) = e^{\alpha\tau + \beta x} u(x,\tau) = e^{-r\tau - \frac{1}{2}\left(\frac{\mu}{\sigma} - \frac{1}{2}c\sigma\right)^2 \tau + \left(\frac{1}{2} - \frac{\mu}{c\sigma^2}\right)} \qquad u(x,\tau) = I_1 + I_2
$$

Let us now first calculate I_1 as follows:

$$I_1 = \int_{\ln \bar{I}}^{+\infty} \frac{1}{c\sigma\sqrt{2\pi\tau}} \exp\left\{ \begin{array}{l} -r\tau - \dfrac{1}{2}\left(\dfrac{\mu}{\sigma} - \dfrac{1}{2}c\sigma\right)^2 \tau + \left(\dfrac{1}{2} - \dfrac{\mu}{c\sigma^2}\right)x \\ -\dfrac{(x-\xi)^2}{2c^2\sigma^2\tau} + \left(\dfrac{1}{2} + \dfrac{\mu}{c\sigma^2}\right)\xi \end{array} \right\} d\xi$$

$$= \int_{\ln \bar{I}}^{+\infty} \frac{1}{c\sigma\sqrt{2\pi\tau}} \exp\left\{ -\frac{1}{2c^2\sigma^2\tau}\left[(x-\xi) + \left(\mu + \frac{c\sigma^2}{2}\right)c\tau\right]^2 + (c\mu - r)\tau + x \right\} d\xi$$

$$= e^{x+(c\mu-r)\tau} \int_{-\infty}^{\frac{x-\ln\bar{I}+\left(\mu+\frac{c\sigma^2}{2}\right)c\tau}{c\sigma\sqrt{\tau}}} e^{-(w^2/2)} dw$$

$$= e^{x+(c\mu-r)\tau} \cdot N(d_1)$$

where

$$d_1 = \frac{\ln\left(I/\bar{I}\right) + \left(\mu + \left(c\sigma^2/2\right)\right)c(T-t)}{c\sigma\sqrt{T-t}}$$

Next, we calculate I_2 as follows:

$$I_2 = -\bar{I}\int_{\ln \bar{I}}^{+\infty} \frac{1}{c\sigma\sqrt{2\pi\tau}} \exp\left\{ \begin{array}{l} -r\tau - \dfrac{1}{2}\left(\dfrac{\mu}{\sigma} - \dfrac{1}{2}c\sigma\right)^2 \tau + \left(\dfrac{1}{2} - \dfrac{\mu}{c\sigma^2}\right)x - \dfrac{(x-\xi)^2}{2c^2\sigma^2\tau} \\ + \left(-\dfrac{1}{2} + \dfrac{\mu}{c\sigma^2}\right)\xi \end{array} \right\} d\xi$$

$$= -e^{-r\tau}\bar{I} \cdot N(d_2)$$

where

$$d_2 = \frac{\ln\left(I/\bar{I}\right) + \left(\mu - \left(c\sigma^2/2\right)\right)c(T-t)}{c\sigma\sqrt{T-t}}$$

Finally, we obtain the analytical solution of the country's potential expenditure for its carbon emission:

$$C(I,t) = aIe^{(c\mu-r)(T-t)}N(d_1) - a\bar{I}e^{-r(T-t)}N(d_2) \tag{3.29}$$

Next, we calculate the partial derivative of (3.20) with respect to I:

$$\frac{\partial C(I,t)}{\partial I} = ae^{(q\mu-r)(T-t)}N(d_1(t)) + \frac{a}{c\sigma\sqrt{T-t}}\left[\begin{array}{l} e^{(q\mu-r)(T-t)} \cdot \frac{1}{\sqrt{2\pi}} \cdot e^{-\left(d_1^2/2\right)} \\ \\ -\frac{\overline{I}}{I}e^{-r(T-t)} \cdot \frac{1}{\sqrt{2\pi}}e^{-\left(d_2^2/2\right)} \end{array} \right]$$

$$= ae^{(q\mu-r)(T-t)}N(d_1(t)) > 0 \qquad (3.30)$$

This inequality shows that when other factors remain unchanged, the cost C is an increasing function of the carbon emission I. The greater the carbon emission I is, the higher the cost will be. Meanwhile, from the principles of economics, it follows that $(\partial C(I,t)/\partial I)$ is the marginal cost when the CO_2 emission is increased by a unit. Considering the marginal cost can provide a basis for when the country invests in carbon reduction projects. If the marginal cost of carbon emission reduction projects is less than the marginal cost of carbon emissions, one can invest in emission reduction projects; on the contrary, one does not invest.

By calculating the second-order partial derivative of C with respect to I, we have

$$\frac{\partial^2 C}{\partial I^2} = ae^{(q\mu-r)(T-t)}\frac{\partial N(d_1)}{\partial I} = \frac{a}{Ic\sigma\sqrt{T-t}}e^{(q\mu-r)(T-t)-((1/2)d_1^2(t))} > 0$$

From the inequality $(\partial^2 C(I,t)/\partial I^2) > 0$, it follows that the marginal cost of carbon emission is an increasing function of emissions. This means that the greater the amount of carbon emission is, the higher marginal cost is. Therefore, investing early in carbon reduction projects is necessary. As long as the marginal cost of reduction is less than the marginal cost of the initial time $(\partial C(I,0)/\partial I)$, one can continue to invest in projects to reduce carbon emission.

By taking the partial derivative with respect to σ based on (3.30), we get the following relationship between the cost of carbon emission and GDP growth volatility:

$$\frac{\partial C}{\partial \sigma} = aIe^{(q\mu-r)(T-t)} \cdot \frac{1}{\sqrt{2\pi}} \cdot e^{-((1/2)d_1^2)} \cdot \frac{\partial d_1}{\partial \sigma} = \frac{ac\overline{I}\sqrt{T-t}}{\sqrt{2\pi}}e^{(q\mu-r)(T-t)-((1/2)d_1^2(t))} > 0$$

$$(3.31)$$

This inequality shows that the cost function of carbon reduction $C(I,t)$ is an increasing function of the volatility σ. That is, when the volatility in the GDP growth increases, the risk that the carbon emission I exceeds a predetermined threshold $\overline{I}$ increases at moment T so that the potential spending will increase.

2. *Case 2*: When $t \to 0$, $\lim\limits_{t \to \infty} r(t) = \tilde{r}(0)$. That is, since the beginning, we have to consider the potential cost of the carbon emission problem at the present moment. Since $\tilde{r}(0) = \left(1 - \left(P_0/P_m\right)\right)b\rho + c\mu$, we denote it as r_0; in this situation, (3.26) implies

$$\begin{cases} \alpha = \dfrac{r_0}{2} - \dfrac{1}{8}(c\sigma)^2 - \dfrac{1}{2}\dfrac{r_0^2}{c^2\sigma^2} - r_0 \\[4mm] \beta = \dfrac{1}{2} - \dfrac{r_0}{c^2\sigma^2} \end{cases} \tag{3.32}$$

Therefore, we obtain the following formula:

$$C(I,t) = aIe^{(r_0 - r)(T-t)}N(d_1) - a\bar{I}e^{-r(T-t)}N(d_2) \tag{3.33}$$

By calculating the partial derivative with respect to I based on (3.33), we get

$$\frac{\partial C(I,t)}{\partial I} = ae^{(r_0 - r)(T-t)}N(d_1(t)) + \frac{a}{c\sigma\sqrt{T-t}}\left[\begin{array}{l} e^{(c\mu - r)(T-t)} \cdot \dfrac{1}{\sqrt{2\pi}} \cdot e^{-\left(d_1^2/2\right)} \\[3mm] -\dfrac{\bar{I}}{I}e^{-r(T-t)} \cdot \dfrac{1}{\sqrt{2\pi}}e^{-\left(d_2^2/2\right)} \end{array}\right]$$

$$= ae^{(r_0 - r)(T-t)}N(d_1(t)) > 0 \tag{3.34}$$

$$\frac{\partial^2 C}{\partial I^2} = ae^{(r_0 - r)(T-t)}\frac{\partial N(d_1)}{\partial I} = \frac{a}{Ic\sigma\sqrt{T-t}}e^{(r_0 - r)(T-t)-\left((1/2)d_1^2(t)\right)} > 0 \tag{3.35}$$

By calculating the partial derivative with respect to σ based on (3.33), we get the following relationship between the cost of carbon emission and the volatility of the GDP growth:

$$\frac{\partial C}{\partial \sigma} = aIe^{(r_0 - r)(T-t)} \cdot \frac{1}{\sqrt{2\pi}} \cdot e^{-\frac{1}{2}d_1^2} \cdot \frac{\partial d_1}{\partial \sigma} = \frac{ac\bar{I}\sqrt{T-t}}{\sqrt{2\pi}}e^{(r_0 - r)(T-t)-\frac{1}{2}d_1^2(t)} > 0 \tag{3.36}$$

By using comparisons, we find that the potential short-term expenditure is similar to that of the long term. Note the fact that when $t \to 0$, $r(t) \to r_0$, $r_0 = \left(1 - \dfrac{P_0}{P_m}\right)b\rho + c\mu > c\mu, (t \to \infty)$ indicates that the marginal cost of carbon emission increases with the amount of emission, and that the impact of the volatility of GDP growth on the potential expense is larger in the short term than that in the long term. So, it is possible to wait for quite a while before starting to think about the problem of carbon emission.

3.2.4 Conclusions

Undoubtedly, urbanization represents a significant factor for China's recent increasing carbon emission. The rapid and excessive urbanization have caused some irreversible consequences. The push of urbanization is not a natural development for the purpose of meeting market needs, but a consequence of the strong planned economy system and strong administrative measures. Recently, the proposal to accelerate the pace of urbanization represents a typical phenomenon of a planned economy. The urbanization is a complex process that is hundred times more complex than any single product and any single economic indicator. It involves population distribution, resource allocation, and various aspects of the environment and the society. Whether or not the Chinese growth rate of carbon emission can be controlled by appropriately regulating the speed of urbanization remains to be seen and be further demonstrated.

In terms of the policy instruments that might be effective for carbon emission reduction, there are generally two kinds of emission trading and taxation. Existing research has different views on which policy instruments is more suitable for China. From our point of view, the carbon tax might be more effective than the carbon emission market in China. It also requires further use of a dynamic CGE model of Chinese economy to make long-term analysis on carbon tax policies taken in the future and their impact on the economy.

3.3 Pricing of Foreign Currency Option: Linear Parabolic PDEs (II)

3.3.1 Introduction

With the development of economic globalization, trade exchanges worldwide have been increasing constantly. Fluctuating exchange rates between national currencies make the transaction process of trades often experience additional risk. Efforts have been made to get rid of the negative impact of the economic crises that are triggered by infection. Because of the movement of globalization, foreign exchanges will become more and more widespread. That will naturally encounter additional risks arising from exchange rate fluctuations. Especially since the reform of the RMB exchange rate mechanism, the exchange rate fluctuation of RMB has been expanded in magnitude. Effectively avoiding and controlling the exchange rate risk have become an important task for all kinds of foreign economic entities. Rate option is an effective tool to avoid the foreign exchange risk. It is of not only a characteristic of nonlinear gain, but also taking into account the investment, speculation, hedging, and risk aversion function. Therefore, the study of exchange rate options, especially for the reasonable pricing of the options, has important theoretical and practical significance for the development of exchange rate options market of China. Currently, the pricing method of exchange rate options is primarily derived by Black–Scholes model. Garman and Kohlhagen (1983) explicitly provided the first formula (G-K model) for pricing European exchange rate options. Subsequent studies could at most be considered as amendments and extensions of the G-K model. This model assumes that the domestic rate is constant; but in reality, both domestic and foreign interest rates are random variables. Grabbe (1983) studied the problem of modeling the pricing of short-term interest rate options, but did not give an explicit pricing formula. Amin and Jarrow (1991) considered the pricing model for European exchange options, where the national interest and foreign interest rates obey a forward rate model, while the exchange rate follows a geometric Brownian motion, and established an explicit pricing formula. Because G-K model assumes that the exchange rate follows a geometric Brownian motion, while in reality, the rate often experiences the phenomenon of random jumps, Bernard et al. (1995) introduced the Merton jump diffusion model to analyze option pricing rates.

Chinese scholars also have a strong interest in the study of the pricing of exchange rate options. For example, Lin and Xu (2006) have analyzed several specific pricings of financing products related to foreign exchanges. Shen (2008) has studied the pricing problem of European exchange rate options under fractional Brownian motion (FBM). Zhang et al. (2008) connected jumps with FBM, and derived a pricing formula for European exchange rate options under jump fractal processes. On this basis, we establish a kind of trigger rate option pricing model, and derive a condensed option pricing formula. This formula facilitates investors to analyze the influence of various parameters on the value of options.

3.3.2 The Model

In order to establish a simple pricing model, we make the following assumptions:

1. There are two kinds of currencies, A and B; the exchange rate S_t between the two currencies obey the following geometry Brown motion:

$$\frac{dS_t}{S_t} = (r_f - r_d)dt + \sigma dW_t \qquad (3.37)$$

where
r_d and r_f are the riskless rates of currencies of A and B, respectively
σ is the market volatility
r_d, r_f, and σ are some positive constants
W_t is the standard Brown motion

2. The market provides no arbitrage opportunity.
3. No transaction cost and revenue are considered.

Let $V = V(S, t)$ be the product value of 1 unit savings in currency A. By using hedging techniques, we construct the following PDE for the currency option price:

$$\frac{\partial V}{\partial t} + \frac{1}{2}\sigma^2 S^2 \frac{\partial^2 V}{\partial S^2} + (r_d - r_f)S\frac{\partial V}{\partial S} - r_d V = 0 \tag{3.38}$$

According to the specific content of structured deposit product, we can establish the boundary conditions:

$$V(S_a, t) = (1 + r_0 T + \rho T)e^{-r_d(T-t)}, \quad 0 < t < T \tag{3.39}$$

which can be rewritten as follows:

$$V(S,T) = (1 + r_1 T + \rho T)\left[1 - \frac{1}{K}(K - S)^+\right], \quad S < S_a \tag{3.40}$$

Therefore, the value $V(S,t)$ of this kind of structured product is the solution that satisfies the following problem with its solution in the region $\Omega = \{(S,t) | 0 < S < +\infty, 0 < t < T\}$:

$$\begin{cases} \dfrac{\partial V}{\partial t} + \dfrac{1}{2}\sigma^2 S^2 \dfrac{\partial^2 V}{\partial S^2} + (r_d - r_f)S\dfrac{\partial V}{\partial S} - r_d V = 0 \\ V(S_a, t) = (1 + r_0 T + \rho T)e^{-r_d(T-t)} \\ V(S,T) = (1 + r_1 T + \rho T)\left[1 - \dfrac{1}{K}(K - S)^+\right] \end{cases} \tag{3.41}$$

3.3.3 The Solution

Denote $1 + r_0 T + \rho T = \bar{A}$ and $1 + r_1 T + \rho T = \bar{B}$. Let $U = V - \bar{A}e^{-r_d(T-t)}$. Then the problem of solving (3.41) in the region Ω can be converted to

$$\begin{cases} \dfrac{\partial U}{\partial t} + \dfrac{1}{2}\sigma^2 S^2 \dfrac{\partial^2 U}{\partial S^2} + (r_d - r_f)S\dfrac{\partial U}{\partial S} - r_d U = 0 \\ U(S_a, t) = 0 \\ U(S,T) = (\bar{B} - \bar{A}) - \dfrac{\bar{B}}{K}(K - S)^+ \end{cases} \tag{3.42}$$

So, the problem of determining the solution can be decomposed into two parts:

$$U = U_1 - \frac{\bar{B}}{K}U_2$$

where U_1 satisfies the following boundary problem in the region Ω:

$$\begin{cases} \dfrac{\partial U_1}{\partial t} + \dfrac{1}{2}\sigma^2 S^2 \dfrac{\partial^2 U_1}{\partial S^2} + (r_d - r_f)S \dfrac{\partial U_1}{\partial S} - r_d U_1 = 0 \\ U_1(S_a, t) = 0 \\ U_1(S, T) = (\bar{B} - \bar{A}) \end{cases} \tag{3.43}$$

and U_2 satisfies the following boundary problem in the region Ω:

$$\begin{cases} \dfrac{\partial U_2}{\partial t} + \dfrac{1}{2}\sigma^2 S^2 \dfrac{\partial^2 U_2}{\partial S^2} + (r_d - r_f)S \dfrac{\partial U_2}{\partial S} - r_d U_2 = 0 \\ U_2(S_a, t) = 0 \\ U_2(S, T) = (K - S)^+ \end{cases} \tag{3.44}$$

Lemma 3.1 The system (3.44) has an analytical solution in the following form:

$$U(S,t) = p_{down}^{out}(S,t)$$

$$= Se^{-r_f(T-t)}N(d_1) - Ke^{-r_d(T-t)}N(\hat{d}_1) - S_a \left(\frac{S}{S_a}\right)^{-2((r_d - r_f)/\sigma^2)} e^{-r_f(T-t)}N(d_2)$$

$$+ K\left(\frac{S}{S_a}\right)^{1-2((r_d - r_f)/\sigma^2)} e^{-r_d(T-t)}N(\hat{d}_2) - Se^{-r_f(T-t)}N(d_3) + Ke^{-r_f(T-t)}N(\hat{d}_3)$$

$$+ \left(\frac{S}{S_a}\right)^{1-2((r_d - r_f)/\sigma^2)} \left[\frac{S_a^2}{S} e^{-r_f(T-t)}N(d_4) - Ke^{-r_d(T-t)}N(\hat{d}_4)\right] \tag{3.45}$$

where

$$d_1 = \frac{\ln\dfrac{S}{K} + \left(r_d - r_f + \left(\sigma^2/2\right)\right)(T-t)}{\sigma\sqrt{T-t}}$$

$$d_2 = \frac{\ln \frac{S_a^2}{SK} + (r_d - r_f + \frac{\sigma^2}{2})(T - t)}{\sigma\sqrt{T-t}}$$

$$d_3 = \frac{\ln \frac{S}{S_a} + \left(r_d - r_f + \left(\sigma^2/2\right)\right)(T-t)}{\sigma\sqrt{T-t}}$$

$$d_4 = \frac{\ln \frac{S_a}{S} + (r_d - r_f + \frac{\sigma^2}{2})(T-t)}{\sigma\sqrt{T-t}}$$

$$\hat{d}_i = d_i - \sigma\sqrt{T-t}, \quad i = 1,2,3,4$$

By solving (3.43), we obtain

$$U_1(S,t) = (\bar{B} - \bar{A})e^{-r_d(T-t)} \left[N(\hat{d}_3) - \left(\frac{S}{S_a}\right)^{1-2\left((r_d - r_f)/\sigma^2\right)} N(\hat{d}_4) \right]$$

Finally, we can derive the following result:

$$V(S,t) = \bar{A}e^{-r_d(T-t)} - \frac{\bar{B}}{K} p_{down}^{out}(S,t)$$

$$+ (\bar{B} - \bar{A})e^{-r_d(T-t)} \left[N(\hat{d}_3) - \left(\frac{S}{S_a}\right)^{1-2((r_d - r_f/\sigma^2))} N(\hat{d}_4) \right] \quad (3.46)$$

In light of the density function of normal distributions and the integral mean value theorem, it is ready to prove the following result.

Lemma 3.2 There exist constants $0 < \theta_i < 1$, $i = 1,2,3,4$, such that

$$N(\hat{d}_i) = N(d_i) - \sigma\sqrt{T-t} \cdot \exp\left\{-\frac{1}{2}(d_i - \theta_i\sigma\sqrt{T-t})^2\right\}$$

By using both Lemmas 3.1 and 3.2, we can derive the following result.

Theorem 3.1 The system (3.42) has an analytical solution in the following form:

$$V(S,t) = a_0 + \sum_{i=1}^{4} a_i N(d_i) \tag{3.47}$$

where

$$a_0 = \bar{A}e^{-r_d(T-t)}$$

$$a_1 = -\bar{B}Se^{-r_f(T-t)} + \bar{B}e^{-r_d(T-t)} + \bar{B}\sigma\sqrt{T-t}\,e^{-((1/2)(d_1-\theta_1\sigma\sqrt{T-t})^2)}$$

$$a_2 = \bar{B}S_a\left(\frac{S}{S_a}\right)^{-2((d-r_f)/\sigma^2)} e^{-r_f(T-t)} - \bar{B}\left(\frac{S}{S_a}\right)^{1-2((r_d-r_f)/\sigma^2)} e^{-r_d(T-t)}$$

$$+ \bar{B}\sigma\sqrt{T-t}\left(\frac{S}{S_a}\right)^{1-2((r_d-r_f)/\sigma^2)} e^{-r_d(T-t)-((1/2)(d_2-\theta_2\sigma\sqrt{T-t})^2)}$$

$$a_3 = \bar{B}Se^{-r_f(T-t)} - \bar{B}\left(1 + \frac{\bar{B}-\bar{A}}{K}\right)e^{-r_d(T-t)}$$

$$- \bar{B}\left(1 + \frac{\bar{B}-\bar{A}}{K}\right)\sigma\sqrt{T-t}\,e^{r_d(T-t)-((1/2)(d_3-\theta_3\sigma\sqrt{T-t})^2)}$$

$$a_4 = -\frac{\bar{B}S_a}{K}\left(\frac{S}{S_a}\right)^{-2((r_d-r_f)/\sigma^2)} e^{-r_f(T-t)} + \bar{B}\left(1 + \frac{\bar{B}-\bar{A}}{K}e^{-r_d(T-t)}\right)\left(\frac{S}{S_a}\right)^{1-2((r_d-r_f)/\sigma^2)}$$

3.4 Pricing of Credit Default Swaps: Linear Parabolic PDEs (III)

3.4.1 Introduction

Credit default swaps (CDSs) are currently very popular in the international market of credit derivatives, which is in the form of traditional insurance. The default risk of one or more underlying investments is transferred from the buyer of credit protection to the seller of a contractual agreement. In a CDS contract, the buyer pay a *premium* to the seller (this fee is usually calculated as a percentage of the nominal value of the bonds or loans each quarter, each year to represent the spread in basis points), and if a credit event occurs before the contract expiration, the buyer will be compensated for his seller. Because banks, insurance companies, and hedge funds buy CDS for their increasing exposure to credit risks, by the end of 2007, the gross notional value of CDSs had reached \$62 trillion. This startling development has eliminated the need for a better understanding of credit risk assessment. However, the following questions still puzzle many scholars: Which factor of the implied volatility time series can indeed be used to explain the magnitude of CDS fluctuation? Is it better to forecast the future volatility or is it more effective to capture the risk premium of volatility?

Implied volatility is an important explanatory variable that measures the extent of volatility in the CDS spreads of time series. This relies on the basic principles of implied volatility as shown in Figure 3.1, where implied volatility is applied to the calculation of AT&T CDS spreads, the parameters use the historical volatility of 252 days or put option implied volatility is calculated as the industry standard for credit risk model Credit Grades. From Figure 3.1, we can learn about the 2002 midterm contraction phase of the telecommunications industry when AT&T CDS spreads rose to 700 basis points from 200 basis points. And this use of the implied volatility results was shown to be consistent with the market trend.

Currently, there are two main methods employed in the research on how to price credit risk: a structured approach and a reduction approach. The structured approach was first proposed by Merton, etc., through using structure variable indicators of the company (such as the value of assets, liabilities) to describe changes in the process of breaching the contract. A breach of contract will occur when a company's assets fall below a certain level. In this way, the corporate bonds can be considered as an option value of the assets, the strike price of the debt amount.

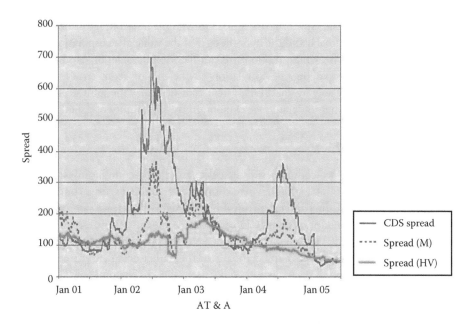

Figure 3.1   CDS spread.

The Merton model (Shiller, 1986) assumes that if the value of the company's assets is worth less than the debt at the time of maturity for the company to repay the debt, then default occurs. Black and Cox (1976) considered that the company could default before the time of maturity. In terms of development, the Merton model concludes that at any time t, if the asset value of the company falls to a certain level or below, default will occur. The reduction approach was used first by Duffie et al. (2003). This method considers default as an exogenous process that is caused by some kinds of unpredictable external factors. Both of these methods in the study of CDS pricing issues have been widely used, and the event of default is described by using a Poisson process.

In the real market place, a certain correlation between different companies and between the events of default of companies often exists. With the development of derivatives markets, an increasing number of companies have defaulted. Among such companies, a common product can be a basket of credit default swaps (basket CDS). A basket of CDSs is in fact an insurance contract for the occurrence of default by a number of companies.

The key to the pricing of a basket of CDSs is about issues related to a basket of corporate defaults. As of the writing of this book, many scholars have studied the pricing by using the following three main methods: conditional independence of default method, infection models, and Copula method. In recent years, Chinese scholars have done some research against the pricing issues of a basket of CDSs.

In this section, let us consider the default correlation under the framework of structural approach, and discuss the problem of pricing the product of a barker of CDSs. On the basis of the value of a barker of CDSs and the no-arbitrage principle, by applying structural approach and the hypothesis that the value of a corporation satisfies the 2D geometric Brownian motion, we can describe the relationship between companies through employing the correlation coefficients of geometric Brownian motions. Therefore, by solving the PDE system that describes the joint probability for two related companies to default, we can obtain the probability of default.

3.4.2 The Model

Assume that there are two companies in the basket, and their contract involves four companies, denoted A, B, C, and D, where companies A and B signed a contract for companies C and D of the basket. Starting from the date of contract signing, company A pays a fixed amount of costs (CDS spread) to company B until such a time that a company defaults. If no company is in default, the premium will be paid until the contractual maturity date. If any company in the basket defaulted, company B, which is the CDS disposal party, would compensate the loss to company A, which is the CDS's buyer, according to the signed contract. Next, let us analyze the CDS pricing problem of first default.

Assume that the event of default debt recovery and riskless rate are independent, and the premium is paid continuously. For the sake of simplicity and clarity, we first introduce some definitions. Assume that T is the term of CDS contract; R is the recovery after the occurrence of an event of default; r is the riskless rate; L is the nominal value of the bond; c_{st} is the CDS buyer (company A) who needs to continuously pay the insurance premium for this contract; τ_1 and τ_2 are the dates of default by company A and B, respectively. Let τ_{st} be the first date a default occurs. Then $\tau_{st} = \min\{\tau_1, \tau_2\}$. The expected present value of company A's premium payments is

$$c_{st} L \int_0^T e^{-rt} P(\tau_{st} > t) dt$$

where $P(\tau_{st} > t)$ is the probability that a default has not occurred at time t.

When an event of default occurs, the expected present value of payments that company A received from company B is

$$(1-R)L\int_0^T e^{-rt} p(\tau_{st},t) dt = (1-R)L\int_0^T e^{-rt} \frac{\partial}{\partial t} P(\tau_{st} \le t) dt$$

$$= (1-R)L\left[1 - e^{-rT} P(\tau_{st} > T) - r\int_0^T e^{-rt} P(\tau_{st} > t) dt\right]$$

where $P(\tau_{st} \le t)$ stands for the probability that a default event has occurred at time t, and $p(\tau_{st}, t)$ is the corresponding default probability density function.

According to the no-arbitrage principle, we have

$$c_{st} = \frac{(1-R)L\left[1 - e^{-rT} P(\tau_{st} > T) - r\int_0^T e^{-rt} P(\tau_{st} > t) dt\right]}{\int_0^T e^{-rt} P(\tau_{st} > t) dt} \tag{3.48}$$

The following formula will calculate the probability of default based on Equation 3.48. Let $p(\tau_{st}, t)$ be the probability that two companies are not in default. From the knowledge of the structured approach and Kolmogorov theorem, we know that $p(x_1, x_2, t)$ must satisfy the following PDE and boundary conditions:

$$\begin{cases} \dfrac{\partial p}{\partial t} + \dfrac{\sigma_1^2}{2} x_1^2 \dfrac{\partial^2 p}{\partial x_1^2} + \dfrac{\sigma_2^2}{2} x_2^2 \dfrac{\partial^2 p}{\partial x_2^2} + \rho\sigma_1\sigma_2 \dfrac{\partial^2 p}{\partial x_1 \partial x_2} + \mu_1 x_1 \dfrac{\partial p}{\partial x_1} + \mu_2 x_2 \dfrac{\partial p}{\partial x_2} = 0, \\ \qquad\qquad\qquad\qquad\qquad m_1 < x_1 < \infty, m_2 < x_2 < \infty, 0 \le t < T \\ p(m_1, x_2, t) = 0 \\ p(x_1, m_2, t) = 0 \\ p(x_1, x_2, T) = 1 \end{cases} \tag{3.49}$$

By taking the independent variable substitution and function substitution as shown in Section 3.1,

$$z_1 = \frac{1}{\sqrt{1-\rho^2}}\left[\left(\frac{x_1 - m_1}{\sigma_1}\right) - \rho\left(\frac{x_2 - m_2}{\sigma_2}\right)\right], z_2 = \frac{x_2 - m_2}{\sigma_2}$$

$$p(x_1, x_2, \tau) = e^{a_1 x_1 + a_2 x_2 + b\tau} W(z_1, z_2, \tau)$$

the problem of solving (3.49) is converted to the following standard form:

$$\begin{cases} \dfrac{\partial W}{\partial \tau} = \dfrac{1}{2}\left(\dfrac{\partial^2 W}{\partial z_1^2} + \dfrac{\partial^2 W}{\partial z_2^2}\right) \\[2mm] W(L_1, \tau) = 0 \\[2mm] W(L_2, \tau) = 0 \\[2mm] p(z_1, z_2, 0) = e^{-\left[a_1\sigma_1\sqrt{1-\rho^2}\, z_1 + (a_1\sigma_1\rho + a_2\sigma_2) z_2 + a_1 m_1 + a_2 m_2\right]} \end{cases} \qquad (3.50)$$

where

$$L_1 = \left\{(z_1, z_2)\big|\, z_2 = 0\right\}$$

$$L_2 = \left\{(z_1, z_2)\,\Big|\, z_2 = -\frac{\sqrt{1-\rho^2}}{\rho}\, z_1\right\}$$

$$a_1 = \frac{\rho\mu_2\sigma_1 - \mu_1\sigma_2}{(1-\rho^2)\sigma_1^2\sigma_2}$$

$$a_2 = \frac{\rho\mu_1\sigma_2 - \mu_2\sigma_1}{(1-\rho^2)\sigma_1^2\sigma_2}$$

$$b = \mu_1 a_1 + \mu_2 a_2 + \frac{1}{2}\sigma_1^2\mu_1^2 + \frac{1}{2}\sigma_2^2\mu_2^2 + \rho\sigma_1\sigma_2\mu_1\mu_2$$

$$\tau = T - t$$

3.4.3 The Solution

Lemma 3.3 (Decomposition principle). If the following problems

$$
\begin{cases}
\dfrac{\partial v}{\partial t} - a^2 \dfrac{\partial^2 v}{\partial x^2} = 0 & x \in R, t > 0 \\[2mm]
v(x,0) = \varphi_1(x)
\end{cases}
\tag{3.51}
$$

and

$$
\begin{cases}
\dfrac{\partial w}{\partial t} - a^2 \dfrac{\partial^2 w}{\partial y^2} = 0 & y \in R, t > 0 \\[2mm]
w(y,0) = \varphi_2(y)
\end{cases}
\tag{3.52}
$$

have solutions $v(x, t)$ and $w(y, t)$, respectively, then $u(x, y, t) = v(x, t)w(y, t)$ is a solution of the following problem:

$$
\begin{cases}
\dfrac{\partial u}{\partial t} - a^2 \left(\dfrac{\partial^2 u}{\partial x^2} + \dfrac{\partial^2 u}{\partial y^2} \right) = 0 & (x, y) \in R^2, t > 0 \\[2mm]
u(x, y, 0) = \varphi_1(x)\varphi_2(y)
\end{cases}
\tag{3.53}
$$

Proof. From $u(x, y, t) = v(x, t)w(y, t)$, we have

$$
\frac{\partial u}{\partial t} = w\frac{\partial v}{\partial t} + v\frac{\partial w}{\partial t}, \quad
\frac{\partial^2 u}{\partial x^2} = w\frac{\partial^2 v}{\partial x^2}, \quad
\frac{\partial^2 u}{\partial y^2} = v\frac{\partial^2 w}{\partial y^2}
$$

By plugging Equations 3.51 and 3.52 into (3.53), one can complete the proof of this result.

Lemma 3.4 If the problem

$$
\begin{cases}
\dfrac{\partial u_i}{\partial t} - a^2 \dfrac{\partial^2 u_i}{\partial x_i^2} = 0 & x_i \in R, t > 0 \\[2mm]
u_i(x_i,0) = \varphi_i(x_i)
\end{cases}
\tag{3.54}
$$

has a solution $u_i(x_1, t), i = 1,2,\ldots,n$, then $u(x_1,x_2,\ldots,x_n,t) = \prod_{i=1}^{n} u_i(x_i)$ is a solution of the problem

$$
\begin{cases}
\dfrac{\partial u}{\partial t} - a^2 \displaystyle\sum_{i=1}^{n} \dfrac{\partial^2 u}{\partial x_i^2} = 0, \quad (x, y) \in R^2, t > 0 \\[3mm]
u(x_1,x_2,\ldots,x_n,0) = \displaystyle\prod_{i=1}^{n} \varphi_i(x_i)
\end{cases}
\tag{3.55}
$$

From Lemma 3.3, we know that the solution of (3.50) is

$$
W(z_1,z_2,\tau) = W_1(z_1,\tau)W_2(z_2,\tau)
$$

where $W_1(z_1, \tau)$ satisfies

$$
\begin{cases}
\dfrac{\partial W_1}{\partial \tau} = \dfrac{1}{2}\dfrac{\partial^2 W_1}{\partial z_1^2} \\[3mm]
W_1(L_1,\tau) = 0 \\[3mm]
W_1(z_1,0) = e^{-\left[a_1\sigma_1\sqrt{1-\rho^2}\,z_1 + a_1 m_1\right]}
\end{cases}
\tag{3.56}
$$

and $W_2(z_2, \tau)$ satisfies

$$
\begin{cases}
\dfrac{\partial W_2}{\partial \tau} = \dfrac{1}{2}\dfrac{\partial^2 W_2}{\partial z_2^2} \\[3mm]
W_2(L_2,\tau) = 0 \\[3mm]
W_2(z_2,0) = e^{-\left[(a_1\sigma_1\rho + a_2\sigma_2)z_2 + a_2 m_2\right]}
\end{cases}
\tag{3.57}
$$

After applying odd expansions to Equations 3.56 and 3.57, and then by using the Poisson formula, we obtain

$$
W_1(z_1,\tau) = \frac{1}{\sqrt{2\pi\tau}} \int_{0}^{+\infty} \left[e^{-((z_2-\xi)^2/2\tau)} - e^{-((z_2+\xi)^2/2\tau)} \right] e^{-\left[a_1\sigma_1\sqrt{1-\rho^2}\,\xi + a_1 m_1\right]} d\xi
$$

$$
= e^{-c_1 z_1 + (c_1^2/2)\tau - a_1 m_1} \left[N(d_1) - e^{2c_1 z_1} \cdot N(\tilde{d}_1) \right]
$$

where

$$c_1 = a_1\sigma_1\sqrt{1-\rho}$$

$$d_1 = \frac{-c_1\tau + z_1}{\sqrt{\tau}}$$

$$\tilde{d}_1 = \frac{-c_1\tau - z_1}{\sqrt{\tau}}$$

$$N(x) = \int_{-\infty}^{x} \frac{1}{\sqrt{2\pi}} e^{-(y^2/2)} dy$$

Meanwhile, we have

$$W_2(z_2,\tau) = \frac{1}{\sqrt{2\pi\tau}} \int_0^{+\infty} \left[e^{-\left((z_2-\xi)^2/2\tau\right)} - e^{-\left((z_2+\xi)^2/2\tau\right)} \right] e^{-\left[(a_1\sigma_1\rho + a_2\sigma_2)\xi + a_2 m_2\right]} d\xi$$

$$= e^{-c_2 z_2 + (c_2^2/2)\tau - a_2 m_2} \left[N(d_2) - e^{2c_2 z_2} \cdot N(\tilde{d}_2) \right]$$

where

$$c_2 = a_1\sigma_1\rho + a_2\sigma_2$$

$$d_2 = \frac{-c_2\tau + z_2}{\sqrt{\tau}}$$

$$\tilde{d}_2 = \frac{-c_2\tau - z_2}{\sqrt{\tau}}$$

Therefore, we derive the following formula.

Theorem 3.2 Problem 3.50 has the following solution:

$$W(z_1,z_2,\tau) = e^{-c_1 z_1 - c_2 z_2 + ((c_1^2 + c_2^2)/2)\tau - a_1 m_1 - a_2 m_2} \left[N(d_1) - e^{2c_1 z_1} \cdot N(\tilde{d}_1) \right]$$

$$\times \left[N(d_2) - e^{2c_2 z_2} \cdot N(\tilde{d}_2) \right] \qquad (3.58)$$

where

$$p(x_1,x_2,t) = e^{((c_1^2+c_2^2+2b)/2)(T-t)} \left[N(d_1) - e^{2a_1\sigma_1\left[((x_1-m_1)/\sigma_1)-\rho((x_2-m_2)/\sigma_2)\right]} \cdot N(\tilde{d}_1) \right]$$

$$\cdot \left[N(d_2) - e^{2(a_1\sigma_1\rho+a_2\sigma_2)((x_2-m_2)/\sigma_2)} \cdot N(\tilde{d}_2) \right]$$

Notice that $p(x_1, x_2, t)$ is the density function of the distribution function $P(\tau_{st} > t)$. By substituting for $p(x_1, x_2, t)$, we get the value of the two companies' CDS in the market. For the CDS pricing problem with N companies in the market, it can be analyzed in the same way as by using Lemma 3.4.

3.5 Pricing of Forward Exchange Rate: Linear Parabolic PDEs (IV)

In this section, we consider a simple situation, where there are three state variables (s is the spot exchange rate, r_d is the domestic interest rate, r_f is the foreign interest rate) in the international market of forward exchange rates. Then the price of forward exchange rate $F(t,T)$ satisfies the following PDE:

$$\begin{cases} \dfrac{1}{2}\dfrac{\partial^2 F}{\partial s^2}s^2\sigma_s^2 + \dfrac{1}{2}\dfrac{\partial^2 F}{\partial r^2}\sigma_r^2 + \dfrac{1}{2}\dfrac{\partial^2 F}{\partial f^2}\sigma_f^2 + \dfrac{\partial^2 F}{\partial s\partial f}s\rho_{sf}\sigma_s\sigma_f + \dfrac{\partial^2 F}{\partial s\partial r}\rho_{sr}\sigma_s\sigma_r \\[2mm] + \dfrac{\partial^2 F}{\partial f\partial r}\rho_{fr}\sigma_f\sigma_r + \dfrac{\partial F}{\partial s}s(r_d - r_f) + \dfrac{\partial F}{\partial f}(\alpha_2(\beta_2 - r_f) - \lambda_f) \\[2mm] + \dfrac{\partial F}{\partial r}(\alpha_1(\beta_1 - r_d) - \lambda_r) - \dfrac{\partial F}{\partial \tau} = r_d F \\[2mm] S.L \quad F(s,r_d,r_f,0) = S_0 \end{cases} \qquad (3.59)$$

where u_s, σ_s are the spot rates on the mean and standard deviation of the return of property, respectively; $d\omega_s$, $d\omega_r$, $d\omega_f$ standards are the Wiener processes corresponding to the spot exchange rate, domestic interest rate, and foreign interest rate, respectively; α_1, α_2 are the restitution coefficients of the mean of the domestic and foreign interest rates, respectively; β_1, β_2 are the long-term averages of domestic and foreign short-term interest rates, respectively; σ_r, σ_f are the instant variances of short-term interest rate volatility corresponding to domestic interest rate market and foreign interest rate market, respectively; J is the jumping random process, satisfying the normal distribution, $J \sim N(\mu,\sigma^2)$.

Theorem 3.3 The price of the forward exchange rate of the solution of (3.59) is given by

$$
F_t(t,T) = \frac{\hat{E}_t\left[e^{-\int_t^T r_d(s)ds} S_T\right]}{\hat{E}_t\left[e^{-\int_t^T r_d(s)ds}\right]} = \frac{\hat{E}_t\left[e^{-\int_t^T r_d(s)ds} S_T\right]}{P(t,T)}
$$

$$
= \frac{S(t)F_1(\tau)e^{-H_f r_f(t)}}{P(t,T)} \tag{3.60}
$$

where

$$
F_1(\tau) = \exp\left[\frac{(H_f - \tau)(\hat{\beta}_2\alpha_2^2 - \sigma_f^2/2 + \alpha_2\rho_{fs}\sigma_f\sigma_s)}{\alpha_2^2} - \frac{\sigma_f^2 H_f^2}{4\alpha_2}\right], H_f(\tau) = \frac{1-e^{-\alpha_2\tau}}{\alpha_2}
$$

$$
P(r,t,\tau) = \exp\left[-H_r(\tau)r(t) + I(\tau)\right]
$$

$$
H_r(\tau) = \frac{1-e^{-\alpha_1\tau}}{\alpha_1}
$$

$$
I(\tau) = \frac{(\hat{\beta}_1 + \lambda u)((1-e^{-\alpha_1\tau}) - \alpha_1\tau)}{\alpha_1}
$$

$$
- \frac{\left[\sigma_r^2 + \lambda(u^2 + \sigma^2)\right](4(1-e^{-\alpha_1\tau}) - (1-e^{-2\alpha_1\tau}) - 2\alpha_1\tau)}{4\alpha_1^3}
$$

$$
\hat{\beta}_1 = \beta_1 - \frac{\lambda_r}{\alpha_1}
$$

Proof. Suppose

$$
X = \ln S - \int_t^T r_d(u)du \tag{3.61}
$$

By using Ito lemma, X's risk-neutral process can be shown to be

$$
d\hat{X} = (r_d - f_f - 0.5\sigma_s^2)dt + \sigma_s dW_s - d\left(\int_t^T r_d(u)du\right) \tag{3.62}
$$

By integrating both sides of Equation 3.62, we get

$$X(T) = \ln S(t) - 0.5\sigma_s^2 \tau - \int_t^T r_f(v)dv + \sigma_s \int_t^T dW_s(u) \qquad (3.63)$$

X is normally distributed so that its expectation is

$$\hat{E}_t[X(T)] = \ln S(t) - 0.5\sigma_s^2 \tau - \int_t^T \hat{E}_t[r_f(v)]dv \qquad (3.64)$$

In order to compute the integration in Equation 3.64, let us consider the following stochastic differential equation of the risk-neutral foreign interest:

$$dr_f = \alpha_2\left(\hat{\beta}_2 - r_f\right)dt + \sigma_f dW_f^* \qquad (3.65)$$

Hence,

$$r_f(v) = e^{\alpha_2(t-v)}e^{\alpha_2(t-v)}r_f(t) + \hat{\beta}_2(1 - e^{\alpha_2(t-v)}) + \sigma_f e^{-\alpha_2 v}\int_t^v e^{\alpha_2 u}dW_f(u) \quad (3.66)$$

Therefore, we obtain the expectation of (3.66) as follows:

$$\hat{E}_t(r_f) = e^{\alpha_2(t-v)}r_f(t) + \hat{\beta}_2(1 - e^{\alpha_2(t-v)}) \qquad (3.67)$$

By plugging (3.67) into (3.64) and integrating the result, we have

$$\hat{E}_t[X(T)] = \ln S(t) - 0.5\sigma_s^2 \tau - r_f(t)H_f - \hat{\beta}_2(\tau - H_f) \qquad (3.68)$$

where

$$H_r = \frac{1 - e^{-\alpha_1 \tau}}{\alpha_1}, H_f = \frac{1 - e^{-\alpha_2 \tau}}{\alpha_2}$$

The variance of $X(T)$ can be written as

$$\text{var}[X(T)] = \text{var}\left[\int_t^T r_f(v)dv\right] + \sigma_s^2 \tau - 2\,\text{cov}\left(\int_t^T r_f(v)dv, \sigma_s \int_t^T dW_s(v)\right) \quad (3.69)$$

Furthermore, integrating the stochastic term of (3.66) gives us

$$\int_t^T \sigma_f e^{-\alpha_2 v} \int_t^v e^{\alpha_2 u} dW_f(u) dt = \int_t^T [e^{\alpha_2 u} \int_u^T \sigma_f e^{-\alpha_2 v}] dW_f(u) \qquad (3.70)$$

Finally, we obtain the variance and covariance of Equation 3.69, respectively, as follows:

$$\text{var}\left[\int_t^T r_f(v) dv\right] = \int_t^T \left[e^{\alpha_2 u} \int_u^T \sigma_f e^{-\alpha_2 v} dv\right]^2 du$$

$$= \int_t^T e^{\alpha_2 u} \frac{\sigma_f^2}{\alpha_2^2} (e^{-\alpha_2 T} - e^{-\alpha_2 u})^2 du = -(H_f - \tau) \frac{\sigma_f^2}{\alpha_2^2} - \frac{\sigma_f^2 H_f^2}{2\alpha_2^2} \qquad (3.71)$$

$$\text{cov}\left(\int_t^T r_f(v) dv, \sigma_s \int_t^T dW_s(v)\right) = \text{cov}\left(\left[\int_t^T e^{\alpha_2 s} \int_s^T \sigma_f e^{-\alpha_2 v} dv\right] dW_f(s), \sigma_s \int_s^T dW_s(v)\right)$$

$$= \int_t^T \left[e^{\alpha_2 s} \int_s^T \sigma_f e^{-\alpha_2 v} dv\right] \sigma_s \rho_{fs} dt = -\frac{\rho_{fs} \sigma_s \sigma_f}{\alpha_1} (H_f - \tau)$$

$$(3.72)$$

Plugging (3.71) and (3.72) into (3.69) yields

$$\text{var}[X(T)] = -(H_f - \tau) \frac{\sigma_f^2}{\alpha_2^2} - \frac{\sigma_f^2 H_f^2}{2\alpha_2} + \sigma_s^2 \tau + \frac{2\rho_{fs}\sigma_s\sigma_f}{\alpha_1}(H_f - \tau) \quad (3.73)$$

So, the risk-neutral expectation of $e^{X(T)}$ can be calculated as follows:

$$\hat{E}_t\left[e^{-\int_t^T r_d(s) ds} S_T\right] = \hat{E}_t\left[e^{X(T)}\right] = e^{\hat{E}_t[X(T)] + (1/2)\text{var}[X(T)]} \qquad (3.74)$$

The price of the forward exchange rate is

$$F(t,T) = \frac{\hat{E}_t\left[e^{-\int_t^T r_d(s)ds} S_T\right]}{\hat{E}_t\left[e^{-\int_t^T r_d(s)ds}\right]} = \frac{\hat{E}_t\left[e^{X(T)}\right]}{P(t,T)} = \frac{e^{\hat{E}_t[X(T)]+(1/2)\operatorname{var}[X(T)]}}{P(t,T)} \tag{3.75}$$

By using Feynman–Kac pricing of the forward exchange rate and plugging Equations 3.64, 3.73, and 3.74 into 3.75, we at last get the price of the forward exchange rate.

Nest, we consider the volatility term structure of the forward exchange rate of model (3.72), which represents the second moment of the forward exchange rate.

Theorem 3.4 The unit volatility term structure of the forward exchange rate for the problem of solving (3.59) is

$$\sigma_F^2(\tau) = \sigma_S^2 + H_f^2 \sigma_f^2 + H_r^2\left[\sigma_r^2 + \lambda(u^2 + \sigma^2)\right] + 2H_r\rho_{sr}\sigma_s\sigma_r$$

$$- 2H_f\rho_{sf}\sigma_s\sigma_f - 2H_rH_f\rho_{rf}\sigma_r\sigma_f \tag{3.76}$$

Proof. Because $EJ^2 = [E(J)]^2 + \operatorname{var}(J) = u^2 + \sigma^2$, by using Ito lemma to expand Equation 3.62, one can get readily the unit volatility term structure of the forward exchange rate (3.76).

Expression (3.76) stands for the volatility term structure of the forward interest rate, and the corresponding curve is referred to as the fluctuation curve. The volatility of the forward exchange rate is related with the spot exchange rate, the volatility of interest rate at home and abroad, the mean regression coefficient of interest rate at home and abroad, the jump density of interest rate at home and abroad, the magnitudes of jump in the mean and standard deviation of the interest rates, and the date of expiration. At the same time, on the date of expiration of the forward exchange rate, Equation 3.76 is equal to σ_s^2, because the forward exchange rate is equal to the spot exchange rate, the volatility of the forward exchange rate is equal to the volatility spot exchange rate. Furthermore, the jump of interest rate strengthens the volatility of the forward exchange rate. Because $\partial\sigma^2/\partial\lambda\tau = 2(u^2 + \sigma^2)H_r e^{-\alpha_1 \tau} > 0$, the jump random process density has a positive influence on volatility and will be increased with the date of expiration.

With an infinite date of expiration, the volatility of the forward exchange rate gets really close to

$$\sigma_F^2(\tau) = \sigma_S^2 + \frac{\sigma_f^2}{\alpha_2^2} + \frac{\left[\sigma_r^2 + \lambda(u^2 + \sigma^2)\right]}{\alpha_1^2} + \frac{2\rho_{sr}\sigma_s\sigma_r}{\alpha_1} - \frac{2\rho_{sf}\sigma_s\sigma_f}{\alpha_2} - \frac{2\rho_{rf}\sigma_r\sigma_f}{\alpha_1\alpha_2} \tag{3.77}$$

3.6 Pricing of Arithmetic Average Asian Option: Nonlinear Parabolic PDEs(I)

3.6.1 Introduction

As a new financial product, Asian options, also known as average options, can be seen as innovative European options. The commonality with European options is that investors are only allowed to exercise their option contracts on the maturity dates, while the difference is that investors of Asian options decide whether or not to exercise their option contracts based on the price level of the average share price during the contract term. Because the value of European options on the maturity date has nothing to do with the price path and depends only on the maturity date of the stock price, it is difficult to prevent speculators from manipulating the maturity price and consequently from arbitrage. On the other hand, because Asian options are associated with the price path, they can be applied to ease the volatility of market behaviors. They represent the exotic options most actively traded in the financial derivatives market today. Their difference with the usual-sense stock options is the implementation of price limits so that the exercise price is the average price of the stock price of the secondary market over a period of 6 months prior to exercise. Their difference with the standard options is that the options' maturity-date benefits are determined not by the prevailing market price of the underlying asset, but by the average price of the underlying asset over some time period during the option contract period. This time period is known as an average period, over which either the arithmetic or geometric mean is applied. Asian options can be well designed to avoid stock price manipulation and damage of the interests of the underlying company that might be caused by insider trading. Therefore, they are welcomed by both investors and issuers. Additionally, compared to the standard options, Asian options also possess advantages such as lower prices and applicability to hedge the risk of a specified time period. Along with baskets, spreads, as well as other strategies, Asian options have been extensively studied by many scholars. Even so, few of published works deal with the problem of pricing general Asian options.

Asian options are exotic, and have been widely used in the financial derivatives market. Their earnings depend on the average price of the underlying assets during the lifespan of the options. According to different representations of the average, they can be categorized into geometric Asian options and arithmetic Asian options. The former is more common than the latter in the financial market. Although the arithmetic Asian options have been widely used in the market, there has not been any analytical expression that can represent their prices, and most studies only provide numerical solutions or approximate analytical solutions. So the pricing study about these options becomes an important topic of the option theory.

In recent years, some scholars have obtained the semiexplicit solution about arithmetic Asian options by using different methods. A semianalytical method is first proposed by Zhang (2001). He decomposes the value of arithmetic Asian options into an approximation value and the relevant error term. The approximation value is given by a modified equation. Zhang derives an analytical expression of this approximation value and establishes the PDE that the error term satisfies. Zhang (2003) furthered this method. By using the perturbation method and the singularity removal techniques, the error term of the second-order approximation is given, which satisfies a new PDE that can be solved by using the same technique. Zhang presents the analytical results up to the fourth order and points out that the process converges quickly. Based on their research, Zhang et al. (2004) found an approximate analytical expression by applying the Fichera theory of PDEs and the theory of fixed solution of boundary value problems. Numerical simulations were carried out, and a more satisfactory approximation was obtained. A different approach is suggested by Cruz-Baez and Gonzalez-Rodriguez (2008). According to the PDE theory and integral transforms, and by using mathematical approaches, they produced a semiexplicit solution about arithmetic Asian options. Through the use of Laplace transforms and power series expansion, the progressive solutions of the Asian options are given by Shaw (2008) and Dewynne and Shaw (2008), where they also obtained the solution of a fifth-order problem in the case of $r=q$ and the result of a third-order problem in the general case of $r \neq q$. Kim (2009) estimated the solution of a one-dimensional (1D) PDE that Asian options satisfy, and gave the scope of solution under certain conditions. Xu et al. (2007) established a pricing model for a class of foreign exchange deposit product of option style. By using the convex analysis and filtering technique, a Black–Scholes style formula for option pricing with dividends under partial information is obtained by Wu and Wang (2007). The optimal investment strategy for maximizing the terminal wealth problem under partial information is also obtained. Recently, Wang (2010) discussed a class of problems about discrete time option pricing. He used a mean self-financing delta-hedging argument in a discrete time setting in the fractional Black–Scholes model with transaction costs, and obtained a new European call option pricing formula (Wang, 2010).

In this section, arithmetic Asian options are priced by using the decomposition approach and an upper bound of error is given on the basis of Xu's study. After several lemmas are introduced, we put forward a method of decomposing the value of Asian options (Ying et al., 2010). Finally, the error term is estimated by employing the comparison principle.

3.6.2 The Lemma

Lemma 3.5 Let $\Sigma = \{(x, t) | -T < x < +\infty, 0 < t \le T\}$ and consider the following Cauchy problems:

$$\begin{cases} u_t - a^2 u_{xx} = f(x,t) & (x,t) \in \Sigma \\ u(-T,0) = 0 & -T < x < +\infty \\ u(x_0,t) = 0 & 0 < t \le T \end{cases} \tag{3.78}$$

and

$$\begin{cases} v_t - a^2 v_{xx} = g(x,t) & (x,t) \in \Sigma \\ v(-T,0) = 0 & -T < x < +\infty \\ v(x_0,t) = 0 & 0 < t \leq T \end{cases} \qquad (3.79)$$

If $f(x, t) \leq g(x, t)$ holds true in the region Σ, then $u(x, t) \leq v(x, t)$ also holds true in the region Σ.

Proof. Denote $w = \left(m/x_0^2 \right)(x^2 + Kt)e^{\lambda t} + (v - u)$, where $K > 2a^2$, $\lambda > 0$, $x_0 > 0$, and $m > 0$, then we have

$$a^2 w_{xx} - w_t = \frac{m}{x_0^2} e^{\lambda t} \left[2a^2 - K - \lambda(x^2 + Kt) \right] + (a^2 v_{xx} - v_t) - (a^2 u_{xx} - u_t)$$

$$\leq \frac{m}{x_0^2} e^{\lambda t} \left[2a^2 - K - \lambda(x^2 + Kt) \right] < 0$$

By considering the function $w(x,t)$ in the region $Q_{x_0} = \{(x,t); x \leq x_0, 0 \leq t \leq T\}$, we have

$$w(x,0) \geq v(x,0) - u(x,0) = 0, \quad w(-T,t) \geq v(-T,t) - u(-T,t) = 0$$

Hence, $w(x,t) \geq 0$ in the region Q_{x_0}. For any point $P \in \Sigma$, let Q_{x_0} contain the point P and x_0 be big enough so that $w = \left(m/x_0^2 \right)(x^2 + Kt)e^{\lambda t} + (v - u) \geq 0$ holds true at P. When $x_0 \to +\infty$, we are sure that $u(x,t) \leq v(x,t)$.

Lemma 3.6 There exists a number $x_0 > 0$ such that the following inequality holds true:

$$\frac{(x - \tau)^2}{4\eta} \geq a + kx + \frac{2}{\sigma^2 x} \quad \text{for } x > x_0 \, (a < 0 \text{ and } k > 0)$$

Proof. Let φ_1 be a function defined by $\varphi_1(x) = \frac{(x - \tau)^2}{4\eta} - \left(a + kx + \frac{2}{\sigma^2 x} \right)$, for $x > 0$.

For the constants $a < 0$ and $k > 0$, it is clear that $\varphi_1(0^+) < 0$ and $\varphi_1(+\infty) < +\infty$. Hence, by using the intermediate value theorem, we know that there is $x_0 > 0$ such that $\varphi_1(x_0) = 0$. Let $x_0 = \inf\{x; \varphi_1(x) > 0\}$ such that the conclusion of Lemma 3.6 holds true.

Lemma 3.7 There exists a constant $b>0$ such that the inequality $((x-\tau)^2/4\eta) \geq b - (2/\sigma^2 x)(x>0)$ holds true.

Proof. Let $\varphi_2(x) = \left(x(x-\tau)^2/4\eta\right) - \left(bx + \left(2/\sigma^2\right)\right)$, where $x>0$. It is clear that $\varphi_2'(x) = \left(3x^2 - 4\tau x + \tau^2/4\eta\right) - b$ is a quadratic polynomial. Choose such a $b>0$ that makes the discriminant $\Delta \geq 0$ so that we have $\varphi_2'(x) \geq 0$. Due to $\varphi_2(0) = (2/\sigma^2) > 0$, we have $\varphi_2(x) \geq 0$ $(x>0)$, and therefore, $(\varphi_2(x)/x) \geq 0$ $(x>0)$. It means the inequality of Lemma 3.7 is correct.

Lemma 3.8 If $0<x<x_0$, then there exists a constant $N>0$ such that $1-\beta e^{-\lambda x} \leq (N/x^2)$.

Proof. Define $\varphi_3(x) = (N/x^2) - (1 - \beta e^{-\lambda x}), (x>0)$.

If $\lambda>0$, $\beta>0$, then there is a constant $N>0$ such that the inequalities $\varphi_3(0^+)>0$ and $\varphi_3(+\infty) = -1 <0$ hold true. Therefore, according to the intermediate value theorem of calculus, there exists $x_0>0$ such that $\varphi_3(x_0) = 0$. Note that $N = x_0^2(1 - \beta e^{-\lambda x_0})$, where x_0 is the minimum zero point of $\varphi_1(x)$ in Lemma 3.6, so we have $\varphi_3(x)>0$ in the interval $0<x<x_0$. The proof is completed.

3.6.3 Decomposition of the Solution

Assume the price S_t of the underlying asset is subject to the following geometric Brown motion:

$$dS_t = (r - q)S_t dt + \sigma S_t dW_t \tag{3.80}$$

where
r and q are risk-free interest rate and the dividend of the underlying asset
σ is the asset volatility
W_t is the standard Brown motion

Then an arithmetic Asian call option, whose strike price is K and maturity date T, satisfies $V_T = (J_T - K)^+$, where $J_T = (1/T)\int_0^T S_t dt$.

Using the Δ-hedging and no-arbitrage principle, we can see that the price of the Asian option satisfies the following PDE:

$$\begin{cases} \dfrac{\partial V}{\partial t} + \dfrac{1}{2}\sigma^2 S^2 \dfrac{\partial^2 V}{\partial S^2} + \dfrac{S - J}{t}\dfrac{\partial V}{\partial J} + (r - q)S\dfrac{\partial V}{\partial S} - rV = 0 \quad S \in R^+, J \in R^+, t \in [0,T) \\ V(S,J,T) = (J - K)^+ \end{cases} \tag{3.81}$$

By using the following translations,

$$x = \frac{TK - tJ}{S} e^{-(r-q)\tau} - \frac{1}{r-q} \left[1 - e^{-(r-q)\tau}\right], \quad \tau = T - t, \quad V(S, J, t) = \frac{S}{T} e^{-q\tau} U(x, \tau)$$

(3.82)

Equation 3.81 can be simplified as follows:

$$\begin{cases} \dfrac{\partial u}{\partial \tau} - \dfrac{1}{2}\sigma^2 \left[x + \dfrac{1}{r-q}\left(1 - e^{-(r-q)\tau}\right)\right]^2 \dfrac{\partial^2 U}{\partial x^2} = 0 & x \in R, 0 < \tau \le T \\ U(x, 0) = (-x)^+ \end{cases}$$

(3.83)

Let us just consider the case $r \approx q$ in the actual financial markets and $(1/(r-q)) \times (1 - e^{-(r-q)\tau}) = \tau + o(|r-q|\tau)$, then Equation 3.83 becomes

$$\begin{cases} \dfrac{\partial u}{\partial \tau} - \dfrac{1}{2}\sigma^2 \left[x + \tau\right]^2 \dfrac{\partial^2 U}{\partial x^2} = 0 & x \in R, 0 < \tau \le T \\ U(x, 0) = (-x)^+ \end{cases}$$

(3.84)

According to the Fichera theory of PDE, we know $U(x, \tau) = -x(x < -\tau)$. Hence, we just need to look for an approximate solution of $U(x, \tau)$ for when $x > -\tau$.

Now we solve for $U(x, \tau)$ that satisfies the following equation:

$$\begin{cases} \dfrac{\partial U}{\partial \tau} - \dfrac{1}{2}\sigma^2 \left[x + \tau\right]^2 \dfrac{\partial^2 U}{\partial x^2} = 0 & x > -T, 0 < \tau \le T \\ U(-T, \tau) = T \\ U(x, 0) = (-x)^+ \end{cases}$$

(3.85)

Suppose that $U = U_0 + U_1$, where U_0 is an approximate solution of U, and U_1 is an error term. By substituting it into formula (3.85), we can see that U_0 satisfies the following initial boundary value problem of homogeneous parabolic equation:

$$\begin{cases} \dfrac{\partial U_0}{\partial \tau} - \dfrac{1}{2}\sigma^2 \tau^2 \dfrac{\partial^2 U_0}{\partial x^2} = 0 & x > -T, 0 < \tau \le T \\ U_0(-T, \tau) = T \\ U_0(x, 0) = (-x)^+ \end{cases}$$

(3.86)

Theorem 3.5 The solution of Equation 3.86 can be represented as follows:

$$U_0(x,\tau) = -xN\left(-\frac{x}{\sqrt{2\eta}}\right) + (x+2T)N\left(-\frac{x+2T}{\sqrt{2\eta}}\right) + \sqrt{\frac{\eta}{\pi}}\left[e^{-(x^2/4\eta)} - e^{-((x+2T)^2/4\eta)}\right]$$

(3.87)

where $N(\cdot)$ is the probability function of the standard normal distribution and $\eta = (\sigma^2\tau^3/6)$.

Proof. Let us choose the following transformation for initial problem (3.86):

$$y = x + T, \eta = \frac{\sigma^2\tau^3}{6}, \quad u(y,\eta) = U_0(x,\tau) - T$$

After applying an odd prolongation for the function u, we have

$$\begin{cases} \dfrac{\partial u}{\partial \eta} - \dfrac{\partial^2 u}{\partial y^2} = 0, & y \in R, y > 0 \\ u(y,0) = \phi(y), & y \in R \end{cases}$$

(3.88)

where

$$\phi(y) = \begin{cases} -T, & y > T \\ -y, & -T \le y \le T \\ T, & y < T \end{cases}$$

In light of the integral formula in Lemma 3.5, we get the solution (3.87).

Theorem 3.6 The error term U_1 satisfies the following initial-value problem of nonhomogeneous parabolic PDE:

$$\begin{cases} \dfrac{\partial U_1}{\partial \tau} - \dfrac{1}{2}\sigma^2(x+\tau)^2\dfrac{\partial^2 U_1}{\partial x^2} = \dfrac{\sigma^2}{4\sqrt{\pi\eta}}x(x+2\tau) \\ \qquad \times\left[e^{-(x^2/4\eta)} - e^{-((x+2T)^2/4\eta)}\right] \qquad x > -T, 0 < \tau \le T \\ U_1(-T,\tau) = 0 \\ U_1(x,0) = 0 \end{cases}$$

(3.89)

3.6.4 Estimation for Error

Theorem 3.7 There exists a constant M_1 such that one of the upper bounds of the error term U_1 is

$$U_{21} = \int_{x_0}^{\xi} e^{-(2/\sigma^2 \zeta)} (M_1 e^{-k\zeta} + C) d\zeta \tag{3.90}$$

Proof. As $x > x_0$, according to Lemma 3.7, we have the following inequality:

$$\frac{\sigma^2}{4\sqrt{\pi\eta}} x(x + 2\tau) \left[e^{-(x^2/4\eta)} - e^{-((x+2T)^2/4\eta)} \right] \leq \frac{\sigma^2}{4\sqrt{\pi\eta}} (x + \tau)^2 e^{-\left[a + k(x+\tau) + (2/(\sigma^2(x+\tau))) \right]} \tag{3.91}$$

In light of Lemma 3.6, for $x > x_0$, we have $U_1(x,\tau) \leq U_{21}(x,\tau)$, where $U_{21}(x,\tau)$ satisfies the following initial-value problem:

$$\begin{cases} \dfrac{\partial U_{21}}{\partial \tau} - \dfrac{1}{2} \sigma^2 (x + \tau)^2 \dfrac{\partial^2 U_{21}}{\partial x^2} = \dfrac{\sigma^2}{4\sqrt{\pi\eta}} (x + \tau)^2 \times e^{-\left[a + k(x+\tau) + (2/(\sigma^2(x+\tau))) \right]} \\ \\ \hspace{6cm} x > x_0, 0 < \tau \leq T \quad (3.92) \\ U_2(-T, \tau) = 0 \\ U_2(x, 0) = 0 \end{cases}$$

Let $\xi = x + \tau$. Then Equation 3.92 can be converted to the following initial-value problem of ordinary differential equation:

$$\begin{cases} U_{21}' - \dfrac{1}{2} \sigma^2 \xi^2 U_{21}'' = \dfrac{\sigma^2}{4\sqrt{\pi\eta}} \xi^2 e^{-\left[a + k\xi + (2/(\sigma^2 \xi)) \right]} & x > x_0, 0 < \tau \leq T \\ \\ U_{21}' |_{\xi = x_0} = 0. \end{cases} \tag{3.93}$$

Solving this initial problem (3.93) leads to

$$U_{21} = \int_{x_0}^{\xi} e^{-(2/(\sigma^2 \zeta))} \left(\frac{2M}{k\sigma^2} e^{-k\zeta} + C \right) d\zeta \tag{3.94}$$

where

M is the maximum of $\left(\sigma^2/4\sqrt{\pi\eta} \right) e^{-a}$ for $\tau \in [0, T]$
C is an integral constant

By letting $M_1 = (2M/k\sigma^2)$, the proof is finished.

Theorem 3.8 There exists a constant M_2 such that one of upper bounds of the error term U_1 is

$$U_{22} = \int_0^\xi M_2 \cosh\left(\frac{2}{\sigma^2 \zeta}\right) d\zeta \tag{3.95}$$

Proof. If $0 < x < x_0$, according to Lemma 3.8, we know

$$\frac{\sigma^2}{4\sqrt{\pi\eta}} x(x + 2\tau)\left[e^{-(x^2/4\eta)} - e^{-((x+2T)^2/4\eta)}\right] \le \frac{\sigma^2}{4\sqrt{\pi\eta}} Ne^{-\left[b - (2/(\sigma^2(x+\tau)))\right]} \tag{3.96}$$

By using Lemma 3.6, we get $U_1(x, \tau) \le U_{22}(x, \tau)$ for $0 < x < x_0$, where $U_{22}(x, \tau)$ satisfies the following equation:

$$\frac{\partial U_{22}}{\partial \tau} - \frac{1}{2}\sigma^2(x + \tau)^2 \frac{\partial^2 U_{22}}{\partial x^2} = \frac{\sigma^2}{4\sqrt{\pi\eta}} Ne^{-\left[b - (2/(\sigma^2(x+\tau)))\right]} \quad 0 < x < x_0, 0 < \tau \le T \tag{3.97}$$

From the proof of Theorem 3.7, we have

$$U_{22} = \int_0^\xi e^{-(1/(\sigma^2\zeta))}\left(\frac{M'N}{2} e^{(4/(\sigma^2\zeta))} + C'\right) d\zeta \tag{3.98}$$

where

M' is the maximum of $\left(\sigma^2/4\sqrt{\pi\eta}\right)e^{-b}$ for $\tau \in [0, T]$
C' is an integral constant

If we let $M_2 = M'N$ and $C' = (M_2/2)$, expression (3.95) is obtained and the proof is completed.

3.6.5 Estimation of the Error Term

In his study of a degenerated parabolic PDE in the pricing of Asian options, Kim (2009) obtained an inequality that the price of Asian options satisfies. In this section, we try to improve this particular Kim's inequality. Our inequality has an advantage over Kim's, for our estimation has an upper bound as R vanishes. Therefore, our newly established estimation can help us determine the solution error of a second-order PDE with variable coefficients.

In terms of the pricing of Asian options, many scholars (Ehrlich, 1971; York, 2003; Wang, 2005; Chunbo, 2008) represent some of the most recent results, while Shaw (2008) provided a simplified means of pricing Asian options by using parabolic PDEs. A progressive solution is established by using Laplace transforms and power series expansions. Equations of the Kolmogorov type have also turned out to be relevant in option pricing in the setting of certain models with stochastic volatility and in the pricing of Asian options. Frentz et al. (2010) numerically solved the Cauchy problem of a general class of second-order degenerated parabolic differential operators of the Kolmogorov type with variable coefficients by using posteriori error estimates and an algorithm developed for adaptive weak approximation of stochastic differential equations. On the basis of these works, we show how to apply the relevant results in the context of mathematical finance and option pricing. The approach outlined in this section circumvents many of the difficulties confronted by any deterministic approach based on, for example, a finite-difference discretization of PDEs. Meanwhile, Frentz (2010) also analyzed the second-order PDE operators arising in the pricing of Asian options, while proving the optimal interior regularity. Dai et al. (2010) presented a lattice algorithm for pricing both European- and American-style moving average barrier options (MABOs). They developed a finite-dimensional PDE model for discretely monitoring MABOs and solved it numerically by using a forward shooting grid method. However, their modeling PDE for continuously monitored MABOs is of infinite dimensions and cannot be solved directly by using existing numerical methods. Recently, Bayraktar and Xing (2011) constructed a sequence of functions based on the value of Asian options. As a result, each term of the sequence is the unique classical solution of a parabolic PDE so that they provide the relevant numerical approximation.

Deelstra et al. (2010) obtained an approximation formula by using comonotonic bounds, leading to four different approximations: the upper, the improved upper, the lower, and the intermediary bounds. In this way, they improved the traditional hybrid moment matching method. Their methods have the advantage that they can be applied in other frameworks, for example, in Lévy settings, as well. These results can be adapted to deal with options written in a foreign currency, such as compo and quanto options. Kim (2009) studied a simple 1D parabolic PDE that the price of Asian options satisfies. The result indicates that the generalized solution is a classical solution and cannot be obtained. Additionally, Kim's inequality is shown to be unbounded as the parameter R vanishes. By improving Kim's inequality, our estimation has an upper bound as R vanishes

Kim considered the following parabolic type equation:

$$u_t + \frac{1}{2}\left[x - e^{-\int_0^t dv(s)} q(t)\right]^2 \sigma^2 u_{xx} = 0 \tag{3.99}$$

along with the boundary condition

$$u(T,x) = \max(x - K_1, 0) \tag{3.100}$$

where
   $v(t)$ stands for dividend yield
   σ is the volatility of the underlying asset
   $q(t)$ is the trading strategy

First let us introduce a lemma about Gauss estimation. Suppose that $g(x)$ is a continuous function in the interval $[-R, R]$ satisfying

$$\frac{1}{2} \le g(x) \le \frac{3}{2}, \quad x \in [-R, R]$$

Denote

$$Q := \{(t,x) \in R^2 : 0 < t < 2, |x| < R\}$$

$$\Omega := \{(t,x) \in Q : t > g(x)\}, \Sigma := \{(t,x) \in Q : t = g(x)\}$$

Lemma 3.9 (Kim, 2009) Let Ω and Σ be defined as previously presented and $a(t,x)$ a function satisfying

$$\begin{cases} Lu := u_t - a(t,x)u_{xx} = 0, & (t,x) \in \Omega \\ u = 0, & (t,x) \in \Sigma \end{cases}$$

Assume that $u \in C_{loc}^{1,2}(\Omega) \cap C(\bar{\Omega})$ and $a(t,x)$ satisfy

$$0 \le a(t,x) \le 1, \quad \forall (t,x) \in \Omega$$

Then, the following estimate holds true:

$$|u|_{0;\Omega'} \le \left(\frac{16}{\sqrt{2\pi}}\right) R^{-1} e^{(-R^2/32)} |u|_{0;\Omega}, \quad \Omega' := \left\{(t,x) \in \Omega : |x| < \frac{R}{2}\right\} \tag{3.101}$$

Theorem 3.9 If $|x| < R$, then the following holds true:

$$\int_E \phi(2, x - y) dy < 2 \int_0^{2R} \phi(2, y) dy + \frac{1}{\sqrt{2\pi}R} e^{-2R^2} \qquad (3.102)$$

where $E = \bigcup_{j \in Z} ((4j+1)R, (4j+3)R)$.

Proof. First we show that the following equality holds true:

$$\int_E \phi(2, x - y) dy = \sum_{j=1}^{\infty} \int_{(4j-3)R+x}^{(4j-1)R+x} \phi(2, y) dy + \sum_{j=0}^{\infty} \int_{(4j+1)R-x}^{(4j+3)R-x} \phi(2, y) dy$$

In fact,

$$\int_E \phi(2, x - y) dy = \int_{\bigcup_{j=1}^{\infty} ((-4j+1)R, (-4j+3)R)} \phi(2, x - y) dy + \int_{\bigcup_{j=0}^{\infty} ((4j+1)R, (4j+3)R)} \phi(2, x - y) dy$$

$$= \sum_{j=1}^{\infty} \int_{(-4j+1)R}^{(-4j+3)R} \phi(2, x - y) dy + \sum_{j=0}^{\infty} \int_{(4j+1)R}^{(4j+3)R} \phi(2, x - y) dy$$

$$= \sum_{j=1}^{\infty} \int_{x-(-4j+1)R}^{x-(-4j+3)R} -\phi(2, y) dy + \sum_{j=0}^{\infty} \int_{x-(4j+1)R}^{x-(4j+3)R} \phi(2, y) dy$$

Because $\phi(2, y) = \left(1/\sqrt{8\pi}\right) e^{-(y^2/8)}$ is an even function, we have

$$\int_E \phi(2, x - y) dy = \sum_{j=1}^{\infty} \int_{(-4j+1)R-x}^{(-4j+3)R-x} \phi(2, y) dy + \sum_{j=0}^{\infty} \int_{(4j+1)R-x}^{(4j+3)R-x} \phi(2, y) dy$$

$$= \sum_{j=1}^{\infty} \int_{(4j-3)R+x}^{(4j-1)R+x} \phi(2, y) dy + \sum_{j=0}^{\infty} \int_{(4j+1)R-x}^{(4j+3)R-x} \phi(2, y) dy$$

Secondly, we show that when $-R < x < 0$, the following inequality holds true:

$$\sum_{j=1}^{\infty} \int_{(4j-3)R+x}^{(4j-1)R+x} \phi(2, y)dy \leq \int_{0}^{2R} \phi(2, y)dy + \frac{1}{2\sqrt{2\pi R}} e^{-2R^2} \tag{3.103}$$

In fact, when $-R < x < 0$, we have

$$\sum_{j=1}^{\infty} \int_{(4j-3)R+x}^{(4j-1)R+x} \phi(2, y)dy < \sum_{j=0}^{\infty} \int_{4jR}^{(4j+2)R} \phi(2, y)dy$$

$$= \int_{0}^{2R} \phi(2, y)dy + \sum_{j=1}^{\infty} \int_{4jR}^{(4j+2)R} \phi(2, y)dy$$

So, we have

$$\sum_{j=1}^{\infty} \int_{4jR}^{(4j+2)R} \phi(2, y)dy < \int_{4R}^{\infty} \phi(2, y)dy = \int_{4R}^{\infty} \frac{1}{\sqrt{8\pi}} e^{-(y^2/8)} dy$$

$$< \frac{1}{\sqrt{8\pi}} \int_{4R}^{\infty} \frac{y}{4R} e^{-(y^2/8)} dy = \frac{1}{2\sqrt{2\pi R}} e^{-2R^2}$$

and

$$\sum_{j=1}^{\infty} \int_{(4j-3)R+x}^{(4j-1)R+x} \phi(2, y)dy \leq \int_{0}^{2R} \phi(2, y)dy + \frac{1}{2\sqrt{2\pi R}} e^{-2R^2}$$

Thirdly, we show that for $0 < x < R$, we have

$$\sum_{j=0}^{\infty} \int_{(4j+1)R-x}^{(4j+3)R-x} \phi(2, y)dy \leq \int_{0}^{2R} \phi(2, y)dy + \frac{1}{2\sqrt{2\pi R}} e^{-2R^2} \tag{3.104}$$

In fact, if $0 < x < R$, then we have

$$\sum_{j=0}^{\infty} \int_{(4j+1)R-x}^{(4j+3)R-x} \phi(2, y)dy \le \sum_{j=0}^{\infty} \int_{4jR}^{(4j+2)R} \phi(2, y)dy$$

$$= \int_{0}^{2R} \phi(2, y)dy + \sum_{j=1}^{\infty} \int_{4jR}^{(4j+2)R} \phi(2, y)dy$$

$$= \int_{0}^{2R} \phi(2, y)dy + \sum_{j=1}^{\infty} \int_{4jR}^{(4j+2)R} \phi(2, y)dy$$

By putting (3.103) and (3.104) together, the proof is completed.

Theorem 3.10 If $|x| < R$, then the following holds true:

$$\int_{E} \phi(2, x - y)dy < \sqrt{1 - e^{-R^2}} + \frac{1}{\sqrt{2\pi R}} e^{-2R^2} \tag{3.105}$$

Proof. For an arbitrary $R > 0$, we have the following inequality:

$$\int_{0}^{2R} \phi(2, y)dy < \frac{1}{2} \sqrt{1 - e^{-R^2}}$$

In fact, we have

$$\left(\int_{0}^{2R} \phi(2, y)dy \right)^2 = \int_{0}^{2R} \int_{0}^{2R} \frac{1}{8\pi} e^{-((x^2+y^2)/8)} dx dy$$

$$\le \iint_{D} \frac{1}{8\pi} e^{-((x^2+y^2)/8)} dx dy$$

$$= \frac{1}{4} \left(1 - e^{-R^2} \right)$$

where $D = \{(x, y) | x^2 + y^2 \le 8R^2, x \ge 0, y \ge 0\}$. Hence, $\int_{0}^{2R} \phi(2, y)dy < \frac{1}{2} \sqrt{1 - e^{-R^2}}$.

By substituting the aforementioned inequality in Theorem 3.9, inequality (3.105) is obtained at once.

Theorem 3.11 If $k > 1$ and $R^2 < \left(2\ln 2/(16k - 9)\right)$, then the following holds true:

$$\int\limits_{4jR}^{(4j+2)R} \phi(2, y)dy \leq \int\limits_{(8j-4)R}^{8jR} \phi(2, \sqrt{ky})dy \tag{3.106}$$

Proof. After changing the variable $t = \left((y + 4R)/2\right)$, we have

$$\int\limits_{(8j-4)R}^{8jR} \phi(2, \sqrt{ky})dy = \frac{1}{\sqrt{8\pi}} \int\limits_{(8j-4)R}^{8jR} e^{-(ky^2/8)}\, dy = \frac{1}{\sqrt{8\pi}} \int\limits_{4jR}^{(4j+2)R} e^{\ln 2 - ((k(y-2R)^2)/2)}\, dy$$

For $R^2 < \left(2\ln 2/(16k - 9)\right)$, $4jR < y < (4j + 2)R$, $j = 1, 2, \ldots$, the following inequality holds true:

$$\ln 2 - \frac{k(y - 2R)^2}{2} + \frac{y^2}{8} \geq 0$$

Hence, we have

$$\frac{1}{\sqrt{8\pi}} \int\limits_{4jR}^{(4j+2)R} e^{-(y^2/8)}\, dy \leq \frac{1}{\sqrt{8\pi}} \int\limits_{4jR}^{(4j+2)R} e^{\ln 2 - ((k(y-2R)^2)/2)}\, dy$$

It means that

$$\int\limits_{4jR}^{(4j+2)R} \phi(2, y)dy \leq \int\limits_{(8j-4)R}^{8jR} \phi(2, \sqrt{ky})dy$$

Theorem 3.12 If $k > 1$ and $R^2 < \left(2\ln 2/(16k - 9)\right)$, then the following holds true:

$$\int\limits_{E} \phi(2, x - y)dy < 2\int\limits_{0}^{2R} \phi(2, y)dy + \frac{1}{\sqrt{2\pi kR}} e^{-2kR^2} \tag{3.107}$$

Proof. Combining the results of Theorems 3.10 and 3.11 leads to

$$\int_E \phi(2,x-y)dy \le 2\int_0^{2R} \phi(2,y)dy + 2\sum_{j=1}^\infty \int_{4jR}^{(4j+2)R} \phi(2,y)dy$$

$$\le 2\int_0^{2R} \phi(2,y)dy + 2\sum_{j=1}^\infty \int_{(8j-4)R}^{8jR} \phi(2,\sqrt{ky})dy$$

$$\le 2\int_0^{2R} \phi(2,y)dy + 2\int_{4R}^\infty \phi(2,\sqrt{ky})dy$$

$$\le 2\int_0^{2R} \phi(2,y)dy + \frac{1}{\sqrt{2\pi}}\int_{4R}^\infty \frac{y}{4R}e^{-(ky^2/8)}dy$$

$$= 2\int_0^{2R} \phi(2,y)dy + \frac{1}{\sqrt{2\pi kR}}e^{-2kR^2}$$

Theorem 3.13 If $k>1$ and $R^2 < \left(2\ln 2/(16k-9)\right)$, then the following holds true:

$$\int_E \phi(2,x-y)dy < \sqrt{1-e^{-R^2}} + \frac{1}{\sqrt{2\pi kR}}e^{-2kR^2}$$

Proof. By imitating the proof procedure of Theorem 3.10, one knows that inequality (3.107) is true.

3.6.6 Conclusions

With the increasing globalization of investment in recent years, a variety of Asian options have obtained a wide range of applications. With the rapid development of international trades and the globalization of the financial industry, this type of option will definitely attract more attention. An example is the moving average call option. It has been often used to design a poison pill, a business strategy used to increase the likelihood of negative results over positive ones against a party that attempts a take-over. The moving average calls, issued to existing shareholders, would be triggered by the event of a hostile takeover. The French investment bank Compagnie Financiere Indosuez and the French construction company Bouygues have successfully issued such options/warrants to protect themselves against potentially unfriendly investors.

In this section, we propose an estimation approach for pricing arithmetic average Asian options by using the decomposition methodology. After setting up a comparison theorem for a class of parabolic PDEs, the problem of finding the upper bound of the error term of the Asian options can be reduced to that of looking for a solution of an ordinary differential equation. Finally, we obtain the integral form of an upper bound of the error. From the conclusion of Theorem 3.6 and the second mean value theorem of integrals, one can readily find that U_{21} converges to zero very quickly as the independent variable approaches infinity. From the conclusion of Theorem 3.13, one may argue that the obtained error estimation suffers from some minor precision disadvantage as the independent variable vanishes sufficiently quickly. However, considering the facts that the error estimation possesses simplicity in the form of hyperbolic cosine for the upper bound of the error and that the length of its domain can be adjusted to sufficiently small, the conclusions of this section can surely be regarded as an improvement of previously published results.

Kim's inequality is very useful for Ying et al. (2009). However, Kim's estimation is possibly unbounded as R vanishes. Our inequality, developed in this section, has an advantage over Kim's inequality, because our estimation is bounded on the top as R vanishes. Therefore, the results in this section may help us to look for the solution error of a second-order PDE with variable coefficients.

3.7 Pricing of European Exchange Options: Nonlinear Parabolic PDEs (II)

3.7.1 Introduction

The concept *Fractal* was produced by the French mathematician Mandelbrot in 1967 in his article *How Long Is the British Coast*, published in the academic magazine *Science*. In 1982, Mandelbort's monograph *The Fractal of Nature* came out. Along with further research development, systematology, synergics, and dissipative structures sprung up in succession, and the concept *self-similarity of fractal* was enriched and enlarged to contain the self-similarity with regard to information, function, and time. So, if it has something to do with the self-similarity displayed in form, function, and information, the research object is referred to as *general fractal* as a joint name.

Fractal aims to study such a geometrical object that is not described by the traditional mathematics. In fact, mathematicians have long found out that there is a type of nondifferentiable function, which jeopardizes classical mathematics. Therefore, fractal geometry came into being. Along similar lines of study, scholars have found evidence for the exchange rate to exhibit self-similarity. In terms of the same exchange rate, it is found that the low-frequency data of changes in currency exchange have demonstrated the property of self-similarity. This kind of self-similarity feature exists in certain scope, while restricted from some of its characteristic scale at both ends. Therefore, the use of the fractal theory in the study of the pricing of foreign exchange options can be expected to bring forward some new progress in the study of self-similarity.

By relying on the development and application of nonlinear dynamical systems, many scholars have recently used the concepts of chaos and fractal as their research tools to study the in-depth statistical characteristics of the changes in exchange rates. Meese and Rogoff (1983) and Frankel and Froot (1986) were the first to discover that there are nonlinear characteristics in the exchange rate fluctuation. They were then followed by other scholars (Garman and Kohlhagen, 1983; Amin and Jarrow, 1991; Bernard and Bertill, 1995). Bershadskii (1999) believed that changes in the foreign exchange rate have the properties of fractal. Sim (1989) used the Dutting-Holmes equation to test the money supply fluctuation in the United States financial market covering the time period from January 1959 to November 1987. He successfully demonstrated the existence of chaos. Chen (1988) proved that the dimension of the monetary strange attracter is 1.5. Cecen and Eerkal (1996) studied the probability for hourly exchange rate of the low-dimensional chaotic attractor to experience major fluctuations.

Through demonstration, it is found that the estimated correlation dimension is not stable, but the estimated maximum Lyapunov index is positive in short time intervals. Martites (1999) described the chaotic behavior of the foreign exchange in Philippine in the framework of dynamical systems and proposed the exchange rate forecasting models based on the chaos theory. At the end, he calculated the fractal dimension and the maximum Lyapunov index of the time series from Philippine foreign exchange market. Schwartz and Yousefi (2003) studied the correlation dimension of the wide range fluctuation of currency exchange rate. By analyzing 6500 figures, it was shown that the exchange rate, such as German mark/dollar, pound/Japanese yen, and pound/dollar, displayed the low sequence fractal dimension. Wieland (2002) suggested that the central bank effectively controls the foreign exchange market by making use of chart analyses to verify the chaos control model. Vandewalle and Ausloos (1998) used the method of multiaffine characteristics to prove the existence of fractal changes in foreign exchange market. By using empirical study, Richard (2000) found that there are fractal structures in the foreign exchange market in many different countries. Ghashghaie et al. (1996) studied the scale of the exchange rate action by using 1,472,241 figures on the exchange of the dollar against the mark. They concluded that there are information cascades in the foreign exchange market. So, they advocated the need to study the movement of exchange rates.

In recent years, Chinese scholars began to apply the chaos theory to the study of the movement of exchange rates. By making use of partition function, Yang and Xie (2008) determined the existence of multifractal characteristics of the exchange rate series. Meanwhile, Xie and Yang (2008) systematically discussed the chaotic behavior of the exchange rate and the problem of how to determine the chaotic behavior by using quantitative methods. They concluded that applying the chaos theory to research, the exchange rate represents an important branch of research.

A large number of academic papers have shown that the behavior of exchange rates displays a long-term memory, and does not comply with normal distribution. Since the changes in an exchange rate have a chaotic fractal mechanism, it is essential to amend the traditional formula of Black–Scholes option (Hull, 1999). Continuing this line of thought, in this section, we will derive an option pricing formula under the FBM condition.

Zhong and Daye (2003) thought that the international capital market is subject to a nonlinear process, showing the properties of a sharp peak and heavy tail. Therefore, it is assumed that the underlying asset of financial derivatives follows a geometric FBM. The properties of self-similarity and long-term dependency of the geometric FBM can fit very well the fractal characteristics of the underlying asset of the financial derivatives. Shen (2008) selected the general foreign exchange options as the object to study the option pricing formula under the hypothesis that the prices of foreign currencies are subject to a geometric FBM. By assuming that both domestic and foreign risk-free interest rates vary with time, he derived the pricing of European foreign exchange options. By using physical probabilistic measure of pricing process and the principle of fair premium, Zhang (2006) dealt with the pricing formula of foreign currency options under the assumption that the pricing process of foreign currency options is driven by an FBM process. By using the basic building blocks of the derivatives theory, such as delta hedging, no arbitrage principle, and the consequent standard argument, Peng (2007) established the pricing formula for European style calls and puts with a fractional Ornstein–Uhlenbeck process. Potgieter (2009) analyzed the problem of how to price options from the perspective of arbitrage. By assuming that the underlying asset (the exchange rate) follows the fractal geometry for parameter H ($1/2 < H < 1$), he attempted to prove that there is a solution similar to the geometric Brownian motion, resulting in pricing formulas for foreign exchange call options and put options.

3.7.2 Foreign Exchange Option with Fractional Brownian Motion

FBM is a random process, which deals with the phenomenon of long-term dependency. This kind of random process manifests a continuous Gaussian process $\{W_H(t), t \in R\}$, which fits

$$W_H(0) = 0, \quad E[W_H(t)] = 0$$

and covariance is equal to

$$C_H(t,s) = \left(\frac{1}{2}\right)\left\{|t|^{2H} + |s|^{2H} + |t-s|^{2H}\right\}$$

From the covariance function, we can see that the correction coefficient for the increment over time is $C(t) = 2^{2H-1} - 1$. Different values of H give the form of time series of movement. As the change in the capital market indicates strong positive serial correction, the range of the parameter H is located within (0.5, 1). And W_t^H shows the similarity, meaning that for any $\alpha > 0$, $W_{\alpha t}^H$ and $\alpha^H W_t^H$ obey the same distribution.

Changes in exchange rate are not an independent random-walk process, but a state of a persistently biased random-walk process. From Table 3.1 (Hou, 2002), one can see that, for the three types of foreign exchange rates, as mentioned in the table, their yield kurtosis of the sample is much larger than 3, the skewness is not 0, and their distribution is not consistent with the normal distribution. It can be seen that changes in currency exchange rate obviously have a *long-term memory* effect.

Because the distribution possesses the characteristics of narrow-peaks and fat-tails, currency fluctuation can be described by a fractal distribution. For details, see Table 3.2 (Hou, 2002).

Assume that the price of the underlying asset follows the following FBM:

$$dS_t = \mu_t S_t dt + \sigma_t S_t dW_t^H, \quad \ln S_0 = x \tag{3.108}$$

where
  S_t is the underlying asset price of the financial derivatives of concern
  W^H is the FBM with Hurst index $H \in (1/2, 1)$
  μ_t is the average recovery rate
  σ_t is the underlying asset price volatility

Table 3.1   Statistical Results on the Earnings of Foreign Exchange

Classification	Standard Deviation	Kurtosis	Skewness
Yen/dollar	0.9632E02	−0.9354	89.63
Pound/dollar	0.2359E01	0.3162	264.59
Mark/dollar	0.1634E01	−0.2135	369.42

Table 3.2   Regression Result of Exchange Hurst Index *H*

Classification	P Value	R^2	Correlation Scale
Yen/dollar	0.001	0.994	0.1887
Pound/dollar	0.001	0.986	0.1430
Mark/dollar	0.001	0.986	0.2753

Lemma 3.10 Suppose that μ_t and σ_t are identified bounded functions. Then the solution of Equation 3.108 can be expressed as

$$\ln S_t = x + \int_0^t \mu_s ds + \int_0^t \sigma_s dW_s^H \tag{3.109}$$

Lemma 3.11 (*Itô Lemma*) Given certain functions $b:R_+ \to R$ and $a:R_+ \to R$, b_t is a boundary function in $[0, \infty)$ and $\int_0^\infty \int_0^\infty a_s a_t \phi(s,t) ds dt < \infty, \phi(s,t) = H(2H-1)$ $|s-t|^{2H-2}$, $H \in (1/2, 1)$, and

$$\eta_t = \xi + \int_0^t b_u du + \int_0^t a_u \, dW_u^H \tag{3.110}$$

If a function $f \in C^{1,2}(R_+ \times R)$ has bounded partial derivatives, then it can be expressed as

$$f(t, \eta_0) = f(0, \xi) + \int_0^t \frac{\partial f}{\partial s} ds + \int_0^t \frac{\partial f}{\partial \eta_s} b_s d_s + \int_0^t \frac{\partial f}{\partial \eta_s} a_s dW_s^H$$

$$+ \int_0^t \frac{\partial^2 f}{\partial \eta_s^2} a_s \left(\int_0^s a_u \phi(s,u) du \right) ds \tag{3.111}$$

Lemma 3.12 If the underlying asset price is subject to the FBM type Equation 3.108, then the price of a European style call satisfies the following boundary value problem:

$$\begin{cases} \dfrac{\partial V}{\partial t} + rs\dfrac{\partial V}{\partial S} + \sigma_t \left(\int_0^t \sigma_s \phi(t,s) ds \right) \dfrac{\partial^2 V}{\partial S^2} - rv = 0 \\[2mm] V(T,S) = (S-K)^+ \end{cases} \tag{3.112}$$

in which r is the risk-free interest rate and $(t,s) \in \Sigma_1$.

Theorem 3.14 If the underlying asset price is subject to the FBM type Equation 3.108, then a pricing formula for a European style call option is

$$V_c(t,S) = e^{-r(T-t)} \sqrt{\frac{\bar{\sigma}_1(t)}{2\pi}} \exp\left(-\frac{e^{r(T-t)}S - K}{2\bar{\sigma}(t)} \right) + e^{-r(T-t)}(e^{r(T-t)}S - K)N(d) \tag{3.113}$$

where

$$
d_1 = \frac{e^{r(T-t)}S - K}{\sqrt{\bar{\sigma}_1(t)}}, \bar{\sigma}_1(t) = 2\int_0^t \sigma_u e^{2r(T-u)} \left(\int_0^u \sigma_s \phi(u,s)ds \right) du
$$

Theorem 3.15 Under the same conditions as in Theorem 3.14, a pricing formula for a European style put option is

$$
V_p(t,S) = e^{-r(T-t)} \sqrt{\frac{\bar{\sigma}_2(t)}{2\pi}} \exp\left(-\frac{e^{r(T-t)}S - K}{2\bar{\sigma}(t)} \right) - e^{-r(T-t)}(e^{r(T-t)}S - K)N(-d_2)
$$

(3.114)

where

$$
d_2 = \frac{e^{r(T-t)}S - K}{\sqrt{\bar{\sigma}_2(t)}} \text{ and } \bar{\sigma}_2(t) = 2\int_t^T \sigma_u e^{2r(t-u)} \left(\int_0^u \sigma_s \phi(u,s)ds \right) du
$$

3.7.3 Conclusions

Nonlinear changes in foreign exchange are caused by the traders' reaction to market expectations. So when it intervenes in the foreign exchange market, the central bank should pay more attention to the role of the microstructure in the exchange rate fluctuation. Only by fully taking into account the main microinvestment psychology, it will enable the policy adopted for intervening in the foreign exchange market to obtain better results. On the other hand, since many countries employ the floating exchange rate system, central banks can intervene in the foreign exchange market to keep the exchange rate fluctuate within the appropriate scope. This kind of *long-term memory* inherent in exchange rate changes causes the central bank's intervention to affect not only the short-term movement of the exchange rate, but also the long-term movement. Therefore, when the central bank intervenes in the foreign exchange market, it must take into consideration the changes in the effect of the *long-term memory*. At the same time, the central bank should take corresponding measures to achieve better results.

In this section, on the basis of the fractal distribution and Hurst index, we derived our pricing formula for foreign exchange options. To some extent, this work can make the pricing of foreign exchange options more in line with economic realities. But, it is still difficult to a degree to price foreign exchange options by using the method provided in this section. The reason lies in the fact that the central bank tends to implement intervention policies in the foreign exchange market, so that the foreign exchange rate is not entirely decided by the market forces. Generally speaking, polices of economic intervention can all lead to external impacts, which would have a linear influence to exchange rate. To some extent, this influences the statistical distribution characteristics of the exchange rate. In the economic reality, not only do fluctuations in the foreign exchange rate show the statistic law of chaotic systems, but also external disturbances cause fluctuations. Therefore, it is very important to combine *chaos* and *external disturbance* in our understanding of the statistical characteristics of the underlying asset. Therefore, deducing the pricing equation for foreign exchange options in line with the economic reality should be the direction of future research.

Chapter 4

From Financial Crises to Currency Substitution: Limit Cycle Theory for Capital Account Liberation

4.1 General Theory of Limit Cycles

Consider the following nonlinear differential equations of the first order:

$$\frac{dx}{dt} = F(x,t) \tag{4.1}$$

where $x \in R^n$, $t \in R$.

Assume that $F \in C(R \times G)(G \subset R^n)$ satisfies the existence and uniqueness conditions of solution. If the right-hand function of Equation 4.1 obviously contains the independent variable t, then it is called nonautonomous system of differential equations; correspondingly, if the right-hand function of Equation 4.1 does not contain the independent variable t, that is,

$$\frac{dx}{dt} = F(x) \tag{4.2}$$

then it is called autonomous system of differential equations.

As for the phase space, which is the Euclidean plane R^2, some important results were established by H. Poincaré and I. Bendixson. For example, the given differential equations

$$\begin{cases} \dfrac{dx}{dt} = P(x, y) \\ \dfrac{dy}{dt} = Q(x, y) \end{cases} \tag{4.3}$$

where $P(x, y)$, $Q(x, y)$ are continuous in a connected region $D \subset R^2$ and must satisfy some conditions in order to keep any initial-value problem have a unique solution. This kind of system appears in the studies of physics and finance frequently. For example, there is the following differential equation with damping in the oscillating financial market:

$$\frac{d^2 x}{dt^2} + \varepsilon(1 - x^2)\frac{dx}{dt} + x = 0 \tag{4.4}$$

where $\varepsilon > 0$. Let $y = (dx/dt)$. Then, (4.4) becomes the following quadratic system:

$$\begin{cases} \dfrac{dx}{dt} = y \\ \dfrac{dy}{dt} = -x - \varepsilon(1 - x^2)y \end{cases} \tag{4.5}$$

In this section, we introduce a few important theorems.

Theorem 4.1 (Jordan theorem). A simple curve J in the plane R^2 divides R^2 into two parts, that is, the open set $R^2 - J$ has exactly two connected components D_1, D_2 such that D_1 is bounded (it is called the interior of J), and D_2 is unbounded (it is called the exterior of J).

It is obvious that a continuous path starting out at an arbitrary point in the region D_1 to another arbitrary point in the region D_2 must intersect with the curve J.

Theorem 4.2 (Poincaré–Bendixson theorem). Let the functions X and Y have continuous first partial derivatives in a domain D of the xy-plane. Assume that D_1 is a bounded subdomain in D and $\bar{D_1}$ is the region that consists of D_1 plus its boundary. Suppose that $\bar{D_1}$ contains a critical point of the system (4.3). If there exists a constant t_0 such that $x = \varphi(t)$, $y = \psi(t)$ is a solution of the system (4.3) that exists and stays in $\overline{D}_1$ for all $t \geq t_0$, then either $x = \varphi(t)$, $y = \psi(t)$ is a periodic solution, or $x = \varphi(t)$, $y = \psi(t)$ spirals toward a closed trajectory as $t \to \infty$. In either case, the system (4.3) has a periodic solution in $\overline{D}_1$.

The concept of singular points plays an important role in the study of the local structure of dynamic systems, while the concept of limit cycles occupies a specially important position in investigating the global structure of dynamic systems. Now, we introduce a few important theorems from the limit cycle theory.

Theorem 4.3 (Bendixson theorem). There exists at least one singular point in the interior region enveloped by an arbitrary closed trajectory.

Theorem 4.4 (Poincaré–Bendixson ring domain theorem). Suppose that a ring region R is surrounded by two simple curves Γ_1 and Γ_2, where Γ_1 lies entirely within the interior of Γ_2, all the trajectories that intersect with both Γ_1 and Γ_2 enter the ring region R as t increases (or decreases), and there does not exist any singular point in the interior of R. Then, the system (4.3) has at least one limit cycle in the ring region R.

In general, the curves Γ_1 and Γ_2 are called the inside boundary curve and outside boundary curve, respectively.

Theorem 4.5 (Bendixson). Let the functions P and Q have continuous first partial derivatives in a domain D of the xy-plane. If

$$\frac{\partial P}{\partial x} + \frac{\partial Q}{\partial y} \geq 0 \quad (\leq 0)$$

then there is not any closed trajectory such that the system (4.3) lies entirely within the region D.

Theorem 4.6 (Bendixson–Dulac theorem). Let the functions P, Q, and B have continuous first partial derivatives in a domain D of the xy-plane. If

$$\frac{\partial(BP)}{\partial x} + \frac{\partial(BQ)}{\partial y} \geq 0 \quad (\leq 0)$$

then there is not any closed trajectory of the system (4.3) that lies entirely within the region D.

4.2 Poincaré Problem: Quadratic Polynomial Differential Systems (I)

4.2.1 Introduction

We consider the general planar polynomial differential system of the form

$$
\begin{cases}
\dot{x} = \displaystyle\sum_{i+j=0}^{2} a_{ij} x^i y^j = P_2(x, y) \\[4mm]
\dot{y} = \displaystyle\sum_{i+j=0}^{2} b_{ij} x^i y^j = Q_2(x, y)
\end{cases}
\tag{4.6}
$$

where the dot stands for the derivative with respect to the independent variable t. Here, both $P_2(x, y)$ and $Q_2(x, y)$ are coprime polynomials such that max{deg P_2, deg Q_2} = 2. As usual, the symbol R^2 denotes the ring of the polynomials in two variables. In what follows, system (4.6) will simply be referred to as a quadratic system.

A point (x_0, y_0) is called a finite singular point of system (4.6) if $P_2(x_0, y_0) = Q_2(x_0, y_0) = 0$.

Let DJ be the Jacobian matrix associated with the plane R^2, that is,

$$
DJ = \begin{pmatrix}
\dfrac{\partial P_2}{\partial x} & \dfrac{\partial P_2}{\partial y} \\[4mm]
\dfrac{\partial Q_2}{\partial x} & \dfrac{\partial Q_2}{\partial y}
\end{pmatrix}
\tag{4.7}
$$

The flow of (4.6) in the neighborhood of the singular point (x_0, y_0) is classified according to the eigenvalues of the matrix $DJ(x_0, y_0)$. In particular, if the eigenvalues are zero, then the singular point is called degenerating. Otherwise, it is termed as nondegenerating. If $DJ(x_0, y_0)$ has exactly one eigenvalue that is equal to zero, then the singular point (x_0, y_0) is called elementarily degenerating. Finally, if the Jacobian matrix $DJ(x_0, y_0)$ is not identically null and it possesses two zero eigenvalues, we say that (x_0, y_0) is a nilpotent point.

Knowing the concept of invariant algebraic curves is really important for the understanding of the dynamics of a differential system. Moreover, these curves play an important role even when they are complex. Finding an upper bound for the degree of the irreducible invariant algebraic curves of a family of polynomial differential systems, if it exists, is known as the Poincaré problem. Already Poincaré in his monograph, entitled *Theory of Limit Cycles*, studied such a problem in connection with the problem of the algebraic integrability of these families (Ye, 1986). The famous works on this topic can be found in Poincaré (1891), Qin (1966),

Seidenberg (1968), Odani (1995), Li et al. (1996), Walcher (2000), Chavarriga et al. (2003), Chavarriga and Garcia (2004), and Chavarriga et al. (2004).

A limit cycle for (4.6) is defined as an isolated periodic orbit. When a limit cycle is an oval of an invariant algebraic curve, it is called an algebraic limit cycle. Quadratic systems that can have limit cycles are classified into the following three families:

$$\begin{cases} \dot{x} = \delta x - y + lx^2 + mxy + ny^2 \\ \dot{y} = x(1 + ax + by) \end{cases} \tag{4.8}$$

according to the following conditions:

Family (I), if $a = b = 0$
Family (II), if $a \neq 0$ and $b = 0$
Family (III), if $b \neq 0$

For details, see, for instance, the classical textbook (Ye, 1986).

It is well known, see, for instance, Ye (1986), that when $d = 0$ or $m(l + n) = 0$, any system in family (I) does not have any limit cycle. If $l = 0$, then each system in family (I) is a quadratic Leonard system with linear damping without invariant algebraic curves as shown in Zoladek (1998). Here, let us introduce some detailed results regarding family (I) by solving the Poincaré problem and proving the nonexistence of any algebraic limit cycle.

In this section, we summarize some results about formal differential equations and their solutions. Concerning the Poincaré problem for family (I), we have the following result.

Theorem 4.7 Any irreducible invariant algebraic curve for family (I) has at most degree 3.

Let us consider the families of quadratic systems with algebraic limit cycles known as of the writing of this section. These families contain all the algebraic limit cycles defined by polynomials of degrees 2 and 4 for a quadratic system, as it is proved by Chavarriga et al. (2005). It is well known from Evdokimenko's work (Evdokimenko, 2011) that no cubic limit cycles exist for quadratic systems.

Theorem 4.8 If a quadratic system has an algebraic limit cycle of degree 2, then after an affine change of variables, the limit cycle becomes the circle: $x^2 + y^2 = 1$.

Theorem 4.9 (Yablonskii system). If $abc \neq 0$, $a \neq b$, $ab > 0$, $4c^2(a - b)^2 + (3a - b)$ $(a - 3b) < 0$, and $l = 4abc^2 - (3/2)(a + b)^2 + 4ab$, then the Yablonskii system

$$\begin{cases} \dot{x} = -abcx - (a+b)y + 3(a+b)cx^2 + 4xy \\ \dot{y} = (a+b)abx - 4abcy + lx^2 + 8(a+b)cxy + 8y^2 \end{cases} \tag{4.9}$$

has the following invariant algebraic curve:

$$(y + cx^2)^2 + x^2(x - a)(x - b) = 0 \tag{4.10}$$

whose oval is a limit cycle for system (4.9).

Theorem 4.10 (Filipstov system). If $0 < a < (3/13)$, then the Filipstov system

$$\begin{cases} \dot{x} = 6(1+a)x + 2y - 6(2+a)x^2 + 12xy \\ \dot{y} = 15(1+a)y + 3ax(1+a)x^2 - 2(9+5a)xy + 16y^2 \end{cases} \tag{4.11}$$

has the following invariant algebraic curve:

$$3(1+a)(ax^2 + y)^2 + 2y^2(2y - 3(1+a)x) = 0 \tag{4.12}$$

whose oval is a limit cycle for system (4.11).

Theorem 4.11 (Chavarriga system). If $\dfrac{-71 + 17\sqrt{17}}{32} < a < 0$, then the Chavarriga system

$$\begin{cases} \dot{x} = 5x + 6x^2 + 2y + 4(1+a)xy + ay^2 \\ \dot{y} = x + 2y + 4xy + (2+3a)y^2 \end{cases} \tag{4.13}$$

has the following invariant algebraic curve:

$$x^2 + x^3 + x^2y + 2axy^2 + 2axy^3 + a^2y^4 = 0 \tag{4.14}$$

whose oval is a limit cycle for system (4.13).

Theorem 4.12 (Llibre and Sorolla system). If $0 < a < (1/4)$, then the Llibre and Sorolla system

$$\begin{cases} \dot{x} = 2(1 + 2x - 2ax^2 + 6xy) \\ \dot{y} = 8 - 3a - 14ax - 2axy - 8y^2 \end{cases} \tag{4.15}$$

has the following invariant algebraic curve:

$$\frac{1}{4} + x - x^2 + ax^3 + xy + x^2 y^2 = 0 \qquad (4.16)$$

whose oval is a limit cycle for system (4.15).

4.3 Foreign Assets and Foreign Liabilities: Quadratic Polynomial Differential Systems (II)

4.3.1 Introduction

By foreign assets, it means the value of asset claims that a nation's residents own against foreigners, and by foreign liabilities, it stands for the value of liabilities that they owe to foreigners. A nation's net foreign asset (NFA) position is measured by foreign assets minus foreign liabilities. Although the NFA position is important in macroeconomic study, surprisingly, little is known about the respective role that each of the two accounting sources of NFA plays in determining the dynamics and comparative statics of a nation's wealth. Both an increase in foreign assets and a decrease in foreign liabilities lead to an increase in NFA; nevertheless, they have different transmission mechanisms to affect macroeconomic conditions. An increase in foreign assets or a decrease in foreign liabilities raises aggregate real wealth of the nation and therefore increases consumption expenditure and the holding of real money balances. That constitutes the wealth effect that increases the aggregate demand. On the other hand, a decrease in foreign liabilities (foreign physical capital, in particular) also reduces the aggregate real capital domiciled in the home economy or at least the loanable funds supplied for investment. Therefore, the aggregate supply will decrease, and the consequent income effect further weakens the aggregate demand for goods and services and the derived demand for real balances. In this section, we compare and differentiate the asset and liability components of NFA in terms of their respective macroeconomic effects.

4.3.2 Macroeconomic Model

The macroeconomic model in this section has four characteristics:

1. Foreign capital inflow adds to the total capital stock domiciled in the domestic economy, thus increasing the aggregate supply.
2. The domestic capital outflow increases the real wealth of the nation and thus leads to increases in consumption expenditure and the demand for real money balances.

3. The demand for the domestic currency depends on the exchange rate and the household wealth.
4. The exchange rate that clears the money market changes over time.

The total capital stock domiciled in the domestic economy (K) consists of the capital stock owned both by domestic residents (K_d) and by foreigners (K_f). The home economy's NFA position can be described as

$$\text{NFA} = K_d^* - K_f \tag{4.17}$$

where K_d^* is the domestically owned capital stock domiciled in the rest of the world.

Thus, total real wealth of the nation can be identified as the sum of the domestic capital assets and the NFA position.

Suppose that the home economy has an aggregate production function and produces a composite tradable good at the level of $Y = F(K)$. Interest rates and prices are flexible, and output is at its natural rate or full-employment level, and the natural rate level of output increases as the aggregate capital stock (K) increases. The price level P is determined in the world market and subject to the purchasing power parity. A simplifying assumption is that the foreign price level (P^*) is standardized at unity so that the domestic price level (P) is equivalent to the exchange rate of domestic currency for foreign currency (E), and hence an inflation rate is identified as the currency depreciation rate.

On the expenditure side of the goods market, the amount of private wealth and the level of income determine the consumption spending. The private wealth is equal to the total national capital wealth plus real money balances, that is, $K_d + NFA + m$, where m is defined as the ratio of money stock (M) over the exchange rate (E). The difference between the output and spending determines the trade balance, which translates into the net foreign investment (NFI). So, the time derivative of NFA is given as follows:

$$NFI \equiv \dot{NFA} = \dot{K}_d^* - \dot{K}_f$$
$$= F(K_d + K_f) - C[K_d + NFA + m, F(K_d + K_f)] - I(r) - G \tag{4.18}$$

where
  r is the real interest rate
  $C(\cdot)$ is the consumption function, which is subject to the wealth effect (the first argument in the consumption function) and income effect (the second argument in the function)
  $I(\cdot)$ is the investment function
  G is the government expenditure, which is assumed not to be financed by seigniorage

The market-clearing condition for real money balances is

$$\dot{m} = L[\pi^e + r, F(K_d + K_f)](K_d + NFA + m) \tag{4.19}$$

where
   $L(\cdot)$ is the liquidity-preference function
   π^e is the expected inflation rate

In contrast to the income effect where the real money demand rises with income (liquidity preference), the wealth effect in the money market suggests that people increase their holdings of real money balances as private wealth increases. This interpretation of the wealth effect broadly follows the portfolio approach suggested by Tobin in viewing real balances as a direct portfolio substitute for holdings of interest-yielding assets. Particularly, the real private wealth enters the real money demand function (4.19) multiplicatively so that real money balances are subject to unitary wealth elasticity, implying that an increase in NFA amplifies the marginal effect of liquidity preference on the real money demand in an exact proportionate pattern, but the wealth effect of NFA does not enter and therefore does not determine liquidity preference per se.

Inflation expectation is assumed to follow rational expectation—perfect foresight—and it adjusts instantaneously with any change in the wealth and income variables (Krugman, 1979). Therefore, the actual inflation rate can be expressed as the expected rate of inflation determined by the money–market equilibrium condition.

$$\pi = L^{-1}\left(\frac{m}{K_d + NFA + m}, F(K_d + K_f)\right) - r \equiv h(m, NFA) \tag{4.20}$$

Suppose that the monetary authority sets a constant money growth rate g_M. The exchange rate that clears the domestic money market changes over time. Then, the evolution of the real balance can be presented as

$$\dot{m} = (g_M - \pi)m = [g_M - h(m, NFA)]m \tag{4.21}$$

4.3.3 Dynamics Analysis

Equations 4.18 and 4.21 constitute a dynamic equation system of the two wealth variables, NFA and m. Depending on whether changes in NFA stem from the asset component K_d^* or the liability component K_f, the economy exhibits different dynamics because of the different conduits for them to influence the economy.

In this section, we consider only a special case of Equations 4.18 and 4.21, that is, a quadratic polynomial differential system in the neighborhood of $(\overline{NFA}, \bar{m})$, where $\overline{NFA}$ and $\bar{m}$ denote the steady-state values of NFA and m, respectively. Let us take the linear approximation for the functions $C(\cdot)$ and $h(\cdot)$ as follows:

$$C(NFA, m) = c_0 - a_1(NFA - \overline{NFA}) + b_1(m - \bar{m}) \tag{4.22}$$

$$h(NFA, m) = h_0 + a_2(NFA - \overline{NFA}) + b_2(m - \bar{m}) \tag{4.23}$$

By letting

$$x = NFA - \overline{NFA}, \ y = m - \bar{m} \quad \text{and} \quad c_1 = F - I - G - c_0, c_2 = g_M - h_0$$

we obtain the following quadratic polynomial differential system:

$$\begin{cases} \dot{x} = a_1 x + b_1 y + c_1 = P_2(x, y) \\ \dot{y} = (a_2 x + b_2 y + c_2)y = Q_2(x, y) \end{cases} \tag{4.24}$$

Notice that the characteristic equation responding to the linear part of quadratic system (4.24) is

$$\lambda^2 - (a_1 + c_2)\lambda + a_1 c_2 = 0 \tag{4.25}$$

Therefore, it is ready to obtain the following results.

Theorem 4.13 The following hold true.

(i) If $a_1 + c_2 < 0$, $a_1 c_2 > 0$, then the steady-state point of (4.24) is a stable node or focus.

(ii) If $a_1 + c_2 > 0$, $a_1 c_2 > 0$, then the steady-state point of (4.24) is an unstable node or focus.

(iii) If $a_1 c_2 < 0$, then the steady-state point of (4.24) is an unstable saddle point.

Theorem 4.14 If $a_1 + c_2 = b_1 = 0$, then the steady-state point of (4.24) is a center or node.

Theorem 4.15 If $a_1 + c_2 > 0$, $a_2 > 0, b_2 > 0$, then in the right-hand region as defined by the straight line $a_2 x + 2b_2 y + a_1 + c_2 = 0$, the quadratic system (4.24) has no limit cycle.

Proof. In light of Bendixson theorem and the condition of Theorem 4.15, one can calculate and obtain

$$\frac{\partial P_2}{\partial x} + \frac{\partial Q_2}{\partial y} = a_2 x + 2b_2 y + a_1 + c_2 > 0$$

Therefore, there does not exist any limit cycle in the quadratic system (4.24).

4.3.4 Conclusion

Theorem 4.13 shows that the economic system experiences a saddle-point equilibrium when changes in NFA item are exclusively from changes in its asset component so that the wealth effect is the only channel through which NFA affects the real money demand. As depicted in Figure 4.1, the two stationary loci divide the phase-diagram space into four quadrants. The stable branch located in quadrants II and IV exhibits the path for the system to converge to its steady-state equilibrium.

In the present case, any point in quadrants I and III necessarily diverges along unsustainable paths in the quadrants in which they begin, and all the points starting from quadrants II and IV, except those on the stable arm, will eventually cross one of the two stationary schedules and merge onto the unstable branch pointing away from the equilibrium. Starting from a point to the right of the stable arm, the economy will undergo a rapid increase in m and slide into a boom shadowed by its deteriorating NFA position (the U.S. economy in the second half of the 1990s, for instance). On the other hand, starting from a point to the left of the stable arm, the economy will suffer from a continual decline in m and ultimately slide into

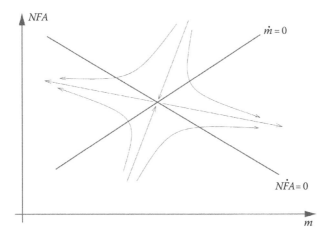

Figure 4.1   The saddle-point equilibrium.

a recession burdened by its external surplus (such as the Japan's economy in the recent decade). Even in the instability case, achieving the steady-state equilibrium calls for correct macroeconomic policies, which will be further discussed later.

4.4 Dynamics of Employment: Cubic Polynomial Differential Systems (I)

4.4.1 Introduction

In this section, we focus on the dynamics of job vacancies and employment after a productivity shock, and then explain the reality that the output follows hump-shaped dynamics when responding to shocks in the economy. Our main research method is the qualitative theory of ordinary differential equations, through which we can judge the stability of the equilibrium point. Then, we analyze the dynamics of job vacancies and employment after permanent productivity shocks. Our results indicate that job vacancies and employment, under certain conditions, can generate nonmonotonic dynamics, which in turn explains the existence of a hump-shaped output dynamics.

According to the definition of international monetary fund (IMF), the so-called capital account liberalization represents some related measures in which there is not any *payment restraint imposed on capital transactions*. The process of capital account liberalization is a gradual relaxation of capital controls, allowing residents and non-residents to hold assets abroad or engage in cross-border transactions, and eventually realizing the goal of a freely convertible currency. There is always a close relationship between capital account liberalization and a nation's economic growth. In the beginning of the related investigations, theorists basically believed that capital account liberalization contributes a lot to a nation's economic growth because capital inflows may bring huge amounts of money needed in the nation-building. However, with the Mexican crisis, the Asian financial crisis, and others, theorists began to question the role of capital account liberalization in promoting a nation's economic growth.

As of this writing, there mainly exist three types of views. First, Quinn (1997) developed the promotion theory represented from an empirical point of view that changes of indicators in capital account liberalization were positively correlated with a nation's economic growth, which was subjected to the statistical significance test. Second, Edison et al. (2002) established the uncertainty theory by initially studying a sample that consisted of 89 countries by means of Quinn's measurement methods and by drawing a conclusion that the relationship between capital account liberalization and economic growth depended on specific circumstances. This relationship was uncertain and influenced by many factors, such as the maturity of the nation's financial system and the nation's economic foundation and legal environment. Rodrik (2008) selected a data sample that included

100 countries for the time period from 1975 to 1989 (both developed countries and developing countries) and found no significant evidence in promoting economic growth by capital account liberalization. Third, Krugman (1998) represented the retrogression theory by pointing out in 1993 that capital account liberalization was not conducive to the long-term economic growth in developing countries and that international financial integration was not the engine of economic growth. In reality, an important fact of business fluctuation is that economic growth is positively correlated in the short run but negatively correlated in the long run with capital account liberalization. This indicates that the output follows a hump-shaped dynamics when shocks in the economy appear. As Cogley and Nason (1995) have shown, most real business cycle models do not have sufficient propagation mechanisms to account for the hump-shaped output dynamics; the key reason for the failure is that employment responds quickly to shocks in those models. Merz (1995) and Andolfatto (1994) thought that it is necessary to integrate labor market search into an intertemporal model. Based on that idea, Shi (1995), Shi and Wen (1997) integrated the said labor market search into an intertemporal utility maximization framework by endogenizing agents' reservation wage as the marginal rate of substitution between leisure and consumption, through which the agents' intertemporal consumption decision directly affects the labor market behavior. As a result, these authors obtained the nonmonotonic dynamics in employment after a permanent productivity shock. That can explain readily the hump-shaped output dynamics.

This section is mainly constructed within the framework of a standard search model. By restricting the parameters according to the nature of the production function and by matching functions, we can solve the equilibrium point. Then, we judge the stability of this point by means of the stability theory of ordinary differential equations. Our results indicate that the dynamics of employment rely on the initial state of job vacancies and employment.

4.4.2 Model

Our model is derived from differential equations and dynamical systems as proposed by Shi and Wen (1997) to describe the dynamics of outputs.

$$
(D)\begin{cases}
\dot{c} = \sigma c(F_1 - \rho - \delta), \\[2mm]
\dot{x} = q\left[(F_1 + \theta - \delta)x - \dfrac{(1-\lambda)m}{B'}\left(F_2 - \dfrac{\beta}{u'} \right) \right], \\[2mm]
\dot{n} = s_0 m(x) - \theta n, \\[2mm]
\dot{k} = F - \delta k - B(s_0 x) - c,
\end{cases}
$$

The function $F(k, n)$ here is a production function, which is assumed to be increasing and concave in each argument, s stands for the unemployment in the economy and is assumed to be s_0 in the calculation in the next part, while n stands for the employment in each certain time and is equal to n_0 at the initial state. F_1 is the marginal product of capital, and F_2 is the marginal product of labor. β/u' is the reservation wage of labor, θ is the turnover rate of employees, and ρ is the discount rate. B is a function of job vacancies v, and B' is the marginal cost of the firm to maintain an additional job vacancy and is a positive number.

The expression $x = (v/s)$ stands for the vacancy–unemployment ratio. A smaller x represents a tighter labor market, and it is also a positive number. The symbol m is the rate at which an unemployed agent can find a job and can be written in terms of a function signified by x, that is,

$$m = m(x) = M_0 x^{\alpha}, \quad \alpha \in (0,1)$$

where M_0 is a positive constant, which means the matching technology exhibits constant returns to scale (Pissarides, 1986; Blanchard and Diamond, 1989; Layard et al., 1991).

Our model, as shown later, comes from the second and third equations of the earlier dynamic systems.

$$\begin{cases} \dot{x} = q\left[(\theta + \rho)x - \dfrac{\alpha m}{B'}\left(F_2 - \dfrac{\beta}{u'} \right) \right] \\ \dot{n} = s_0 m(x) - \theta n \end{cases} \tag{4.26}$$

Since there is no capital in our model, we hypothesize $F = n^{\gamma}$, $\gamma \in (0,1)$. Hence $F_2 = \gamma n^{\gamma-1}$. We will restrict the parameters of α and γ as some specific numbers in the calculations in the following subsection.

4.4.3 Calculation

Case 1: Assume that $\alpha = \gamma = 1/2$; $m = m(x) = M_0 x^{1/2}$; and $F_2 = (1/2)n^{-(1/2)}$.
Equations 4.26 can be rewritten in the general form as follows:

$$\begin{cases} \dfrac{1}{q}\dot{x} = (\theta + \rho)x - \dfrac{M_0 x^{(1/2)}}{2B'}\left(\dfrac{1}{2}n^{-(1/2)} - \dfrac{\beta}{u'} \right) = f_1(x,n) \\ \dot{n} = s_0 M_0 x^{(1/2)} - \theta n = f_2(x,n) \end{cases} \tag{4.27}$$

Equations 4.27 can be converted into a quadratic system after affine changes of variables. Now, we first obtain the equilibrium point $Q_1(x^*, n^*)$ as follows:

$$x^* = \frac{M_0^{(2/3)}\theta^{(2/3)}}{16B'^{(4/3)}s_0^{(2/3)}(\theta+\rho)^{(4/3)}}\delta_3^4, \quad n^* = \frac{M_0^{(4/3)}s_0^{(2/3)}}{4B'^{(2/3)}\theta^{(2/3)}(\theta+\rho)^{(2/3)}}\delta_3^2,$$

where $\delta_3 = (\omega+1)^{(1/3)} - (\omega-1)^{(1/3)}$ and $\omega = \sqrt{1 + \dfrac{8\tau^3}{27(\theta+\rho)^3}}$.

Next, let us discuss the stability of the equilibrium point $Q_1(x^*, n^*)$. At first, let us consider the linear part of the right-hand side of system (4.27):

$$\begin{cases} f_1(x,n) = A_3(x - x^*) + B_3(n - n^*) \\ f_2(x,n) = C_3(x - x^*) + D_3(n - n^*) \end{cases}$$

The characteristic equation corresponding to the linear part of system (4.27) is

$$\lambda^2 - (A_3 + D_3)\lambda + (A_3D_3 - B_3C_3) = 0 \tag{4.28}$$

where

$$A_3D_3 - B_3C_3 = -\theta(\theta+\rho) - \theta\tau\delta_3^{-2} + \theta(\theta+\rho)\delta_3^{-3} - \frac{\theta(\theta+\rho)}{2}\delta_3^{-3}$$

$$= -\frac{\theta(\theta+\rho)}{2} - \frac{\theta(\theta+\rho)}{2}\delta_3^{-3} < 0$$

Therefore, the characteristic Equation 4.28 must have one positive and one negative eigenvalue. According to the qualitative theory of ordinary differential equations, we obtain the following results.

Theorem 4.16 If $\alpha = \gamma = (1/2)$, then the equilibrium point $Q_1(x^*, n^*)$ is an unstable saddle point.

Case 2: Assume that $\alpha = (1/4)$, $\gamma = 1$; and $m = m(x) = M_0x^{(1/4)}$. This is a degenerating case, where $F_2 = 1$ means that the marginal product of labor is 1 unit.

$$\begin{cases} \dfrac{1}{q}\dot{x} = (\theta+\rho)x - \dfrac{3M_0x^{(3/4)}}{4B'}\left(\dfrac{2}{3}n^{-(1/3)} - \dfrac{\beta}{u'}\right) = f_1(x,n) \\ \dot{n} = s_0M_0x^{(3/4)} - \theta n = f_2(x,n) \end{cases} \tag{4.29}$$

Equations 4.27 can be converted into a cubic system after affine changes of variables. Now, we obtain the equilibrium point $Q_2(x^*, n^*)$ as follows:

$$x^* = \delta_1^{(4/3)}, \quad n^* = \frac{s_0 M_0}{\theta} \delta_1^{(1/3)}$$

where

$$\delta_1 = \frac{M_0\left(1 - \dfrac{\beta}{u'}\right)}{4B'(\theta + \rho)}$$

Next let us discuss the stability of the equilibrium point $Q_2(x^*, n^*)$. To this end, we consider the linear part of the right-hand side of system (4.29):

$$\begin{cases} f_1(x,n) = A_1(x - x^*) + B_1(n - n^*) \\ f_2(x,n) = C_1(x - x^*) + D_1(n - n^*) \end{cases}$$

The characteristic equation corresponding to the linear part of system (4.29) is

$$\lambda^2 - (A_1 + D_1)\lambda + (A_1 D_1 - B_1 C_1) = 0 \tag{4.30}$$

where

$$A_1 D_1 - B_1 C_1 = -\frac{3\theta(\theta + \rho)}{4} < 0$$

Therefore, the characteristic Equation 4.30 must have one positive and one negative eigenvalue. According to the qualitative theory of ordinary differential equations, we obtain the following result.

Theorem 4.17 If $\alpha = (1/4)$, $\gamma = 1$, then the equilibrium point $Q_2(x^*, n^*)$ is an unstable saddle point.

Case 3: Assume that $\alpha = (3/4)$, $\gamma = (2/3)$; $m = m(x) = M_0 x^{(3/4)}$; and $F_2 = (2/3)n^{(1/3)}$.
    Equations 4.26 can be rewritten in the general form as follows:

$$\begin{cases} \dfrac{1}{q}\dot{x} = (\theta + \rho)x - \dfrac{3M_0 x^{(3/4)}}{4B'}\left((2/3)n^{-(1/3)} - \dfrac{\beta}{u'}\right) = f_1(x,n) \\ \dot{n} = s_0 M_0 x^{(3/4)} - \theta n = f_2(x,n) \end{cases} \tag{4.31}$$

Equations 4.31 can be converted into a cubic system after affine changes of variables. Similarly, we first obtain the equilibrium point $Q_3(x^*, n^*)$ as follows:

$$x^* = \frac{16 p_1^4}{\delta_2^4}, \quad n^* = \frac{8 M_0 s_0 q_1^3}{\theta \delta_2^3}$$

where $p_1 = \dfrac{3 M_0 \beta}{4 B'(\theta + \rho) u'}$ and $q_1 = \dfrac{M_0^{(2/3)} \theta^{(1/3)}}{2 B'(\theta + \rho) s_0^{(1/3)}}$.

Next, let us discuss the stability of the equilibrium point $Q_3(x^*, n^*)$. At first, let us consider the linear part of the right-hand side of system (4.31):

$$\begin{cases} f_1(x,n) = A_2(x - x^*) + B_2(n - n^*) \\ f_2(x,n) = C_2(x - x^*) + D_2(n - n^*) \end{cases}$$

The characteristic equation corresponding to the linear part of system (4.31) is

$$\lambda^2 - (A_2 + D_2)\lambda + (A_2 D_2 - B_2 C_2) = 0 \tag{4.32}$$

where

$$A_2 D_2 - B_2 C_2 = -\frac{\theta(\theta + \rho)}{4} - \frac{\theta(\theta + \rho)}{16 q_1} \delta_2^2 < 0$$

Therefore, the characteristic Equation 4.32 must have one positive and one negative eigenvalue. According to the qualitative theory of ordinary differential equations, we have the following.

Theorem 4.18 If $\alpha = 3/4$, $\gamma = 2/3$, then the equilibrium point $Q_3(x^*, n^*)$ is an unstable saddle point.

4.4.4 Conclusions

By restricting the parameters α and γ, we obtained, respectively, the equilibrium point in the three special cases as mentioned earlier and found that the equilibrium point (x^*, n^*) is always an unstable saddle point. As a matter of fact, the equilibrium point is always a saddle point in the general situation, and the method of proof is similar to the earlier three cases. In this section, x represents the job vacancy and unemployment ratio, and n is the employment figure. Since the unemployment rate doesn't change in the macroeconomic background, the increase and decrease in x stand for the rise and fall in the number of job vacancies. According to the features of a saddle point, we know that there exist two possible forms. First, when the initial number of job vacancies is near to its equilibrium position, while the initial employment is far away from its equilibrium position, the employment figure will rise slowly after a short-term decline. At the same time, the number of job vacancies shows a trend that it pulls back and then turns around to rise slowly in the short term. Second, when the initial number of job vacancies is far away from its equilibrium position while the initial employment is near to its equilibrium position, the employment figure will continue to climb upward slowly and so does the number of job vacancies. In another word, it won't appear to be fluctuating between full employment and job vacancies around the equilibrium position. Therefore, we can conclude that if the productivity shock is permanent, the dynamics of employment will be monotonic and won't generate any hump-shaped output dynamics. However, if the shock is temporary, employment will take a nonmonotonic adjustment and thus result in the hump-shaped output dynamics as seen in reality.

4.5 Contagion of Financial Crisis: Cubic Polynomial Differential Systems (II)

4.5.1 Introduction

Since the 1990s, international financial crises have been breaking out frequently, and their contagions have also been greatly enhanced over time. These crises behaved like infectious diseases that often spread quickly from the initial regions to other places after their outbreak. For instance, the 1994 *tequila crisis* in Latin American countries, the 1998 *Russian virus*, and the 1997 Southeast Asian financial crises were all regional financial crises with strong contagion. However, the crisis triggered by the 2007 American subprime crisis was of much stronger contagion and had developed into a global financial crisis. This event has made a good number of countries fall into economic crises, producing a profound impact on the order of global economic development. Therefore, we pay particularly attention to the dynamic transmission problem of this financial crisis within a short period of time after its initial outbreak.

Many researchers have focused on the financial crisis and its contagion test. But they produced different empirical conclusions under different definitions of financial crisis contagion. Sometimes different results may be derived even under the same framework of theory, such as when, for example, the financial crisis contagion is defined as the significantly enhanced correlation among financial markets after the crisis broke out. Some scholars (see King, 1978; King and Wadhwani, 1990; Calvo and Reinhart, 1996; Baig and Goldfaijn, 1998) found that there are significantly enhanced correlations in cross-border capital markets when they studied some of the major financial crises of the past. But Forbes and Rigobon (2002) strictly separated the contagion and the degree of mutual dependence. By taking account the variance of different conditions, these authors did not find any evidence that correlations between various markets are destroyed during those major crises. For the relevant reasons, Embrechts et al. (1999) pointed out that the correlation coefficient is only used to describe a linear relationship between two financial markets and is not suitable for studying nonlinear changes. Boyer et al. (1999) put forward the fact that without considering the conditional heteroskedasticity, the results of correlation tests are biased. Furthermore, Dunger et al. (2004) suggested that the differences in definitions and tests are small; even no difference exists under certain conditions. After classifying the empirical research methods of financial contagion and analyzing their similarities and differences, they explained that some models reflect all of the information while others take only parts of the information. Therefore, we follow the strict distinction between infection and mutual dependence as in Forbes and Rigobon (2002) without distinguishing the amount of information used in the model. After that, we investigate contagions of financial crises between two financial markets by analyzing their changes of mutual impact within a short time period during the crises. Whatever nonlinear effect is there between two financial markets, contagion may occur if those changes are in an unstable state. Otherwise, the markets maintain their relationship of mutual dependence.

As we take on a different approach from the literature as outlined previously to analyze the contagion of financial crises, it is not rational for us to employ directly the relevant nonlinear research methods based on the aforementioned definitions. Although these methods, such as the minimum spanning tree method, Copula functions, seemingly unrelated Probit techniques, GARCH model, symbolic time series analysis, dynamic factor model, and other methods (Ortega and Matesanz, 2006; Fazio, 2007; Rim, 2007, 2008; Rodriguez, 2007; Brida et al., 2009; Cipollini and Kapetanios, 2009), can capture more nonlinear characteristics of the contagions of financial crises, it has been an outstanding issue as for how to reduce the error of testing the financial crisis contagion between two markets. So we try to introduce the method of differential dynamics to set up a nonlinear model for the financial crisis contagion between two markets. Additionally, we use the qualitative theory of ordinary differential equations to describe the state of mutual impact between two countries or two financial markets within a short period of time during the crisis. Then, we discuss the infectious state and path.

4.5.2 Model

The contagion of a financial crisis first affects the financial security of a nation. It may lead to a high level of price volatility of financial assets (such as stocks, bonds, currencies, and real estate), deteriorate the operating conditions of financial institutions, cause capital flight, diminish foreign exchange reserves, and increase the amount of foreign debts. One of the first signs during the crisis is the great volatility of stock prices. So stock prices or the volatility of the stock return rate has been one of the most common means used for analyzing the contagion of a financial crisis. Thus, we put forward our nonlinear dynamics model of financial crisis contagion between two countries by examining the trend of dynamic changes of the stock return rate after the outbreak of the crisis.

On the one hand, by directly observing the changes in stock return rate in the market, we found that when one country experiences a financial crisis, its stock returns may drop drastically and quickly affect other countries' stock markets in a short period of time. Then, a crisis contagion appears among the initial country of crisis and its neighboring countries or countries with close contacts. On the other hand, by summarizing previous studies, we recognize that the degree or efficiency with which the crisis country affects other countries usually depends on the affected country's immunity ability and control force, and the aggregation of investors' confidence. The immunity ability is determined by such fundamental factors as economic strength, economic structure, financial system security, financial market openness, management of the exchange system, etc. The control force depends on timely financial assistance, and the positive attitude and actions to rescue other infected countries. And the growth in the aggregated investors' confidence comes largely from the increase in the first two capabilities. Furthermore, we suggest that before the outbreak of the crisis, one country's average rate of change in stock return depends on its immunity ability and reflects the relationship of interdependence with the country of the initial crisis. However, during the crisis, affected by its own control force and aggregated investors' confidence as well as changes in the stock return of the crisis country, one country's stock return may move away from its original average rate and experience nonlinear fluctuations following the change in the crisis country's stock market in a short period of time. Meanwhile, the stock return of the crisis country continues to drop sharply within a short period of time, while experiencing a nonlinear impact of the control force and aggregated investors' confidence on the market volatility.

Thus, we assume that a nonlinear function of the stock return rate in two markets can be established to reflect the previously described volatility as nonlinear affected by the control force, aggregated investors' confidence, and other hidden factors in the two countries. At the same time, we adopt the power function to describe that the impact on other countries is far greater than that on the crisis countries. So, we construct a nonlinear dynamic model with a minimum power law as follows:

$$\begin{cases} \dfrac{dr_1}{dt} = a - r_1 r_2^2 \\[2mm] \dfrac{dr_2}{dt} = r_2(-b + r_1 r_2) \end{cases} \tag{4.33}$$

where a and b are positive constants with a being the increasing rate of the average stock return of country A under the normal situation, and b the decreasing rate of stock return of country B, and r_1 and r_2 the stock return rates of countries A and B, respectively. From the angle of dynamics, cubic system (4.33) shows the following:

1. Within a short period of time after the crisis breaks out, the stock market of country A cannot evolve with a constant speed a due to the nonlinear effect of country B. Here, r_2^2 is similar to the variable coefficient of stock return in country A. When r_2 is small, the financial crisis is of a smaller impact on the average stock return of country A. But when r_2 is becoming larger, the impact of the financial crisis on the average stock return of country A is going to grow upward at the squared speed of the larger r_2.
2. At the initial onset period of the crisis, the increase rate of the average stock return in country B changes at the speed of $-br_2$. This means that its rate of return may decrease gradually in accordance with the exponential law. When the contagion develops to a certain extent, this speed of fall becomes $-b + r_1 r_2$ under the mutual effect of the stock returns in these two countries. Additionally, the following scenarios can be seen:
   a. When country B increases its control force or other countries step up their controls and boost their aggregated investors' confidence, the speed r_2 of fall may be controlled, namely, $r_1 r_2 > 0$.
   b. When these measures cannot be applied for whatever reason, the stock return of country B may experience an accelerated decreasing trend, namely, $r_1 r_2 < 0$.

4.5.3 Analysis

At first, one can readily find that the singular point in (4.33) is $\bar{o} = \left(b^2 / a, a/b \right)$.

By exchanging r_1 and r_2 in (4.33) and moving the singular point to the origin of the Cartesian coordinate system, (4.33) can be transformed into the following form:

$$
\begin{cases}
\dfrac{dr_1}{dt} = br_1 + a^2 r_2 / b^2 + r_1 \left(ar_1/b + 2ar_2/b + r_1 r_2 \right) \\[2mm]
\dfrac{dr_2}{dt} = -2br_1 - a^2 r_2 / b^2 - b^2 r_1^2 / a^2 + r_2 \left(r_1^2 + 2ar_1/b \right)
\end{cases}
\tag{4.34}
$$

According to the qualitative theory of ordinary differential equations (Li et al., 1996), we can establish the following conclusions.

Theorem 4.19 The characteristics of the singular point $\bar{o} = \left(b^2/a, \, a/b \right)$ in (4.33) are

  (i) If $a^2 > b^3$, then $\bar{o}$ is a stable node or focus.
 (ii) If $a^2 < b^3$, then $\bar{o}$ is an unstable node or focus.
(iii) If $a^2 = b^3$, then $\bar{o}$ is a center point or focus.

Proof. The characteristic equation of the corresponding linear part of (4.33) at point $\bar{o}$ is

$$\lambda^2 + (a^2/b^2 - b)\lambda + a^2/b = 0$$

Notice that $p = a^2/b^2 - b$ and $q = a^2/b$. Thus, if $b < 0$, then $\bar{o}$ is a saddle point. If $p > 0$, namely, $a^2 > b^3$ and $b > 0$, then $\bar{o}$ is a stable node or focus. And if $p < 0$, namely, $a^2 < b^3$, then $\bar{o}$ is an unstable node or focus. And if $p = 0$, namely, $a^2 = b^3$, then $\bar{o}$ may be either a center point or a focus.

Theorem 4.20 Equation 4.33 may be described as the Liénard equation. That is, the following holds true:

$$\begin{cases} \dfrac{dx}{dt} = y - F(x) \\[2mm] \dfrac{dy}{dt} = -g(x) \end{cases} \tag{4.35}$$

where $F(x) = x + (bx + b^2x^2/a)/(a^2/b^2 + 2ax/b + x^2)$ and $g(x) = (a^4x/b)/(a^2 + 2abx + b^2x^2)$.

Proof. Let $x = r_1 - b^2/a$ and $y = r_2 - a/b$. Then, we can convert (4.33) into the following form:

$$\begin{cases} \dfrac{dx}{dt} = -\dfrac{a^2x}{b^2} - 2by - \dfrac{2axy}{b} - \dfrac{b^2y^2}{a} - xy^2 \\[2mm] \dfrac{dy}{dt} = \dfrac{a^2x}{b^2} + by + \dfrac{2axy}{b} + \dfrac{b^2y^2}{a} + xy^2 \end{cases} \tag{4.36}$$

By exchanging x and y in (4.36), and still mark x and y, one can get

$$\begin{cases} \dfrac{dx}{dt} = A_0(x) + A_1(x) \cdot y \\[2mm] \dfrac{dy}{dt} = B_0(x) + B_1(x) \cdot y \end{cases}$$

where

$$A_0(x) = bx + b^2x^2/a, \ B_0(x) = -2bx - b^2x^2/a, \ \text{and} \ A_1(x) = -B_1(x) = a^2/b^2 + 2ax/b + x^2$$

Thus, the result follows from Lemma 6.3 in Li et al. (1996).

Theorem 4.21 If $b > 0$ and $a^2 = b^3$, then the singular point of (4.33) is an unstable focus.

Proof. Introduce the following binary function:

$$H(r_1, r_2) = r_1^2 + r_1 r_2 + \frac{r_2^2}{2} + \varphi_3(r_1, r_2) + \varphi_4(r_1, r_2)$$

where φ_3 and φ_4 are third- and fourth-degree homogeneous polynomials, respectively. When c_0 is a sufficient smaller positive number, the equation $H(r_1, r_2) = c_0$ indicates a cluster of closed curves containing the origin. Let us now consider the following differential dynamic system:

$$\begin{cases} \dfrac{dr_1}{dt} = br_1 + br_2 + \dfrac{r_1^2}{b} + 2\sqrt{b}\, r_1 r_2 + r_1 r_2^2 \\[3mm] \dfrac{dr_2}{dt} = -2br_1 - br_2 - \dfrac{r_1^2}{b} - 2\sqrt{b}\, r_1 r_2 - r_1 r_2^2 \end{cases} \tag{4.37}$$

After some simple calculation, we obtain

$$\left. \frac{dH}{dt} \right|_{(4.37)} = b(r_1^2 + r_2^2)^2 + h_5(r_1, r_2)$$

where $h_5(r_1, r_2)$ is a polynomial with its degree less than 5.

Denote $s = (2 + 9\sqrt{b} + 13b^3)/8b^4$, $t = (1 + 3b^{3/2})/4b^4$, and assume that the specific forms of φ_3 and φ_4 in (4.37) are given as follows:

$$\varphi_3(r_1, r_2) = m_1 r_1^3 + m_2 r_1^2 r_2 + m_3 r_1 r_2^2 + m_4 r_2^3$$

$$\varphi_4(r_1, r_2) = n_1 r_1^4 + n_2 r_1^3 r_2 + n_3 r_1^2 r_2^2 + n_4 r_1 r_2^3 + n_5 r_2^4$$

where the coefficients are, respectively,

$$m_1 = \frac{12b^{3/2} + 5}{3b^2}, \quad m_2 = \frac{6b^{3/2} + 3}{b^2}$$

$$m_3 = \frac{4b^{3/2} + 2}{b^2}, \quad m_4 = \frac{4b^{3/2} + 2}{3b^2}$$

$$n_1 = -1 + 10s + 8t, \quad n_2 = -\frac{5}{2} - 20s + 20t,$$

$$n_3 = -\frac{9}{4} + 18s + 18t, \quad n_4 = -4s + 4t, \quad n_5 = -\frac{5}{16} - s + t$$

Then, due to $b > 0$, it can be proved from (4.37) that the singular point $\bar{o}$ in (4.33) is an unstable focus.

Theorem 4.22 If $a^2 \geq b^3$, then there is no limit cycle in (4.33).

Proof. Denote $\lambda(x, y) = y^2/2 + G(x)$, where

$$G(x) = \int_0^x g(z)dz = \frac{a^4}{b^3} \times \ln\left(\frac{a + bx}{a}\right) - \frac{a^4}{b^2} \cdot \frac{x}{a + bx}$$

$$\because xF(x) = x^2 + \left(\frac{bx^2}{(a/b) + x}\right)^2 \cdot \left(1 + \frac{bx}{a}\right)$$

$$\therefore \text{ if } x > -a/b, \ xF(x) > 0 (x \neq 0)$$

And there is $g(x) = a^4/b(a + bx)^2 \cdot x$. So $d\lambda/dt|_{(4.36)} = -g(x)F(x) < 0$ $(x \neq 0, x > -a/b)$. Therefore, (4.35) has no limit cycles, and there is no limit cycle in (4.33).

Theorem 4.23 If $a^2 < b^3$, then there is a unique limit cycle in (4.33).

Proof. From Theorem 6.6 in Li et al. (1996), it follows that

(i) If $a_{01}\alpha_0 < 0$, then $b^3 - a^2 > 0$, namely, $a^2 < b^3$.
(ii) If $\beta_0 > 0$, then $a^2 > 0$, namely, $a \neq 0$.
(iii) If $a_{21}\alpha_4 > 0$, then $b > 0$.
(iv) If $\beta_3 > 0$, then $a^2 + 2b^3 > 0$.

By summarizing these four conclusions, we have shown that Theorem 4.23 is true.

Based on Theorems 4.21 through 4.23, we conclude that there are three possible cases for financial crisis contagions between two countries: a weak contagion with instability but inhibition, a contagion with limit and controllable oscillation, and a strong contagion without any control in any brief moment of time.

Case 1: When $a^2 < b^3 (b > 0)$, Theorem 4.23 indicates that (4.33) has a unique limit cycle. Its phase plane shows that there occurs an alternating oscillation of stock returns between the two countries. However, this oscillation may not enlarge without any limitation due to the fact that there exists a limit cycle. The immunity ability and self-repair capacity of the economy system in both of these two countries may limit the magnitude of oscillation within some controllable size, which depends on the size of the limit cycle. So, it is a case of contagion with limit and controllable oscillation.

Case 2: When $a^2 = b^3 (b > 0)$, Theorem 4.21 suggests that the singular point of (4.33) is not a center but an unstable focus. Due to the fact that there is only a little difference between a center point and a focus, it shows that during the initial period of financial crisis contagion, the oscillation magnitudes of the rates of stock return of these two countries increase gradually in an imperceptible way. Thus, this stage should be the best time moment to apply control on the financial crisis contagion. We call this state of contagion as a weak contagion with instability but inhibition.

Case 3: When $a^2 \geq b^3 (b > 0)$, Theorem 4.22 dictates that there is not any limit cycle in (4.33). In this case, the financial crisis contagion has evolved into a disaster, and both of these two countries have to endure additional strengthening, severe impacts. The governments must take the firmest monetary policies and fiscal policies to curb the spread of the contagion.

4.5.4 Conclusions

Within a short period of time after a financial crisis breaks out, the stock return of the country that experiences the crisis would drop abruptly and quickly increase the volatility in the stock returns of other countries. And then, there appears a nonlinear contagion of the financial crisis throughout the region. To investigate this phenomenon, we introduce the method of differential dynamics to construct a simple nonlinear dynamic model and establish four theorems in accordance with the qualitative theory of differential equations. Furthermore, based on the discussions about the stability of foci and the existence of limit cycles in the nonlinear volatility equations of stock returns, we analyze the situation of financial crisis contagion between two countries during the crisis and find that there are three possible cases of financial crisis contagions within a short period of time after the crisis breaks out. The first one is that if the average increasing rate of squared stock returns in the infected country is more than the decreasing rate of cubed stock returns in the initial country of crisis, there is no limit cycle in the nonlinear dynamic model of financial crisis contagion. Thus, there is a strong contagion between the two countries that needs to be controlled difficultly within a short period of time. The second one is that if the former rate is less than the latter rate, then there is a unique stable limit cycle in this model. Thus, there is an oscillating contagion between the two countries, whose oscillation magnitude depends on the size of the limit cycle. The third one is that if the former rate is equal to the latter rate, this model has an unstable focus. Thus, there is a weak contagion between the two countries so that their state of interdependence could be more easily adjusted before the breakout of a financial crisis.

Our results of analysis are closer to the actual state of financial crisis contagions within a short term between two countries. For example, Hong Kong Monetary Authority expected that the U.S. subprime mortgage crisis will not create systemic impact on Hong Kong's banking system in its paper submitted to Hong Kong SAR Legislative Council in January 2008. Later, this point of opinion was also proved to be correct. By using Hong Kong's Hang Seng Index and U.S. S&P 500 stock index, we calculated the stock returns by using the logarithmical return method and the corresponding increasing or decreasing rates. Then, we found that before the outbreak of the subprime crisis, the average growth rate of Hong Kong stock returns was about 4% for the time period from 2006 to July 2007, and that the average decreasing rate of U.S. stock returns was 600% or more for the time period from July to August 2007. It implies that the crisis contagion between the United States and Hong Kong could be controlled by taking some measurements even if the prices of Asia-Pacific stock markets greatly dropped downward in August 2007.

Our analysis indicates that the method of differential dynamics is a better way to investigate the state of a financial crisis contagion or the path between two countries. However, as a preliminary discussion, this section introduces a power function with the lowest power times into a simple nonlinear dynamic model of crisis contagion. The simple form may be extended to others for further research of variable contagions among three or more countries. In addition, another direction of further work may be doing some empirical research such as that addressing the calculating methods of those two indicators in our simple nonlinear model, testing the states of contagion by employing the time series data of stock returns, constructing some early warning indicators to timely monitor the volatility of financial markets, and so on.

4.6 Contagion of Currency Crises: Fractional Differential Systems (I)

4.6.1 Introduction

Since the 1970s, worldwide currency crises occurred frequently along the development of financial globalization, the development of liberalization, and the violent international capital flows. Especially, during the past 20 years since the 1990s, the worldwide financial market has experienced several influencing currency crises: the 1992–1993 crisis of the European exchange rate system, the 1994–1995 Mexican crisis, the 1997–1998 crisis of southeast Asia, the Russian rubles crisis in 1998, the 1998–1999 Brazilian currency crisis, the 2000–2001 Turkish lire crisis, and the 2001–2002 peso crisis. The infectious mechanism of Southeast Asian financial crisis can be summed up as follows. The currencies of Thailand and almost all its neighboring countries are linked with the dollar. So, Thailand's export competitiveness was enhanced after the Thai baht was devaluated. In order to maintain the export competitiveness of domestic products, the countries surrounding Thailand with similar export structure also started to devaluate their domestic currencies. So, competitive currency devaluation ensued. As expected, international speculators started to short sell these currencies, leading to mounting pressures to further devaluate the currencies.

As Merton Miller (1998), an American economist, pointed out, Thailand's neighbors and competitors, such as Malaysia, Indonesia, the Philippines, and South Korea, were soon infected with the currency crises because of "dramatic devaluation of baht." The outbreak of the financial crisis first appeared in Thailand and then spread not only to Southeast Asian and East Asian regions but also to industries in European and American countries, and Latin American countries through foreign trade, foreign investments, and offshore banking channels. According to a report of the international monetary fund in 1998 (International Monetary Fund, 1998), 53 countries has gone through 158 currency crises, thereto, industrialized countries 42, emerging market countries 116. And Kaminsky and Reinhart (1999) were convinced that 20 countries they studied underwent through 71 currency crises during the time period of 1970–1995.

According to the facts described previously, we can find one of the characteristics on how a currency crisis spreads rapidly in the territory when it first occurred in some country. In the 1992–1993 European currency crisis, Italian lire was the first that felt market pressure, and then the pressure evolved into a currency crisis of the whole system beginning with the Finnish mark separating itself from the European monetary system. And then Italian lire and British pound withdrew from the European monetary system successively. Spanish pesetas, Swedish kronor, Portugal Escudo, Norwegian kronor, and Irish pound either devalued or floated in succession. Even those currencies that did announce depreciation, such as French franc and Danish kronor, were also under the intense speculative impact, and their exchange rates even dropped to the lower limit of European exchange rate mechanism. The 1994–1995 Mexican peso crisis also made Argentina, Brazil, and Peru's currencies suffer strong impacts. Beyond that, the impact even affected the emerging markets in the Asia-Pacific region and such nations as Canada, Italy, Hungary, and Spain. With the help of the United States, Canada, other trading partners of Mexico, the international monetary fund, and the bank for international settlements, the crisis was eventually stabilized. The Asian financial crisis broke out in 1997 and led to currency devaluations in Thailand,

Indonesia, Malaysia, and the Philippines. Then, this competitive depreciation was transmitted to Singapore, South Korea, and Taiwan. And even Hong Kong was under intense speculative impact. A year later, the crisis was infected over to Russia and the financial markets in Latin America.

Because currency crises appeared frequently throughout the world, the causes, infections, warnings, and other aspects of financial crises have been investigated with increasing extensity and depth. Along with the acceleration of marketable reformation of RMB exchange rate, it is expected that RMB will experience frequent shocks. So, such issues as how to maintain the stability of RMB's exchange rate, how to prevent RMB from excessive shocks in international financial markets, and how to protect RMB from currency crises have become increasingly important questions to consider.

In the theoretical research, there have been various explanations on how the infection mechanism of crises works. In order to clarify the infection mechanism of currency crises simply and clearly, we divide the infection mechanism into three aspects: trade contagion effect, financial contagion effect, and expected contagion effect. The so-called trade contagion effect refers to the situation that a currency crisis in one nation worsens the balance of international payments and economic operation condition of another country that has close relationship with the former nation through trades. The so-called financial contagion effect refers to the situation that the low market liquidity of one nation as caused by macroeconomic fluctuations leads to disappearing liquidity in another country that has close relationship with the former nation in several financial domains, such as direct investments and bank loans, and the capital market, and then triggers the breakout of a financial crisis in the latter nation. The so-called expected contagion effect refers to the situation that an ongoing crisis may also infect various countries even though these nations do not have links in terms of trade and financial transactions. It is because the crisis of one nation may lead to changes in expectation in other similar national markets so that speculative opportunities emerge for speculators to handily attack the currencies of these nations. The three financial crises of the 1990s had such common characteristic that they all possessed evident contagion effect. One country's financial crisis quickly spread to other countries and regions, and then evolved into a regional financial crisis of a much greater scale. Because crises have occurred frequently, much magnified have infectious consequences of damage become, such as monetary malignant devaluation, exhaustion of foreign currency reserves, crashes of stock markets, bankruptcy of financial institutions, negative economic growth, destabilization of the government, etc.

Krugman (1979) and Robert and Garber (1984) pioneered the first-generation theory of monetary crises. In their models, sustained domestic credit expansion is considered as an attack to the exchange rate. The theory maintains that there is an essential conflict between the macroeconomic policy (mainly too excessive monetary policy and fiscal deficit policy) and the stable exchange rate policy (fixed exchange rate system that pegs the U.S. dollar) in the country that suffers from a crisis. A continuous deterioration in the macroeconomic fundamentals behind the fluctuation in exchange rate represents one of the main reasons for currency speculators to launch their attacks on the fundamentals. The basic factors of macroeconomic deterioration are the cause underneath the increasing level of fluctuation in the exchange rate, and they draw speculative attacks of money. Robert and Garber simplified Krugman's (1979) model by using linear method and obtained the explicit solution for when speculative attacks occur.

Obstfeld (1994, 1996) and Meltzer (1994) created the second-generation theory of monetary crises based on a self-fulfilling mechanism. They believe that adopting a fixed exchange rate policy rules can avoid unnecessary inflation. Because the rules do not have the necessary flexibility, the effect of adjusting the policy specifics on timing and scale would be better than that of merely adopting a single ordinary policy or an incidental policy. Giving up fixed exchange rate and restarting the exchange rate regime may be a rational choice for the government after it weighs the relevant loss and profit.

Morris and Shin (2004) emphasized on the uncertainty of speculators' expectation behavior by relaxing the hypothesis of the classical model of the second generation. They also proved that information asymmetry among speculators can eliminate multiple equilibria. And money being attacked is the only result under this condition. This sole equilibrium relies on not only the basic variables, but also financial variables, such as the sum of hot money in circulation and speculative trading costs. The critical moment for speculative attacks to occur can be obtained from the model. The third-generation theory of financial crises began to emphasize on the behaviors of enterprises and financial intermediaries beyond the scope of exchange rate mechanism and macroeconomic analysis. Currently, most literature considers that the third-generation theory of financial crises focuses on the cause of the crises from microscopic behaviors of enterprises, banks, foreign creditors, etc. Moral hazard crisis model, financial panic crisis model, and balance sheet crisis model are considered as the representatives of third-generation theory of financial crises.

de Jong et al. (2009b) estimated a dynamic heterogeneous agents' model for the British pound during the European monetary system crisis and illustrated the chain of events leading to the suspension of the pound from the exchange rate mechanism in terms of switching beliefs from that of a fundamentalist to that of a chartist. Tony (2010) estimated a simple small open macroeconomic model to analyze the effectiveness of monetary policy rules (MPRs) where either the nominal interest rate or the nominal exchange rate is the policy instrument. The model tries to ascertain which of those MPRs is better suited for a selection of inflation targeting economies of Asia. Normally, one would associate inflation targeting with interest rate rules, but it is thought that, due to fear of flotation, exchange rate rules may well be more effective given the degrees of openness of these economies.

It is found that interest rate rules seem to better reflect the prevailing policy regime than exchange rate rules. It is also found that stronger relationships pertaining to the interest rate rules are found in the case of Korea and Thailand than for Indonesia and the Philippines. Exchange rates appear to be very influential in determining the value of the nominal interest rate but not in a policy sense. Michael et al. (2010) extended the currency crises model of Banerjee et al. (1993) in different directions. Their main result is that a tight monetary policy can have adverse effects beyond the short term and can potentially cause a currency crisis in the medium term, even in cases when the interest rate defense is successful and prevented a currency crisis in the short run. In addition, they added a risk premium and found that this increases the likelihood of a crisis, can help explain the contagion of crisis, and prospective capital controls will increase the likelihood that such controls will be needed as an emergency measure. Stix (2007) studied the effects of French interventions during the 1992–1993 European Monetary System crisis. In particular, a Markov switching model was estimated where interventions influence the probabilities of transition between a calm and a turbulent regime. Along these lines, we will analyze the impact of intervention on the expected rate of realignment. On balance, our results are consistent with the view that publicly known interventions (but not secret interventions) increase both the probability of switching to the turbulent regime and the expected realignment rate.

4.6.2 Model and Analysis

Cross-fertilization of nonlinear science and economic research has helped to make mathematical economics to be more mature. The research of economics and finance on the basis of nonlinear science has attracted greater attention in recent years. Especially, the catastrophe theory has helped to make better explanation on the phenomenon that changes in continuous parameters can lead to the appearance of discontinuation, leading to some very important results in the analysis of problems of finance.

In the analysis of the contagion of currency crises, such four variables as exchange rate, interest rate, foreign exchange reserve, and stock index are considered usually. In this section, we will consider exchange rate as our main index in our analysis of the mechanism of contagion underneath monetary crises between two countries by using differential dynamic method.

The following model

$$\frac{dx}{dt} = \frac{\alpha x(k - x)}{k + \beta x} \tag{4.38}$$

has been widely used in the fields of biology, economics, and sociology. When β is equal to -1 and 0, Equation 4.38 is the index model and the logistic model.

Now let us consider the contagion of a currency crisis between two reciprocal countries. The following are the basic channels of financial infection:

1. International trade
2. Lending to the surrounding area (liquidity effect and capital adequacy ratio effect)
3. Carrying trades between countries in the securities market, the stock market, and the commodities markets
4. International monetary supply (international inflationary effects)
5. Investors' learning effect, investors' risk aversion improving
6. Macroeconomic similarity

These channels can be summarized as incidental crisis infection channels and sporadic crisis infection channels.

Now we establish the following currency crisis contagion model between country A and country B:

$$
\begin{cases}
\dfrac{dr_1}{dt} = \beta_1 r_1 \left[1 - \dfrac{r_1}{k_1 + c_1 r_2} \right] \\[3mm]
\dfrac{dr_2}{dt} = \beta_2 r_2 \left[1 - \dfrac{r_2}{k_2 + c_2 r_1} \right]
\end{cases}
\tag{4.39}
$$

where

r_1 and r_2 stand for the exchange rate of country A and the exchange rate of country B, respectively

β_1 and β_2 the intrinsic growth rate of the exchange rate of country A and country B, respectively

k_1 and k_2 the upper limits of the changing exchange rates of country A and country B, respectively

c_1 the crisis infection coefficient from B to A

c_2 the crisis infection coefficient from A to B, satisfying $0 < c_i < 1$, $k_i > 0$, for $i = 1,2$

First, let us discuss the equilibrium position $(\bar{r}_1, \bar{r}_2)$ in the first quadrant, where

$$
\begin{cases}
\bar{r}_1 = \dfrac{k_1 + c_1 k_2}{1 - c_1 c_2} \\[3mm]
\bar{r}_2 = \dfrac{k_2 + c_2 k_1}{1 - c_1 c_2}
\end{cases}
\tag{4.40}
$$

By considering changes of the equilibrium position, we can obtain the following.

Theorem 4.24 If $0 < c_i < 1$, $k_i > 0$, for $i = 1, 2$, then the following hold true:

$$\frac{\partial \overline{r_1}}{\partial c_1} > 0, \frac{\partial \overline{r_1}}{\partial c_2} > 0, \frac{\partial \overline{r_2}}{\partial c_1} > 0, \text{ and } \frac{\partial \overline{r_2}}{\partial c_2} > 0.$$

Proof. From the facts

$$\frac{\partial \overline{r_1}}{\partial c_1} = \frac{k_2 + c_2 k_1}{(1 - c_1 c_2)^2} > 0, \quad \frac{\partial \overline{r_1}}{\partial c_2} = \frac{c_1(k_1 + c_1 k_2)}{(1 - c_1 c_2)^2} > 0, \quad \frac{\partial \overline{r_2}}{\partial c_1} = \frac{c_2(k_2 + c_2 k_1)}{(1 - c_1 c_2)^2} > 0, \text{ and}$$

$$\frac{\partial \overline{r_2}}{\partial c_2} = \frac{k_1 + c_1 k_2}{(1 - c_1 c_2)^2} > 0$$

It follows that Theorem 4.24 is true.

Theorem 4.24 shows that growth in the infection coefficients can make the equilibrium position of the exchange rate of countries A and B be far away from the origin. That is to say, fluctuations in the exchange rate of the two countries tend to be enlarged no matter what values the coefficients are equal to.

Theorem 4.25 If $c_2 > c_1$ and $\frac{c_2}{c_1} k_1 > k_2 > k_1$, then $\overline{r_2} > \overline{r_1}$.

Proof. We can establish Theorem 4.25 readily according to relationship (4.39).

Theorem 4.25 shows that if it is more likely for country A to infect country B than country B to infect country A, then at the equilibrium position of the exchange rates, the magnitude of fluctuation of the exchange rate of country B is higher than that of country A.

Let us define

$$F_1(r_1, r_2) = \beta_1 r_1 \left[1 - \frac{r_1}{k_1 + c_1 r_2} \right] \text{ and } F_2(r_1, r_2) = \beta_2 r_2 \left[1 - \frac{r_2}{k_2 + c_2 r_1} \right]$$

then we can conclude the following.

Theorem 4.26 If $\beta_1 + \beta_2 < 0$, then the equilibrium point $(\overline{r_1}, \overline{r_2})$ is unstable, and if $\beta_1 + \beta_2 > 0$, then the equilibrium point $(\overline{r_1}, \overline{r_2})$ is stable.

Proof. When $\beta_1 + \beta_2 < 0$ holds true, let us consider the following linear equation corresponding to the equilibrium point $(\overline{r_1}, \overline{r_2})$:

$$\begin{cases} \dfrac{dr_1}{dt} = a_{11}(r_1 - \overline{r_1}) + a_{12}(r_2 - \overline{r_2}) \\[3mm] \dfrac{dr_2}{dt} = a_{21}(r_1 - \overline{r_1}) + a_{22}(r_2 - \overline{r_2}) \end{cases} \tag{4.41}$$

From Equation 4.39, we can obtain

$$a_{11} = \left.\frac{\partial F_1}{\partial r_1}\right|_{(\overline{r_1},\overline{r_2})} = -\beta_1, \quad a_{12} = \left.\frac{\partial F_1}{\partial r_2}\right|_{(\overline{r_1},\overline{r_2})} = -c_1\beta_1$$

$$a_{21} = \left.\frac{\partial F_2}{\partial r_1}\right|_{(\overline{r_1},\overline{r_2})} = -c_2\beta_2, \quad a_{22} = \left.\frac{\partial F_2}{\partial r_2}\right|_{(\overline{r_1},\overline{r_2})} = -\beta_2$$

The characteristic equation corresponding to Equation 4.39 is

$$\lambda^2 + (\beta_1 + \beta_2)\lambda + (1 - c_1 c_2)\beta_1\beta_2 = 0$$

According to the qualitative theory of differential equations, it follows that Theorem 4.26 is true.

Theorem 4.26 indicates that if the exchange rates of two countries grow negatively, then the exchange rates of these countries would evolve away from the equilibrium position and start to fluctuate with an increasing magnitude around the equilibrium position; on the other hand, if the exchanges rates grow positively, then the exchange rates of these countries would move away from the equilibrium position and vibrate within a bounded region of the equilibrium position.

4.6.3 Conclusions

As of this writing, scholars have reached a consensus on the contagion of currency crises. The published studies have focused on the infection mechanism of currency crises, in order to address how the crisis of one country infects another one. In this section, we define *currency crisis infection* as the possibility with which the crisis that occurs in country A leads to a crisis of another country B. What we emphasize is that the cause or premise of the crisis occurring in country B is that the crisis of country A first occurred. The more intimate the trade connection between country A that is in a crisis and country B is, the greater the probability for a contagion crisis to appear in country B. In our modern world, developing countries produce and sell similar goods, while these countries rarely trade with each other. So the infection effect of trading partners is limited for these countries. However, there is a complementary trade structure between developing countries and developed countries. So many developing countries tend to export commodities competitively to the developed markets. Hence, *rival-type* infection becomes one of the main channels of contagion for crises to spread. In this section, we constructed a differential dynamic model to describe the contagion of a currency crisis between two countries based on the generalized logistics model. Then, we analyzed the mechanism of currency crisis contagion and obtained the following conclusions.

First, the growth in the infection coefficients can lead to the equilibrium position of the exchange rate of countries A and B to stay far away from the origin. That is to say, the fluctuations in the exchange rate of the two countries tend to be enlarged, no matter how big values the coefficients are equal to. Second, if the equilibrium position of country B's exchange rate is greater than that of country A, and if country A infects country B more than country B on A, then at the equilibrium position of the exchange rates, country B experiences greater fluctuation in its exchange rate than country A does. Third, when the exchange rates of these countries experience negative growth, these rates will deviate away from the original stable states and start to vibrate severely, and when the exchange rates grow positively, although the rates would fluctuate within a certain range, they would maintain their overall relative instability.

4.7 Contagion of Currency Crises with Extra-Absorption: Fractional Differential Systems (II)

4.7.1 Introduction

Since the 1970s, worldwide currency crises have occurred frequently. Especially, in the past 20 years the worldwide financial market has experienced several influential currency crises: The 1992–1993 European exchange rate system crisis, the 1994–1995 Mexican crisis, the 1997–1998 the crisis of southeast Asia, the Russian ruble crisis in 1998, the 1998–1999 Brazilian currency crisis, the 2000–2001 Turkish lire crisis, the 2001–2002 peso crisis, and the worldwide financial crisis because of American subprime loans crisis that broken out in 2007. As for what had caused the crises, the reason varies from unreasonable domestic economic institutions, too much debt, and wrong monetary policy to external shocks created by international financial speculators. Something that is worse than the frequent occurrence of these crises is that the infectiousness of the crises has become stronger and stronger and the scope of contagion of these crisis has been getting wider and wider. From the phenomenon perspective, the simple reason that is underneath the growing infectiousness of currency crises and expanding scope of the crises' contagion is the continuous development of international trade and increasingly closer economic and financial ties worldwide. However, a deeper level scientific question, such as what complex mechanism underneath monetary crises contagion is hidden under the simple reason? Addressing this question is the focus of this section.

Since the third generation of the financial crisis theory was initially generated, many scholars have done a great deal of work on the contagion mechanism of currency crises and the potential contagion paths. The conclusion shows that the more important trade competition is, the more likely self-fulfilling speculative crises appear and the larger the set of multiple equilibria is. It is also concluded that the closeness of coordination between countries decreases the possibility for simultaneous self-fulfilling speculative crises to occur in the region and reduces the set size of multiple equilibria. However, regional coordination between countries, even though it is welfare improving for all involved parties, makes the countries more dependent on the fundamentals of other countries so that such coordination may induce additional contagion. To this end, Stix (2007) studied the effect of French interventions during the 1992–1993 European Monetary System crises. In this work, a Markov switching model was estimated where the interventions influence the probability of transition between a calm and a turbulent regime. Additionally, the author also analyzed the impact of intervention on the expected rate of realignment. Overall, the results obtained there were consistent with the view that publicly known interventions but not secret interventions increase both the probability of switching to the turbulent regime and the expected realignment rate. Li (2004) proposed a global game method in the research of financial crises and basic ideas of application. Michael et al. (2010) introduced a new method to describe dynamical patterns of the real exchange rate comovements time series and to analyze the contagion of currency crises. Their method combines the tools of symbolic time series analysis with the nearest neighbor single linkage clustering algorithm. By using the method of symbolizing data, they obtained a metric distance between two different time series that was used to construct an ultrametric distance. By analyzing the data of various countries, they derived a hierarchical organization, consisting of minimal-spanning and hierarchical trees. From these trees they detected various clusters of countries according to their proximities. So, they concluded that their methodology permits them to construct a structural and dynamic topology that was useful to the study of the interdependence and contagion effects among financial time series. With all these points in place, however, as a matter of fact, this third generation of the financial crisis theory still experienced difficulties when it was used to explain the 1997 financial crisis of Asia. That indicates that this theory still needs to be further developed.

4.7.2 Dynamic Model between Two Countries

Logistic models have been widely applied to many research areas, such as biology, economics, and sociology (Sutradhar and Jowaheer, 2010; Syaiba and Habshah, 2010). By expanding the generalized logistics model, we set up a differential dynamic model for describing the contagion of currency crises between two countries as follows:

$$\begin{cases} \dfrac{dx}{dt} = \gamma_1 x \left[1 - \dfrac{x}{k_1 + \alpha_1 y} \right] \\[4mm] \dfrac{dy}{dt} = \gamma_2 y \left[1 - \dfrac{y}{k_2 + \alpha_2 x} \right] \end{cases} \tag{4.42}$$

The system (4.42) can be readily rewritten in the following general form:

$$\begin{cases} \dfrac{dx}{dt} = \gamma_1 x \dfrac{\dfrac{k_2}{\alpha_1} - x - \alpha_2 y}{k_1 - \alpha_2 y} \\[6mm] \dfrac{dy}{dt} = \gamma_2 y \dfrac{\dfrac{k_1}{\alpha_2} - \alpha_1 x - y}{k_2 - \alpha_1 x} \end{cases} \tag{4.43}$$

In this section, we consider the following more general dynamic model for describing the contagion of financial crises between two countries:

$$\begin{cases} \dfrac{dx}{dt} = \gamma_1 x \dfrac{\dfrac{k_2}{\alpha_1} - x - \alpha_2 y}{k_1 + \beta_1 x - \alpha_2 y} \\[6mm] \dfrac{dy}{dt} = \gamma_2 y \dfrac{\dfrac{k_1}{\alpha_2} - \alpha_1 x - y}{k_2 - \alpha_1 x + \beta_2 y} \end{cases} \tag{4.44}$$

where
  x and y stand for the exchange rate of country A and the exchange rate of country B, respectively
  γ_1 and γ_2 the intrinsic growths of the exchange rate of country A and country B, respectively
  k_1 and k_2 the upper limits of changes in the exchange rates of country A and country B, respectively
  α_1 the crisis infection coefficient from A to B
  α_2 the crisis infection coefficient from B to A, satisfying $0 < \alpha_i < 1$, $\gamma_i > 0$
  $k_i > 0$ $(i = 1, 2)$

And β_1 represents the *extra-absorption* coefficient of country A toward the crisis infection, and β_2 the *extra-absorption* coefficient of country B toward the crisis infection.

For the sake of convenience, we transform (4.44) into

$$\begin{cases} \dfrac{dx}{dt} = xF_1(x, y) \\[3mm] \dfrac{dy}{dt} = yF_2(x, y) \end{cases} \qquad (4.45)$$

where

$$F_1(x, y) = \gamma_1 \frac{\dfrac{k_2}{\alpha_1} - x - \alpha_2 y}{k_1 + \beta_1 x - \alpha_2 y} \quad \text{and} \quad F_2(x, y) = \gamma_2 \frac{\dfrac{k_1}{\alpha_2} - \alpha_1 x - y}{k_2 - \alpha_1 x + \beta_2 y}$$

with the parameters α_i, γ_i, and k_i $(i = 1, 2)$ being all positive constants and $\beta_i > -1$ $(i = 1, 2)$.

Let us define the region $D = \left\{ (x, y) \mid 0 < x < \dfrac{k_2}{\alpha_1}, 0 < y < \dfrac{k_1}{\alpha_2} \right\}$ (see Figure 4.2).

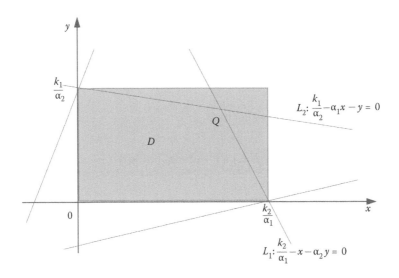

Figure 4.2   Region D and equilibrium point Q.

4.7.3 Stability Analysis

Theorem 4.27 If $\beta_i > -1$ $(i = 1, 2)$, $0 < \alpha_i < 1$, $\alpha_1 k_1 > \alpha_2^2 k_2$, and $\alpha_2 k_2 > \alpha_1^2 k_1$, then there exists a unique equilibrium point $Q(\bar{x}, \bar{y})$ in the region D such that it is a stable node or a focus.

Proof. With some simple calculations, we can obtain an equilibrium point $Q(\bar{x}, \bar{y})$ in addition to the following three equilibrium points $(0,0)$, $(k_1/(\alpha_1\alpha_2, 0))$, and $(0, (k_2/\alpha_1\alpha_2))$, where

$$\bar{x} = \frac{\alpha_2 k_2 - \alpha_1^2 k_1}{(1 - \alpha_1\alpha_2)\alpha_1\alpha_2} \quad \text{and} \quad \bar{y} = \frac{\alpha_1 k_1 - \alpha_2^2 k_2}{(1 - \alpha_1\alpha_2)\alpha_1\alpha_2}$$

If the conditions of Theorem 4.27 are met, then we can obtain

$$0 < \bar{x} < (k_1/\alpha_2) \text{ and } 0 < \bar{y} < (k_2/\alpha_1)$$

So, we can conclude that there is a unique equilibrium point in region D.

By using Taylor formula, we can transform (4.45) to the following form:

$$\begin{cases} \dfrac{dx}{dt} = x\left[\dfrac{\partial F_1}{\partial x}\bigg|_Q \cdot (x - \bar{x}) + \dfrac{\partial F_1}{\partial y}\bigg|_Q \cdot (y - \bar{y})\right] \\[4mm] \dfrac{dy}{dt} = y\left[\dfrac{\partial F_2}{\partial x}\bigg|_Q \cdot (x - \bar{x}) + \dfrac{\partial F_2}{\partial y}\bigg|_Q \cdot (y - \bar{y})\right] \end{cases} \tag{4.46}$$

where the coordinates of point Q are $x_Q = \bar{x} + Q_1(x - \bar{x})$, $y_Q = \bar{y} + Q_2(y - \bar{y})$, and $0 < Q_i < 1$ $(i = 1, 2)$.

An equivalent form for system (4.46) is

$$\begin{cases} \dfrac{dx}{dt} = x\left[a_{11}(x_Q, y_Q) \cdot (x - \bar{x}) + a_{12}(x_Q, y_Q) \cdot (y - \bar{y})\right] \\[4mm] \dfrac{dy}{dt} = y\left[a_{21}(x_Q, y_Q) \cdot (x - \bar{x}) + a_{22}(x_Q, y_Q) \cdot (y - \bar{y})\right] \end{cases} \tag{4.47}$$

With some simple manipulations, we obtain

$$a_{11}(x_Q, y_Q) = -\frac{\gamma_1(\beta_1 + \alpha_1\alpha_2)\bar{x}^2}{A^2} \tag{4.48}$$

$$a_{12}(x_Q, y_Q) = \frac{\gamma_1 \alpha_2 \bar{x} \left[\dfrac{k_2}{\alpha_1} - k_1 - (1 + \beta_1)\bar{x} \right]}{A^2}$$

(4.49)

$$a_{21}(x_Q, y_Q) = \frac{\gamma_2 \alpha_1 \bar{y} \left[\alpha_1 \bar{x} - (1 + \beta_2)\bar{y} \right]}{B^2}$$

(4.50)

$$a_{22}(x_Q, y_Q) = -\frac{\gamma_2 (\beta_2 + \alpha_1 \alpha_2) \bar{y}^2}{B^2}$$

(4.51)

where

$$A = \frac{\beta_1 + \alpha_2^2}{(1 - \alpha_1 \alpha_2)\alpha_1 \alpha_2}(\alpha_2 k_2 - \alpha_1^2 k_1) \text{ and } B = \frac{\beta_2 + \alpha_1^2}{(1 - \alpha_1 \alpha_2)\alpha_1 \alpha_2}\left(\alpha_1 k_1 - \alpha_2^2 k_2\right)$$

In order to judge the state of the equilibrium point Q, let us consider the characteristic equation of (4.45) at point Q:

$$\lambda^2 + p\lambda + q = 0$$

(4.52)

where

$$p = -\left[a_{11}(x_Q, y_Q) + a_{22}(x_Q, y_Q) \right]$$

(4.53)

$$q = a_{11}(x_Q, y_Q)a_{22}(x_Q, y_Q) - a_{12}(x_Q, y_Q)a_{21}(x_Q, y_Q)$$

(4.54)

Now, we can judge the state of the equilibrium point Q by using the theory of ordinary differential equations when we institute Equations 4.48 through 4.51 into Equation 4.53 one by one.

Theorem 4.28 If $\min\{\beta_1, \beta_2\} + \alpha_1 \alpha_2 > 0$, then the point Q is a stable node or a focus.

Theorem 4.29 If $\max\{\beta_1, \beta_2\} + \alpha_1 \alpha_2 < 0$, then the point Q is an unstable node or a focus.

Theorem 4.30 There exists at least one pair of β_1 and β_2 that satisfy $(\beta_1 + \alpha_1 \alpha_2)$ $(\beta_2 + \alpha_1 \alpha_2) < 0$ and $p = 0$.

Theorem 4.31 If $q < 0$, then the point Q is an unstable saddle point.

Theorem 4.32 If $\max\{\beta_1, \beta_2\} < \alpha_1\alpha_2$, then the fractional system (4.45) has no limit cycle.

Proof. In the first quadrant, let us take Dulac function

$$B(x, y) = \frac{(k_1 + \beta_1 x - \alpha_2 y)(k_2 - \alpha_1 x + \beta_2 y)}{xy}$$

By using Bendixson–Dulac theorem, we have

$$\frac{\partial(BxF_1)}{\partial x} + \frac{\partial(ByF_1)}{\partial y} = \frac{\gamma_1}{y}\left[-2k_2 + 2\alpha_1 x + (-\beta_2 + \alpha_1\alpha_2)y\right]$$

$$+ \frac{\gamma_2}{x}\left[-2k_1 + 2\alpha_2 y + (-\beta_1 + \alpha_1\alpha_2)x\right]$$

It is obvious that the equilibrium point Q lies in the region

$$D_1 = \{(x, y) \mid x >, y > 0, -2k_2 + 2\alpha_1 x + (-\beta_2 + \alpha_1\alpha_2)y > 0,$$

$$-2k_1 + 2\alpha_2 y + (-\beta_1 + \alpha_1\alpha_2)x > 0\}$$

Therefore, if the parameters satisfy $\beta_1 < \alpha_1\alpha_2$ and $\beta_2 < \alpha_1\alpha_2$, then the following inequality holds true in region D_1:

$$\frac{\partial(BxF_1)}{\partial x} + \frac{\partial(ByF_1)}{\partial y} > 0.$$

According to Bendixson–Dulac theorem, Theorem 4.32 holds true.

4.7.4 Conclusions

In nature, some vegetables are particularly able to absorb pollutants like phenol. That is known as the phenomenon of *addicted to absorption*. Similarly, some plants have the pronounced tendency of *anti-absorption*. Correspondingly, the financial markets also enjoy a similar phenomenon. When a country suffers from a financial crisis, the investors of both domestic and abroad will withdraw their investments because of the relevant panics. The massive investment withdrawals then cause the financial crisis to spread over into other countries, accelerating the price and return decline of financial assets. That is called a *herding effect*. However, if such a country that would be potentially infected provides investors with confidence in a timely manner or shows a stronger economic and financial support to alleviate the prevailing panic of the investors, and to attract additional investors to pull their money away from the countries that suffer from the crisis into itself, then the country will accelerate the growth of asset prices and investment yields.

Therefore, based on our contagion model (4.42), we introduce a factor $\beta_1 x$ or $\beta_2 y$ to reflect such a phenomenon of acceleration in order to better describe the dynamic behavior of financial contagion between two countries. As inspired by the phenomenon of plants' *addiction to absorption*, we define β_i as the coefficient of *addiction to absorption* of country i's financial market in order to reflect the financial contagion. Intuitively speaking, if $0 > \beta_i > -1$, it means that changes in the exchange rate in country i will experience an increasing volatility, and if $\beta_i > 0$, it stands for that changes in the exchange rate will experience a decreasing volatility.

Furthermore, after analyzing the stability of our model by using the qualitative theory of ordinary differential equations, we found that under the assumption that the upper limit of changes in the exchange rate of a country depends on the ratio between the fluctuation limit of another country's exchange rate and its transmission coefficients, there exist some linear combinations of β_i and α_i that can be useful to reflect when the contagion of a financial crisis is becoming either uncontrollable or controllable. Furthermore, we developed a specific method to approximately calculate β_i and α_i by discretizing our differential dynamic model.

4.8 Thomas Constraint in Currency Substitutions

4.8.1 General Theory

In a closed economy, when a variety of metal money is the means of storing value, there will be such a phenomenon that bad money drives out good money. That is the currency substitution phenomenon as described by *Gresham's law* in the era of metal currency. The modern problem of currency substitution was first proposed by U.S. economist K. V. Chetty. In March 1969, he published an article entitled *On Measuring the Nearness of Near Money* and pointed out that assuming the currency is freely convertible in an open economy, once there is a depreciation of expected exchange rate, economic agents will choose to hold foreign currencies with a relatively stable exchange rates. That makes the modern phenomenon of currency substitution occur. That is the phenomenon of how good money drives out bad money. Seventy years later, with economic integration and globalization, capital controls continue to be relaxed, and international capital flows rapidly increase. So the modern phenomenon of currency substitution is found in a much wider range of economies. As a result, economists have created an extensive theoretical reservoir on the study of currency substitution, where the money demand theory of an open economy is the core underlying the study.

Corresponding to different opinions on the determinants of currency substitution and the money demand function, the following four major theories have been established (Hall and Murphy, 2000; Jiang and Yang, 2009):

1. The production function theory of currency services is established by Miles (1978). It is one of the early theories about currency substitution, and Miles's paper becomes the classic reference to the study of currency substitution. Miles claimed that currency services, such as media of exchange, measurements of value, and storages of value, are provided by either the domestic currency or a foreign currency; people always adjust the proportion of each holding by looking at the respective opportunity costs of different currencies, and that is how the phenomenon of currency substitution occurs. He analyzed the reasons for the emergence of currency substitution from the monetary perspective and introduced the production function theory into the research of monetary problems. Additionally, this research created a formation mechanism of currency substitution.

2. The marginal utility theory of money demand is represented by Bordo and Choudhri (1982). Since the 1980s, this theory has been developed with an emphasis on transaction motivations of money demand. The functional form of Miles's demand is modified, and the currency services in Miles's theory are specifically designated as an assistant utility of holding money in order to facilitate transactions and payments, and money holders maximize their consumer utilities by holding a certain amount of the domestic or a foreign currency.

3. The portfolio theory of money demand is known with King (1978) as the representative. It is another well-recognized theory of currency substitution after the production function of currency services and the marginal utility function of money demand. Based on Miles's research, the portfolio theory takes into account money demand function, which is affected by other economic variables (such as income, wealth, real return of holding money, the opportunity cost of holding money, and risks factors). So, the issue of currency substitution is equivalent to the problem of optimizing portfolio under perfect capital mobility, and methods of optimization can be used in theory to find the optimal form of money demand function. In addition, by considering random fluctuations in bonds and other asset prices, Thomas (1985) used the method developed for solving of random variables to find the best ratio of holding the domestic or a foreign money. So this theory is closer to economic reality than the previous two theories.

4. Poloz (1986) established the prevention demand theory of money. He believed that the previous theories of currency substitution do not consider the transaction cost for obtaining the necessary liquidity and the uncertainty in consumer spending. By using Baumol's *theoretical model of cash*, Poloz founds that to meet the obligation of possible future payments, it is granted

for people to hold a certain amount of the domestic currency and a foreign currency at the same time when they face with uncertainty and liquidity risk. That is, the reason for the occurrence of the phenomenon of currency substitution is to adjust the personal portfolio for preventive motivation.

In short, there are both correlations and inconsistencies among the aforementioned theories. Judging by their respective essentials, the production function theory of money service is to study the problem from the angle of integrated utilities of money; the marginal utility theory of money demand is to research the transaction function of currency; the portfolio theory of money demand is to focus on the storage of value in money; and the preventive demand theory of money emphasizes the principle of caution in order to make payments. Thus, the system, composed of these four theories, covers the common motives of money demand and can explain the formation mechanism of currency substitution from different angles. In addition, these theories have relatively similar research ideas. That is, in a series of constraints of open economies, each of these theories establishes a model with a money demand function and then determines the most moderate demand of the foreign currency through the optimal solution. However, the researchers verify their theoretical perspectives by using empirical analyses with different function models. That creates a lot of differences caused mainly by model assumptions and aspects of testing. Such a problem has also become one of the focuses for the later works.

The portfolio theory occupies a very important position among these theories of currency substitution. As mentioned earlier, the core content of this theory can be refined as follows: as it is similar to other interest-earning assets, currency is also an asset with liquidity in the hands of investors, and the amount of money held by investors is a choice behavior of the asset. The money demand is a stable function of return, gained by holding money by investors, the opportunity cost of holding money, and other risk factors. It emphasizes on the facts that people's asset selection behavior has an impact on currency substitution and that people's motivations to adjust portfolio are to avoid possible risks, to reduce the opportunity cost of holding money, and to maintain a monetary value. Besides King and Thomas, Freitas is also its main representative.

Freitas (2006) explored the implications of the means of payment substitutability and capital mobility on the properties of money demand by using Thomas stochastic dynamic optimization model, where the specific role of money is explicitly considered. At the same time, he extended the complete bond market in Thomas's model to such a case that the consumers cannot hold bonds denominated in a foreign currency and found that there is a channel for portfolio decision that affects the domestic money demand when money is considered as the means of both payment substitutability and storage of value. Contrary to what is suggested by the portfolio balance theory of currency substitution, this research showed that it is a valid test of currency substitution by employing the significance of an expected exchange rate depreciation term in the demand for domestic money. Since the data on foreign bank notes circulating in an economy are not easily available, many empirical studies have measured the extent of currency substitution by the proportion of foreign-currency-denominated deposits (FCDs) in M_2. However, the proportion is sensitive to swaps between foreign bank notes and FCD, and in countries with underdeveloped capital markets, interest-bearing FCD may have a role more comparable to that of foreign bonds than that of foreign currency. Freitas's research also solved the measurability limitation in currency substitution so that it is more convenient to take empirical analyses with the theory.

Recently, Ye (2009) employed Thomas's and Freitas's model to analyze the currency substitution under the capital account liberalization and generalized the model with consumer's budget constraints, where the amount of wealth is dependent on assets and the actual revenue. In discussing the important role of the expected exchange rate in currency substitution, Ye suggested that changes in exchange rates may be appropriate to reflect the changes in currency's purchasing power and its value. And according to the movement of the expected exchange rate and the stability of the real exchange rate, the investor will adjust his portfolio of currency. If there is a sign of depreciation for the domestic currency, the domestic currency will be substituted by a foreign money.

Recently, some scholars introduced cases or model assumptions from a much broader perspective, such as diversification and nondiversification risks, and the geography distribution of different money, to discuss the problem of currency substitution. For example, Groessl and Fritsche (2007) analyzed the effects of money as a storage of value on the decisions of a representative household under diversifiable and nondiversifiable risks given that the central bank successfully stabilizes the inflation rate at a low level. By deriving an explicit relationship from exponential utility between optimal money holdings, the household's desire to smooth and to stabilize the consumption as well as to minimize portfolio risk was established. These authors also showed the impact on the store-of-value function of money from the correlation between the stochastic labor income and stock return. Finally, these authors proved that the store-of-value benefit of money holdings continues to hold even when the riskless alternatives are taken into account. As another example, Coeurdacier and Martin (2009) focused primarily on the impact of the euro for the determinants of trade in bonds, equity, and banking assets by using a set of cross-country data and a set of Swedish data. With the help of a theoretical model, they disentangled the different effects that the euro may have on cross-border asset holdings for both euro zone countries and countries outside of the euro zone. And they found that the euro implies (1) a unilateral financial liberalization, (2) a preferential financial liberalization, and (3) a diversion effect. Their empirical results showed that the elasticity of substitution between bonds from inside the euro zone is three times higher than that between bonds denominated in different currencies.

Yinusa (2008) established an unrestricted portfolio balance model of currency substitution and incorporated the vector error correction technique to analyze the implications of currency substitution and exchange rate volatility for the monetary policy in Nigeria. Results from both impulse response and the forecast error variance decomposition functions suggest that exchange rate volatility and currency substitution respond to monetary policy with some lags, meaning that monetary policy may be effective in dampening exchange rate volatility and currency substitution in the medium horizon but might not be effective in the short horizon. The study concluded that currency substitution was not an instant reaction to the slightest policy mistake; rather, it was a fallout from prolonged period of macroeconomic instability. The major sources of this instability in Nigeria were untamed fiscal deficits that led to high domestic inflation, real parallel market exchange rate volatility, speculative business activities of market agents in the foreign exchange market, and inconsistency or uncertainty in public policies. In terms of policy choice, our result, as developed later, favors exchange rate–based monetary policy as against interest rate–based monetary policy for the stabilization in dollarized economies like Nigeria.

Clearly, these recent literature have analyzed the currency substitution from different perspectives, all including portfolio as an element. At the same time, they also imply a common premise: an open economy or capital account liberalization. Based on these published researches, including the specific references of Thomas, Freitas, and Ye, we will pay more attention to the currency substitution problem during the earlier period of capital account liberalization and discuss a more general application model of currency substitution.

4.8.2 Extended Thomas Model

After opening the capital account, large-scale currency substitutions become possible because the investors can choose the type and quantity of money to hold under an open capital account. And the investors can freely switch between the different currencies or assets due to the facilities provided by the capital account liberation. To this effect, Thomas model is an earlier analysis of the phenomenon of currency substitution under capital account liberalization.

We assume that a country with capital account liberalization has only one kind of consumption good; its domestic price is P, and the direct quotation of the exchange rate is e. Then, P and e obey the following differential equations:

$$\frac{dP}{P} = \pi\,dt + \sigma\,dZ \quad \text{and} \quad \frac{de}{e} = \varepsilon\,dt + \gamma\,dX \tag{4.55}$$

where
$\pi, \sigma, \varepsilon,$ and γ are parameters
dZ and dX the standard Wiener process
ρ the covariance of the two Wiener processes

For the sake of simplicity, assume that consumers hold either bonds or currencies. So there are four assets: the domestic money (M_i), a foreign currency (M_f), bonds in the domestic currency (B_i), and bonds in the foreign currency (B_f). The nominal returns of the domestic bonds and the foreign bonds are, respectively, i and f. Let the actual balance of each asset be

$$m_i = \frac{M_i}{P}, \ m_f = \frac{eM_f}{P}, b_i = \frac{B_i}{P}, \text{ and } b_f = \frac{eB_f}{P} \tag{4.56}$$

After introducing the previous remarks, the actual revenues of the domestic currency, the foreign currency, the domestic bonds, and the foreign bond satisfy the following stochastic differential equations:

$$\frac{dm_i}{m_i} = (\sigma^2 - \pi)dt - \sigma\,dZ$$

$$\frac{dm_f}{m_f} = (\varepsilon + \sigma^2 - \pi - \rho)dt - \sigma\,dZ + \gamma\,dX$$

$$\frac{db_i}{b_i} = (i + \sigma^2 - \pi)dt - \sigma\,dZ \tag{4.57}$$

$$\frac{db_f}{b_f} = (f + \varepsilon - \sigma^2 - \pi - \rho)dt - \sigma\,dZ$$

Now, let us extend Thomas model in order to discuss a more general model of currency substitution. That is,

$$\frac{dm_i}{m_i} = a_{11}dt + a_{12}dZ + a_{13}dX$$

$$\frac{dm_f}{m_f} = a_{21}dt + a_{22}dZ + a_{23}dX$$

$$\frac{db_i}{b_i} = a_{31}dt + a_{32}dZ + a_{33}dX \tag{4.58}$$

$$\frac{db_f}{b_f} = a_{41}dt + a_{42}dZ + a_{43}dX$$

where $a_{11} = \sigma^2 - \pi$, $a_{21} = a_{11} + \varepsilon - \rho$, $a_{23} = \gamma_2$, $a_{31} = a_{11} + i$, $a_{41} = a_{21} + f$, $a_{13} = a_{33} = \gamma_1$, $a_{k2} = -\sigma$ ($k = 1,2,3,4$), $a_{43} = 0$.

Theorem 4.33 There are constants α, β, γ, δ, and C for the Equations 4.58 such that

$$m_i^\alpha m_f^\beta b_i^\gamma b_f^\delta = C \tag{4.59}$$

Proof. In Equations 4.58, there exist constants α, β, γ, and δ such that

$$\alpha \frac{dm_i}{m_i} + \beta \frac{dm_f}{m_f} + \gamma \frac{db_i}{b_i} + \delta \frac{db_f}{b_f} = 0 \qquad (4.60)$$

Equation 4.60 is equal to the following linear equations with existing solutions:

$$a_{11}\alpha + a_{12}\beta + a_{13}\gamma + a_{14}\delta = 0$$
$$a_{21}\alpha + a_{22}\beta + a_{23}\gamma + a_{24}\delta = 0 \qquad (4.61)$$
$$a_{31}\alpha + a_{32}\beta + a_{33}\gamma + a_{34}\delta = 0$$

Let $\delta \neq 0$, denoted by $\tilde{\alpha} = -\alpha/\delta$, $\tilde{\beta} = -\beta/\delta$, and $\tilde{\gamma} = -\gamma/\delta$. Then $\tilde{\alpha}$, $\tilde{\beta}$, and $\tilde{\gamma}$ satisfy the following linear equations:

$$a_{11}\tilde{\alpha} + a_{12}\tilde{\beta} + a_{13}\tilde{\gamma} = a_{14}$$
$$a_{21}\tilde{\alpha} + a_{22}\tilde{\beta} + a_{23}\tilde{\gamma} = a_{24} \qquad (4.62)$$
$$a_{31}\tilde{\alpha} + a_{32}\tilde{\beta} + a_{33}\tilde{\gamma} = a_{34}$$

Because the determinant of the coefficients in (4.62) is det $a_{ij} = -i\sigma\gamma_2 \neq 0$, there exists a unique solutions $\tilde{\alpha}$, $\tilde{\beta}$, and $\tilde{\gamma}$, where

$$\tilde{\alpha} = \frac{\gamma_2(f + \varepsilon - i - \rho) - \gamma_1 f}{i(\gamma_1 - \gamma_2)}$$

$$\tilde{\beta} = \frac{\gamma_1}{\gamma_2 - \gamma_2} \qquad (4.63)$$

$$\tilde{\gamma} = \frac{f(\gamma_1 - \gamma_2) - \gamma_2(\varepsilon - \rho)}{i(\gamma_1 - \gamma_2)}$$

By substituting (4.63) into (4.60), we can obtain

$$\alpha d(\ln m_i) + \beta d(\ln m_f) + \gamma d(\ln b_i) + \delta d(\ln b_f) = 0$$

Hence, by taking integration, we know that for some constant C, the following equation holds true:

$$m_i^\alpha m_f^\beta b_i^\gamma b_f^\delta = C$$

The proof is completed.

To discuss the meaning of this result, let us consider that a country is in the earlier period of capital account liberalization. The actual balance of its foreign bonds remains relatively stable, which is equal to constant b_f. At the same time, the domestic currency and domestic bonds have much lower volatility than the foreign currency, that is, $\gamma_2 \gg \gamma_1$. In such cases, we have

$$\tilde{\beta} = \frac{\gamma_1}{(\gamma_1 - \gamma_2)} < 0 \tag{4.64}$$

According to (4.63), we can readily get

$$\tilde{\alpha} = \frac{\gamma_2(f + \varepsilon - i - \rho) - \gamma_1 f}{i(\gamma_1 - \gamma_2)} \begin{cases} > 0 & f + \varepsilon \leq i + \rho \\ < 0 & f + \varepsilon > i + \rho \end{cases} \tag{4.65}$$

$$\tilde{\gamma} = \frac{f(\gamma_1 - \gamma_2) - \gamma_2(\varepsilon - \rho)}{i(\gamma_1 - \gamma_2)} \begin{cases} > 0 & \varepsilon \geq \rho \\ < 0 & \varepsilon < \rho \end{cases} \tag{4.66}$$

From this analysis, we find four cases of Thomas constraint:

1. $f - i + \varepsilon \leq \rho \leq \varepsilon$
2. $f - i + \varepsilon \leq \rho, \varepsilon < \rho$
3. $f - i + \varepsilon > \rho, \varepsilon \geq \rho$
4. $f - i + \varepsilon > \rho > \varepsilon$

where ε represents the expected exchange rate of inflation. Therefore, the expression $f - i + \varepsilon$ contains the component of speculation. If ε increases, $f - i + \varepsilon$ means greater opportunity for speculation. In addition, from the assumption, we know that in the earlier period of capital account openness, foreign bond balance is stability, and the domestic currency and bonds fluctuate much less than the foreign currency. So, the symbol ρ mainly reflects the difference between the instantaneous fluctuations of the exchange rate (or the foreign currency value). If ρ increases, the hedging opportunity of the foreign currency will grow. Thus, the four cases in the earlier period of capital account openness may correspondingly be described as the following:

Case 1 represents that there are more opportunities for both speculation and hedging.

Case 2 means that there are fewer opportunities for speculation, but more opportunities for hedging transactions.

Cases 3 stands for the fact that there are more opportunities for speculation, but fewer opportunities for hedging transactions.

Case 4 embodies that there are fewer opportunities for speculation and hedging.

4.8.3 Conclusions

Under capital account liberalization, the investors will choose different currency assets by considering the risk-based revenue, costs, and preferences. And the free movement of capital will strengthen the possibility of currency substitution. The traditional theory shows that when there is an expected devaluation of the domestic currency while the relative return of the foreign currency increase or a sign of risk in the exchange rate of the domestic currency appears, the investors will reduce their demand for the domestic currency and be more willing to hold the foreign currency. Ye (2009) employed a utility function with the coefficient of relative risk aversion. By maximizing this function, Ye derived the corresponding first-order conditions, which revealed composition of the optimal portfolio of investors under capital account liberalization.

The difference between what is presented in this section and Ye's research is that we did not use any utility function in our approach. We employ stochastic differential equations to directly describe the real returns of the domestic currency, the foreign currency, the domestic bonds, and the foreign bonds. After that, we find a necessary connection of the aforementioned four factors by obtaining the general integral equation. Furthermore, by assuming under the earlier period of capital account liberalization, the foreign bond balance is stable, and the fluctuations in domestic currency and domestic bonds are much smaller than that of the foreign currency, we discuss the specific meanings about the inside relationships among those four factors. Last, with expected inflation changes of the exchange rates, we obtain four scenarios that may be combined according to the number of how many opportunities for speculation and hedging exists.

This analysis further reveals the implications of the results in our studies. That is, capital account liberalization in fact not only provides for investors a wide range of risk diversification channels, but also opens up additional diversification channels for domestic banks and other financial institutions to select assets from different countries, different areas, and different financial products. So, some current financial risks, which are inherent in the single investment channel available or credit concentration, may be avoided. Hence, in terms of practical applications, it is more beneficial to investigate the problem of currency substitution under capital account liberalization, where more general results can be expected.

4.9 Relative Risk Aversion Coefficient

4.9.1 Introduction

Based on Miles (1978), King (1978) introduced portfolio factors into the money demand function and took the money demand as a function of other economic variables, such as income, wealth, the real value of holding money, the opportunity cost of holding money, and other risk factors. So, the issue of currency substitution is equivalent to the optimal portfolio problem with perfect capital mobility. In general, during the pursuit of maintaining wealth at the maximum level, when the consumers in currency substitution expect increasing returns from a foreign currency or their confidence of holding the domestic currency is deteriorating, they will hold the former instead of the latter in order to balance their asset ratios. And the methods used and resultant composition of their portfolio assets depend on the degree of risk preference and the utility of the holders (Freitas, 2004).

Freitas derived a formula for computing the coefficient of relative risk preference appearing in issues of currency substitution, but he did not analyze the change of the coefficient. In practice, however, it is difficult to measure the risk preference of any asset holder because many factors affected the preference. And holders holding the same assets may very well have different risk preferences in their problem of selecting the portfolio composition. For example, in the classical consumption–investment problem, given a constant relative risk aversion with respect to consumption, the optimal demand of stocks is constant (Merton, 1969, 1971, 1990). But Steffensen (2004) focused on the origins of consumption and investment in his discussion of time-varying risk aversion in consumption. By allowing the curvature of the instantaneous utility function with respect to consumption to vary with time, he studied the relationship between age and risk aversion and found that the relative risk aversion of consumption increases with age. This work provided optimal consumption and investment rules based on life cycle.

In another example, when managers receive equity-based compensation, their decisions on investments and leverage are materially affected by the interaction of risk preference and compensation structure. Lewellen and Nagel (2006) illustrated that holding executable stock options (ESOs) may increase or decrease the managerial risk taking. Ross (2004) derived general conditions under which a compensation schedule concavifies or convexifies the manager's utility functions. And by concentrating on the volatility costs of debt, Lewellen (2006) found that managers holding in the money options are typically worse off with an increase in leverage. As for the problem of ESO valuation, Hall and Murphy (2000) and Hall and Murphy (2002) preferred to use the coefficient of assumed risk preference. And based on the data of ESOs, Pirjetä et al. (2010) employed the method of semiparametric estimation to analyze relative risk preference. Their estimated results show that the relative risk preference is slightly above 1 for a certain stock price range. That might be caused by the typical managers who are wealthy and have low marginal utilities. As for risk aversion, they found that marginal rate of substitution increases considerably in countries with low stock prices.

In addition, some scholars analyzed the risk aversion issue by using the relationship between risk and benefit. After a comprehensive analysis of the relationship between risk and benefit, Berger and De Young (1997) found that cost efficiency has a negative correlation with loan loss in some failed banks. A positive correlation between cost efficiency and loan loss is also possible, at least in the short term.

For example, it seems efficient if banks don't spend sufficient resources to review loan applications. In contrast, those banks with risk-averse managers may be willing to trade off the reduced benefits for reduced risk. It implies higher measured inefficiencies because of taking additional costs for higher-quality loans (Hughes et al., 1995). Meanwhile, the bank's exogenous events, such as a downturn in the economy, can also lead to cost inefficiency (Berger and De Young, 1997). Manlagnit (2010) used stochastic frontier analysis to test the cost efficiency of Philippine commercial banks and found that there are substantial inefficiencies among domestic banks, and that risk and asset quality affect the efficiency of these banks.

On the whole, there is a close relationship among the risk preference of portfolio, the utility function, the cost, and time (Chen and Ying, 2010). Therefore, by considering the time factor, in this section, we derive a more general formula for the risk preference coefficient based on Freitas's study and analyze the change of the risk aversion coefficient under the consumer transaction costs with the common formation of convex function.

4.9.2 Analysis of Relative Risk Aversion Coefficients

First, let us look at the concept of relative risk aversion coefficients. To this end, we assume that a small country with capital account liberalization has only one kind of consumption good with domestic price P; the initial consumption good is fixed as Y, the consumption in period t is C_t, and the direct quotation of the exchange rate is e. Then, the changes of P and e obey the following differential equation system:

$$\begin{cases} \dfrac{dP}{P} = \pi\, dt + \sigma dX \\[3mm] \dfrac{de}{e} = \varepsilon\, dt + \gamma\, dX \end{cases} \tag{4.67}$$

where dt and dX are the standard Wiener processes with π, σ, ε, and γ being parameters, and their covariance is ρ.

Assume that consumers hold either bonds or currencies. So there are four assets: the domestic money (M), a foreign currency (M^*), bonds in the domestic currency (B), and bonds in the foreign currency (B^*). Now, we obtain their actual balances (denoted as m, m^*, b, and b^*, respectively) as follows:

$$m = \frac{M}{P} \qquad m^* = \frac{eM^*}{P} \qquad b = \frac{B}{P} \qquad b^* = \frac{eB^*}{P} \tag{4.68}$$

In addition, the nominal return of the domestic bonds and the foreign bonds are, respectively, i and f. Then, the actual revenues of the domestic currency, the foreign currency, the domestic bonds, and the foreign bonds follow the following general stochastic differential equations:

$$
\begin{cases}
\dfrac{dm}{m} = a_{11}dt + a_{12}dZ + a_{13}dX \\[2ex]
\dfrac{dm^*}{m^*} = a_{21}dt + a_{22}dZ + a_{23}dX \\[2ex]
\dfrac{db}{b} = a_{31}dt + a_{32}dZ + a_{33}dX \\[2ex]
\dfrac{db^*}{b^*} = a_{41}dt + a_{42}dZ + a_{43}dX
\end{cases}
\tag{4.69}
$$

As both the domestic currency and foreign currency can be the media of exchange, Freitas (2004) defined the transaction cost of good purchase as a convex function (dual function).

Definition 4.1 (Freitas, 2004). The transaction cost of good purchase is defined by

$$
\tau = cv\left(\frac{m}{c}, \frac{m^*}{c}\right)
\tag{4.70}
$$

Let us consider the convex functions of specific forms as given in Table 4.1.

Without loss of generality, we select the transaction cost of good purchase as follows:

$$
\tau = ch_1\left(\frac{m}{c}\right)h_2\left(\frac{m^*}{c}\right)
\tag{4.71}
$$

Table 4.1 Convex Function Type Table

Type	I	II	III
$h_1(x)$	$e^{-\alpha x}$	$\dfrac{x^{-\alpha}}{\alpha}$	$\ln\dfrac{1}{x}\ (0<x<1)$
$h_2(x)$	$e^{-\beta x}$	$\dfrac{x^{-\beta}}{\beta}$	$\ln\dfrac{1}{x}\ (0<x<1)$

Table 4.1 shows that there are at least nine common ways to form the combined function. Hence, the total actual wealth held by a consumer is

$$W = m + m^* + b + b^* \tag{4.72}$$

Assume that the budget constraint depends on the changes in the income as caused by the assets and revenue:

$$dW = dm + dm^* + db + db^* + [y - c - \tau]dt \tag{4.73}$$

By substituting (4.69) into (4.73), we can obtain

$$dW = \phi \, dt + (W - b - m)\gamma \, dX + W\sigma \, dZ \tag{4.74}$$

where ϕ is the coefficient of relative risk aversion as defined by Freitas. Therefore, we can derive a general definition of ϕ as follows.

Definition 4.2 The following is known as the coefficient of relative risk aversion:

$$\phi = a_0 + a_1 \times \frac{m}{c} + a_2 \times \frac{m^*}{c} + a_3 \times v\left(\frac{m}{c}, \frac{m^*}{c}\right) \tag{4.75}$$

Second, let us look at changes in relative risk aversion coefficients. For the sake of simplicity, let us analyze changes in the coefficient of relative risk aversion with one of the common forms in (4.75). That is, we let

$$\tau = ce^{-\alpha \frac{m}{c}} \cdot e^{-\beta \frac{m^*}{c}} \tag{4.76}$$

By substituting (4.76) into (4.75), we have

$$\phi = a_0 + a_1 \times \frac{m}{c} + a_2 \times \frac{m^*}{c} + a_3 \exp\left(-\alpha \cdot \frac{m}{c} - \beta \cdot \frac{m^*}{c}\right)$$

where α is the coefficient of the consumer's risk aversion of m and β is the coefficient of the consumer's risk aversion of m^*.

If we assume $a_0 > 0$ and $a_3 < 0$, then we can discuss ϕ's changes as follows:
When $a_1 > 0$ and $a_2 < 0$, the following conclusions are obtained:

1. If the balance m^* of the foreign currency remains unchanged, the coefficient of relative risk preferences of the consumer will increase with the balance of the domestic currency m.

2. If the balance m of the domestic currency remains constant, with the increase in the balance m^* of the foreign currency, the coefficient of relative risk preferences of the consumer will increase at first and then decrease monotonically after reaching a maximum point.

3. If the balances of the foreign currency and the domestic money increase at the same time, there will then be three different forms. Even so, the coefficient of relative risk preference of the consumer will gradually approach a constant through one particular way ($m^* = km$). In fact, if $k = -(a_1/a_2)$, then the coefficient ϕ of the relative risk preference along a straight line $m^* = km$ will take the following form:

$$\phi = a_0 - ce^{-(\alpha - \beta \times (a_1/a_2))(m/c)}, \quad \alpha > 0, \ \beta > 0, \ a_1 > 0, \ a_2 < 0.$$

Obviously, we have $\lim_{m \to \infty} \phi = a_0$. So the coefficient of the relative risk preference will gradually approximate the constant a_0.

When $a_1 < 0$ and $a_2 > 0$, conclusions similar to those in the previous paragraphs can be obtained:

1. If the balance m^* of the foreign currency remains constant, with the increase in the balance m of the domestic money, the coefficient of relative risk preference of the consumer will increase at first, but then decrease monotonically after reaching a maximum point.

2. If the balance m of the domestic currency remains unchanged, the coefficient of relative risk preference of the consumer will increase with the balance m^* of the foreign currency.

3. If the balances of the foreign currency and the domestic currency increase at the same time, there will be three different forms. However, the coefficient of the relative risk preference of the consumer will gradually approach a constant only through a particular way ($m = km^*$). In fact, by letting $k = -(a_2/a_1)$, the coefficient ϕ of the relative risk preference along a straight line $m = km^*$ takes the following form:

$$\phi = a_0 - ce^{-(\alpha - \beta \times (a_2/a_1))(m^*/c)}, \quad \alpha > 0, \ \beta > 0, \ a_2 > 0, \ a_1 > 0$$

Obviously, we have $\lim_{m \to \infty} \phi = a_0$. So the coefficient of relative risk preference will gradually approximate the constant a_0.

When $a_1 > 0$ and $a_2 > 0$, we can conclude that if the balance of one currency remains unchanged, the coefficient of relative risk preference of the consumer will increase with the balance of the other currency. In particular, we have the following:

1. If the balance m^* of the foreign money remains unchanged, with the increase in the balance m of the domestic currency, the coefficient of relative risk

preference of the consumer will increase at first, but then decrease monotonically after reaching a maximum point.

2. If the balance *m* of the domestic currency remains constant, with the increase in the balance *m** of the foreign currency, the coefficient of relative risk preference of the consumer will increase at first, but then decrease monotonically after reaching a maximum point.

4.9.3 Conclusions

When the issue of currency substitution is transformed into the problem of portfolio optimization, the balanced proportion of assets held by a consumer is associated with his risk preference on different currencies, and portfolios of different consumers have their coefficients of relative risk preference change individually. By defining the coefficients of relative risk aversion and by analyzing their changes with varying holding amounts of the foreign or domestic currency, we find that there are three ways for the coefficient of relative risk aversion to change under convex functions of transaction cost. The first way is that ϕ is a stable constant; the second is that ϕ varies monotonically with the change of *m* and *m**; and the third is that ϕ decreases monotonically after an increase at first until reaching a peak with higher values of *m* and *m**.

These three ways for the coefficient ϕ of relative risk aversion to change also reflect three common tendencies in consumers' risk preferences in the issue of currency substitution. For the first case, when the fluctuation of the two currency exchange rate is small, or the exchange rate is stable, the consumers reduce their holdings of the domestic and foreign currencies with the same proportion, reflecting the fact that their risk aversions of the two currencies remain unchanged. For the second case, when the fluctuation of the exchange rate is larger, if the consumers have a clear expectation about the coming trend of change of the exchange rate, they will tend to increase or decrease the holding balance of one of the currencies, reflecting the fact that their risk aversions of the two currencies are totally different from one another. For the third case, where the exchange rates experience large fluctuations, if the consumers increase (or decrease) their balances of currencies to a certain level according to their predetermined principles (or misjudgments), then the coefficients that reflect their risk preferences in the two currencies will take bounded values without any chance for them to evolve in any one direction indefinitely.

Chapter 5

From Normal to Abnormal Flow of Capital: Optimizations for Capital Account Liberation

5.1 Optimization Models in Finance

Capital account liberalization is always accompanied by a large number of international capital flows, and can easily lead to volatility in financial asset prices. Empirical evidence has shown that since the major international financial crises that broke out in the 1990s, the phenomenon of contagion of financial crises, and the frequency for the breakout of financial crises, the depth, breadth and speed of contagion are growing and have everything to do with the increasing amount of the international flows of hot money. In the recent decades, the world economy has experienced a series of financial crises that seemed to be linked by a recurrent pattern: when one country or a sector in the world economy experiences a financial crisis, a massive amount of capital flows in a panic out of the country or sector, while the investors seek for more attractive destinations for their money. At the next destination, capital inflows create a boom that is accompanied by rising indebtedness, rising asset prices, and booming consumption for a time. But all too often, these capital inflows are followed by another crisis at the new destination. Some commentators describe these patterns of capital movements as *hot money* that flows from one sector or country to the next and leaves behind a trail of destruction (Alfaro et al., 2003; Verdier, 2008; Korinek, 2011).

Studies of literature also found that capital account liberalization and financial instability, as well as the financial crises and infection, are closely related. But during the current entry to the postcrisis era, the world economy recovers slowly; the impact of debt problems of some European countries is increasing; the monetary policy of the United States is still loose; and international hot money is flooding into emerging economies. So, the external environment for the economic and financial development of most countries is full of uncertainties, which increases the possibility of a new round of financial crises. With the trend of development of economic and financial globalization, as new financial crises break out frequently, it is reasonable to avoid the rapid spread and movement of hot money flows, which not only have led to greater economic losses along the flow path provided by regional capital account liberalizations but also have and will continue to seriously disrupt the global economic order again and again. In other words, movements of hot money, accompanied by regional capital account liberalizations, have and will continue to create breeding grounds for the future rounds of global financial crises and infections. Therefore, in this section, we introduce a theoretical framework of optimization that can effectively address the phenomenon of contagion of financial crises under the assumption of capital account liberalization and other economic issues.

5.1.1 Hot Money and Serial Financial Crises: Objectives with Recursively Defined Variables

The worldwide availability of investment opportunities will decline, if one region of the world economy experiences a financial crisis. As global investors search for new destinations for their capital, other regions will experience inflows of hot money. However, large capital inflows make the recipient countries more vulnerable to future adverse shocks, creating the risk of serial financial crises. In this section, we introduce a formal model of such flows of hot money and the vulnerability of the recipient country to serial financial crises. This model analyzes the role of macroprudential policies that lean against the wind of such capital flows so as to offset the externalities that occur during financial crises. By summarizing the results of this model in a simple policy rule, it can be found that a 1% point increase in the recipient country's capital inflows/GDP ratio warrants a 0.87% point increase in the optimal level of capital inflow taxation.

Assume that the world economy consists of two types of agents:

1. International investors who represent *hot money* and who hold a large amount of capital that they move to where the opportunity of return is the greatest.
2. Different countries who borrow and who are subject to an endogenous collateral constraint.

We assume that the international investors come in overlapping generations: In each period, a continuum of mass one of investors is born who live for two periods. Denote the variables regarding the investors with the superscript h (as in *hot money* or *households*). Investors value consumption according to a neoclassical period utility function $v(c)$ that satisfies $v'(c) > 0 > v''(c)$ with time discount factor β, which results in a total level of utility

$$v\left(c_t^h\right) + \beta v\left(c_{t+1}^h\right)$$

In this model, we pay our attention to the special case when $v(c) = \log(c)$ so as to obtain analytical solutions. Investors obtain the constant endowments e_1 and e_2 in the first and second periods of their lives. In the first period, they choose how much to consume and how much to save in zero coupon bonds at the gross world interest rate R_{t+1}, where b_{t+1}^h / R_{t+1} denotes the amount saved. In the second period of their lives, they obtain the repayment b_{t+1}^h on their bond holdings, consume all their remaining wealth, and perish. The optimization problem of the investors in generation t (in short notation) is

$$\max_{c_t^h, c_{t+1}^h, b_{t+1}^h} v\left(c_t^h\right) + \beta v\left(c_{t+1}^h\right) \tag{5.1}$$

$$s.t. \quad c_t^h + \frac{b_{t+1}^h}{R_{t+1}} = e_1 \tag{5.2}$$

$$c_{t+1}^h - b_{t+1}^h = e_2 \tag{5.3}$$

which yields the standard Euler equation

$$v'\left(c_t^h\right) = \beta R_{t+1} v'\left(c_{t+1}^h\right) \tag{5.4}$$

For arbitrary utility functions, the response of b_{t+1}^h to changes in the interest rate is

$$\frac{\partial b_{t+1}^h}{\partial R_{t+1}} = \frac{\frac{b_{t+1}^h}{R_{t+1}} v''\left(c_t^h\right) - \beta R_{t+1} v'\left(c_{t+1}^h\right)}{v''\left(c_t^h\right) + \beta R_{t+1}^2 v''\left(c_{t+1}^h\right)} > 0.$$

If $b_{t+1}^h > 0$, the repayment to investors rises with the market interest rate.

Note that the decision-making problems of different generations of investors are not directly linked. This greatly simplifies our analysis—Equation 5.2 defines a time-invariant supply of funds function $b^h(R)$ and it is straightforward to prove that this function is increasing in R.

In the case of log-utility, the previous Euler equation can be solved explicitly for a supply of funds function. We obtain the following expression for the amount of net savings and bond holdings:

$$\frac{b^h(R)}{R} = \frac{\beta e_1 - \dfrac{e_2}{R}}{1+\beta} \qquad (5.5)$$

This expression is increasing in R. That is, investors save more and receive greater repayments when the interest rate is high. Furthermore, the supply of *hot money* is higher when the initial endowment e_1 is larger when compared with the second-period endowment e_2. The inverse demand function is

$$R(b^h) = \frac{(1+\beta)b^h + e_2}{\beta e_1} \qquad (5.6)$$

5.1.2 Dutch Disease and Optimal Taxation: Objectives with Linear Multiple Variables

Tourism is a growing industry in many economies. According to the 2007 Annual Report of the World Tourism Organization (WTO), tourism is a major source of foreign exchange. The number of international tourist arrivals was 69 million in 1960 and reached 922 million in 2008, whereas revenue from tourism was US$6 billion in 1960 and jumped to $944 billion in 2008. The recent *Tourism: 2020 Vision*, issued by WTO, forecasts that 1.6 billion tourists will visit foreign countries annually by the year 2020 (WTO, 1997), generating such a revenue of US$2 trillion per year.

This section develops a dynamic optimization macromodel that sheds light on two stylized facts of tourism:

1. The congestion externalities as caused by tourism expansion
2. The wealth effect generated by the revenues from overseas tourism taxation

Based on the two salient characteristics, the positive analysis indicates that if the tax revenues of tourism are used to provide rebates to local residents, because of the wealth effect, Dutch disease cannot be cured by the consumption tax on tourists. In contrast, if the tax revenues of tourism are used to provide productive government services for the manufacturing sector, Dutch disease can be treated effectively by taxation tailored for tourism. In a normative analysis, we show that to simultaneously correct the distortion caused by the congestion externality of tourism and to generate the revenues of taxation from overseas tourism, the government should not only levy a general tax on tourism consumption, but also discriminate between domestic consumption and overseas tourism consumption so that a positive tax surcharge is imposed on foreign tourists. In addition, the key factors that govern the optimal rates of a general tax and tax surcharge are also examined in this section.

For the sake of simplicity, let us consider an economy with two final goods: manufacturing good X and tourism good Y (which is nontradable). Although the tourist consumption in the receiving country predominantly involves nontraded goods (and services), it contributes to foreign currency earnings when consumers are foreign tourists. There are three types of decision-makers involved here: firms, households, and the government. In both the manufacturing and tourism sectors, firms maximize their profits by producing goods through Cobb–Douglas technology. Domestic households, subject to their budget constraints, seek to maximize their lifetime utility. Domestic residents (households) consume both goods X and Y, whereas foreign tourists consume only the nontraded tourism good Y. As households are also owners of the firms, the economy we consider is a world of a representative household-producer. The government balances its budget in each period; it provides lump-sum transfers to domestic residents by levying taxes on income and consumption. In particular, we assume that the government can discriminate between domestic consumption and overseas tourism consumption so that the latter may be taxed at a higher rate.

The setting of the production environment largely follows (Khan and Mitra, 2007). There are two sectors: the X sector for the manufactured good and the Y sector for the tourism good, which are both perfectly competitive for the sake of simplicity of our discussion. The X good is viewed as the numeraire and, accordingly, the relative price of the tourism good is defined as P.

5.1.2.1 Manufacturing Good Sector X

In the manufacturing sector, each producer employs labor (L_X) and capital (K) to produce good X by using a symmetric technology given as follows:

$$X = f^X(L, K) = AK^{\alpha}(L_X)^{1-\alpha}, \quad 0 < \alpha < 1 \tag{5.7}$$

where A is a constant technology parameter. The term α is the share of capital, and $1 - \alpha$ is the share of labor, implying that the production function exhibits homogeneity of degree 1 in input factors.

Given the production function (5.7), the optimization problem of the manufacturing firm is to choose labor (L_X) and capital (K) so as to maximize its profit, that is,

$$\max \pi_X = X - r_K K - wL_X \tag{5.8}$$

where
   w is the wage rate
   r_K is the rental rate of the physical capital

Solving the optimization problem (5.8) leads to the following first-order conditions:

$$r_K = \frac{\alpha X}{K}, \quad w = \frac{(1-\alpha)X}{L_X} \tag{5.9}$$

Equation 5.9 is referred to as the common MC = MR conditions.

5.1.2.2 Tourism Good Sector Y

In accordance with real-life observations and the common specification in the tourism literature, we also assume that the tourism firms use labor (L_Y) and land (V) to produce the tourism good Y. Land acquisition is apparently important for tourism construction, whereas tourism expansion may deplete the country's natural resource base (Lane, 2001; Gollin, 2002). In this model, V can broadly be referred to as a physical plant: a site, natural resource or facility, such as a waterfall, a wildlife resort, or a hotel. As the tourism resource is limited, we assume that it is supplied inelastically and normalized to $V=1$ for simplicity. Additionally, the physical plant requires the input of *labor services* to make it useful for tourists. Labor services L_Y is referred to as the performance of specific tasks required to meet the needs of tourists. For example, a resort needs management, front desk operation, housekeeping, and maintenance to function as a resort. Given that $\bar{Y}$ is the aggregate output of the tourism good, the production function of a tourism good producer is given by

$$Y = f^Y(L,V,\bar{Y}) = BV^w(L_Y)^{1-w}\bar{Y}^{-\beta} \tag{5.10}$$

where B is a constant technology factor. Compared with the manufacturing good sector, the tourism good sector needs more labor services and is relatively labor intensive. Thus, the utilization of capital in the production of the tourism good is ignored without loss of significant generality. As it is well known, environmental resources are subject to market failure owing to their public good nature and congestion externalities. To incorporate this reality into our model, the term $\bar{Y}^{-\beta}$ (where $\bar{Y}$ denotes the aggregate output of the tourism good) enters into the production function (5.10) and captures negative externalities as caused by tourism expansion, namely, congestion and environmental degradation.

Given $V = 1$ and Equation 5.10, we can express the optimization problem of the tourism producer as

$$\max \pi_Y = PY - r_V - wL_Y \tag{5.11}$$

where
P is the (relative) price of the tourism good
r_V is the rent for the land (the unit cost of using environmental resources)

Thus, the corresponding first-order conditions are

$$w = \frac{(1 - w)PY}{L_Y} \qquad r_V = wPY \tag{5.12}$$

By defining L as the (fixed) labor endowment of the economy, we assume that for the sake of analytical convenience, the level of employment of the X sector is $L_X = uL$ (where $0 < u < 1$ is the share of labor allocated to the manufacturing good sector) and hence, $L_Y = (1-u)L$. Suppose that labor is perfectly mobile between the X and Y sectors (but is not mobile internationally). Consequently, the workforce will move around until wage levels in both of them reaches

$$w = (1 - \alpha)AK^\alpha (uL)^{-\alpha}$$

$$= (1 - \omega)PB((1 - u)L)^{-\omega} \bar{Y}^{-\beta} \tag{5.13}$$

However, given that K and V are specific factors, we can rewrite the optimization conditions concerning capital and land as follows:

$$r_K = \frac{\alpha X}{K} = \alpha AK^{\alpha-1}(uL)^{1-\alpha} \tag{5.14}$$

$$r_V = \omega PY \tag{5.15}$$

Our analysis is confined to a symmetric equilibrium under which $Y = \bar{Y}$ and, as a result, the aggregate production function is

$$Y = \{B[(1 - u)L]^{1-\omega}\}^{1/(1+\beta)} \tag{5.16}$$

Finally, as both sectors X and Y are perfectly competitive, free entry and exit guarantee zero profits for each producer, that is, $\pi_X = \pi_Y = 0$.

5.1.2.3 Households

The economy is populated by a unit measure of identical and infinitely lived households. The representative household derives utility from consuming both manufactured and tourism goods, denoted by C_X and C_Y, respectively. Given that the instantaneous utility is denoted by U, the household facing its budget constraint chooses C_X, C_Y, and K so as to maximize the discounted sum of future instantaneous utilities

$$\max W = \int_0^\infty \frac{\left(C_X^\phi C_Y^{1-\phi}\right)^{1-\sigma} - 1}{1-\sigma} e^{-\rho t} dt, \quad 0 < \phi < 1 \tag{5.17}$$

$$\text{s.t.} \quad \dot{K} = (1-t)(r_K K + r_V + wL) - C_X - (1+\tau)PC_Y + R \tag{5.18}$$

5.1.3 Optimal Growth Rate of Consumable Resource: Objectives with Discrete Variables

In 1928, Ramsey (1928) first addressed the problem of undiscounted optimal growth in which the optimal program of capital accumulation was derived from the maximization of a utility sum over an infinite time period. Samuelson and Solow (1956) extended the framework to a model with many capital goods. Khan and Mitra (2007) utilized geometric techniques to analyze the optimal intertemporal allocation of water resources in a dynamic setup without discounting. The framework features two sectors: the first uses labor to purify water, while the second uses labor and purified water for irrigation to produce an agricultural consumption good. Purified water can also be used as potable water for drinking purposes. The planner allocates the available factors of production between the two sectors every period, and determines the optimal amounts of purified water, potable water, and irrigation water. The geometry characterizes the optimal path depending on whether the irrigation sector is more labor intensive than the purification sector. When the irrigation sector is labor intensive, the optimal path is a nonconverging cycle around the golden rule stock of purified water, while if the purification sector is labor intensive, there is a damped cyclical convergence to the golden rule stock.

To be more specific, let our economy be represented by two sectors: water purification and irrigation. For each date $t \in N$, where N is the set of all nonnegative integers, the production of one unit of purified water requires a unit of labor. One unit of labor and one unit of purified water used in the irrigation sector produce one unit of the consumption good. Moreover, purified water is consumed for drinking purposes. In every period, a planner allocates a given amount of purified water and labor to either sector. There is one technology available to each sector. For each date t, let the technologies be stationary and given by

$$c(t+1) = \min\{w_r(t), l_r(t)\} \tag{5.19}$$

$$z(t+1) = \frac{1}{a} \cdot l_p(t) \tag{5.20}$$

where

$c(t+1)$ and $z(t+1)$ denote the amounts of the consumption good and the investment in water purification in period $t+1$, respectively

$l_r(t)$ and $l_p(t)$ are the labors employed in the irrigation sector for the production of the consumption good and the labor employed in the water purification sector, respectively

$w_r(t)$ is the amount of purified water used in the irrigation sector

The amount of purified water used for drinking purposes in period t, or what is referred to as potable water, is denoted by $w_p(t)$.

We also assume that the stock of purified water evaporates at a rate $e \in (0,1)$. In other words, from a given stock of purified water, a portion is used in both sectors and the leftover evaporates at the given rate. The residual stock plus the purified water produced in the same period form the stock of purified water of the next periodic. If $x(t) \geq 0$ denotes the available stock of purified water in period t, then we have

$$x(t+1) = (1-e) \cdot [x(t) - w_p(t) - w_r(t)] + z(t+1) \tag{5.21}$$

Therefore, the investment in water purification replaces the evaporated water and the amounts used for drinking and irrigation in the previous period.

Labor is normalized to unity in every period of time. The gross increase in purified water stock $z(t+1)$ requires $l_p(t) = a \cdot z(t+1)$ units of labor in period t. The labor required in the irrigation sector is $l_r(t) = w_r(t)$. Therefore, the labor constraint is given by

$$0 \leq a \cdot z(t+1) + w_r(t) = l_p(t) + l_r(t) \leq 1 \tag{5.22}$$

By following Khalifa and Hurcan (2011), we convert this model into its reduced form. The latter is summarized by the transition possibility set Ω as a collection of pairs (x, x'), such that it is possible to have x' amounts of the purified water in the next period from x amounts of purified water available in the current period. Formally, we have

$$\Omega = \{(x, x') \in R_+ \times R_+ : x' - (1-e) \cdot (x - w_r - w_p) \geq 0,$$

$$a(x' - (1-e) \cdot (x - w_r - w_p)) \leq 1\} \tag{5.23}$$

By using this formulation, we can keep track of the transition dynamics of the only state variable $x(t)$ overtime. For any $(x,x') \in \Omega$, let us consider the amounts (w_p, w_r) available for drinking and irrigation purposes, respectively.

Formally, we have a correspondence $\Psi: \Omega \to R_+ \times R_+$, given by

$$\Psi(x,x') = \left\{(w_p; w_r) \in R_+ \times R_+ : 0 \leq w_p + w_r \leq x,\right.$$

$$\left. w_r \leq 1 - a\left(x' - (1-e)(x - w_r - w_p)\right)\right\} \qquad (5.24)$$

Finally, the reduced form utility function $u: \Omega \to R_+$ is defined on Ω such that

$$u(x,x') = \max\{\upsilon(w_p, w_r) : (w_p, w_r) \in \Psi(x,x')\}. \qquad (5.25)$$

In this model, we present a complete characterization of optimal programs, where an optimal program is one that minimizes the aggregate value loss and converges to the golden rule stock. Alternatively, any program that minimizes the aggregate of the sequence of all value losses over the long run is an optimal trajectory. Using cob web diagrams in today–tomorrow plane, every program starting from any initial capital stock can be tracked period by period, and its associated value loss per period and then its aggregate value loss can be calculated. In this way, we can compare the aggregate value losses of two different programs starting from the same initial capital stock and find an optimal program that has the minimum aggregate value loss. Let us consider two cases: when $a \leq 1$ or when $\varepsilon \leq -1$. This is the case when the irrigation sector is more labor intensive than the purification sector. The other case is when $a > 1$ or when $\varepsilon > -1$. This is the case when the purification sector is more labor intensive than the irrigation sector. Irrigation activities occur within the agricultural sector, which is labor intensive, while water purification plants rely on water treatment machinery, which implies it is capital intensive. However, the model is an extension of the Leontief two sector optimal growth model, which analyzes the optimal allocation of capital and labor to a consumption sector and an investment sector. The optimal policy is found to depend upon the factor intensity of the two sectors. Therefore, it is imperative to discuss all possible cases, even in this specific application.

As it is well known, one of consumable capital is water resources. Water can be used in irrigation to produce an agricultural consumption good and can also be used directly for drinking purposes. Therefore, water can serve as capital that is used in the production process of a consumption good and can be consumable as well. In this context, the model features two sectors. The first utilizes labor to purify water. Purified water can be used for either drinking purposes or in an irrigation sector along with labor to produce an agricultural consumption good. The geometric analysis characterizes the optimal path depending on whether the irrigation sector is more labor intensive than the purified sector. In this case, the optimal path is a nonconverging cycle around the golden rule stock of purified water, while in the other case, it exhibits a damped cyclical convergence to the golden rule stock.

This model is not only an extension to a theoretical setup, but also an application to a specific case. The model is motivated by a real-world problem concerning the depletion of the scarce water resource due to the increasing worldwide demand. In addition, the model assumptions are supported by observations from AQUASTAT, the world data extracted by the Food and Agriculture Organization. Finally, the model has policy implications, especially to developing countries that have significant agricultural and irrigation sectors, as it determines the optimal program to be adopted to ensure sustainability of water resources.

5.1.4 Dynamics of Ecosystem Service Provision: Objectives with Bivariate Factors

By using a bioeconomic model of a coral reef–mangrove–seagrass system, Sanchirico and Springborn (2011) analyzed the dynamic path of incentives to achieve an efficient transition to the steady-state levels of fish biomass and mangrove habitat conservation. This model nests different types of species habitat dependency and allows for changes in the extent of habitat to affect the growth rate and the long-run fish level. This model can be described as two-control, two-state nonlinear optimal control problem. Its solution can help us compute the input efficiency frontier characterizing the trade-off between mangrove habitat and fish population. After identifying the optimal locus on the frontier, one can determine the optimal transition path to the frontier from a set of initial conditions to illustrate the necessary investments.

The mathematical approach is most similar to Swallow (1990), who developed a model to investigate the optimal development of coastal habitat (a nonrenewable, nonrestorable resource) that also provides habitat for a biological stock (a renewable resource) that is being optimally harvested. We extend this (Swallow, 1990) by developing an ecological model that nests different species–habitat relationships and by considering the use of restoration (reversible development). Restoration is an important management tool to consider in general and in our system, because worldwide mangroves are being converted at a rate of 1%–2% per year (Duke et al., 2007) and approximately 35%–50% have been cleared.

The infinite horizon optimal control problem of the planner is

$$V = \max_{h_t, D_t} \int_0^\infty e^{-\delta t} [\pi(h_t, N_t) + B(1 - M_t) - C(D_t) + P(M_t)] \, dt \qquad (5.26)$$

$$\text{s.t.} \quad \frac{dN_t}{dt} = \frac{b_1 R_t(N_t, M_t)}{1 + b_2 R_t(N_t, M_t)} - \mu N_t - h_t \qquad (5.27)$$

$$\frac{dM_t}{dt} = F(D_t) \tag{5.28}$$

$$0 \leq M_t \leq 1 \tag{5.29}$$

$$0 \leq N_t, \quad 0 \leq h_t \tag{5.30}$$

$$N\big|_{t=0} = N_0, \quad M\big|_{t=0} = M_0 \tag{5.31}$$

where
   δ is the discount rate
   $\pi(h_t, N_t)$ is the fishing profit in time t
   $B(1-M_t)$ is the benefit from the extent of development given by $1-M_t$
   $C(D_t)$ is the cost of converting mangroves
   $P(M_t)$ is the in situ value of the mangroves that could be due to providing coastal protection (Barbier et al., 2008) or from intrinsic value associated with the habitat

For the sake of simplicity, we will refer to $P(M_t)$ as storm protection for the remainder of the section. Mangroves, therefore, contribute to the value of the system indirectly through the production of fish and directly in their protection of the coastal area. Fishing profit is assumed to be increasing at a decreasing rate in harvest and fish population on the reef ($\pi_h > 0$, $\pi_{hh} \leq 0$; $\pi_N > 0$, $\pi_{NN} \leq 0$).

Because the constraints on the state variables affect the rate of change of N and M with respect to time when the state variables are at the boundaries, one can only derive the current value, Lagrangian rather than Hamiltonian.

Following the steps outlined in Duke (2007), the *Golden rule* equations for the optimal fish stock size and mangroves at the steady state can be derived. By putting aside for now the possibility of corner solutions in the steady state (e.g., all development, all mangroves, and no fishing), the equations that correspond to an interior steady-state solution are

$$\delta = \left(\alpha - 2\beta G(N,M) - \frac{c}{N} \right) G_N + \frac{c}{N} \frac{G(N,M)}{N} \tag{5.32}$$

$$B_M = P_M + \left[\alpha - 2\beta G(N,M) - \frac{c}{N} \right] \left[\frac{b_1 R_M}{(1+b_2 R(N,M))^2} \right] \tag{5.33}$$

where $G(N,M) = \dfrac{b_1 R(N,M)}{1+b_2 R(N,M)} - \mu N.$

This model illustrates that the qualitative nature of the path to the long-run steady state is similar for the obligate and facultative settings, while the steady-state level of mangroves is (intuitively) greater in an obligate relationship. This model also shows that the optimal path can involve temporarily overshooting the long-run mangrove stock. In the case of rebuilding, for example, the over-shoot is optimal, because additional mangroves speed up the rebuilding of fish stocks. The robustness of the optimal overshoot is an interesting area for future research, especially when the assumption that restored habitat is immediately and equally ecologically productive for the fishery is relaxed. Other interesting research questions include measuring the costs of going to other (not optimal) points on the frontier and the economic ecological differences in the transition to these nonoptimal points.

5.1.5 Illiquid Markets with Discrete Order Flows: Objectives with Double Integrals

It is important to study the problem of optimal portfolio selection in an illiquid market with discrete order flow. In this sort of market, bids and asks are not available at any time, but trading occurs more frequently near a terminal horizon. The investor can observe and trade the risky asset only at exogenous random times corresponding to the order flow given by an inhomogeneous Poisson process. By using a direct dynamic programming approach, we first derive and solve the fixed point dynamic programming equation satisfied by the value function, and then perform a verification argument, which provides the existence and characterization of optimal trading strategies. What can be proved is the convergence of the optimal performance, when the deterministic intensity of the order flow approaches infinity at any time, to the optimal expected utility for an investor trading continuously in a perfectly liquid market model with no-short sale constraints.

Let us consider an illiquid market in which an investor can trade a risky asset over a finite horizon. In this market, bids and asks are not available at any time, but trading occurs more frequently near the horizon. This is typically the case in power markets with forward contracts. This market illiquidity feature is modeled by assuming that the arrivals of buy/sell orders occur at the jumps of an inhomogeneous Poisson process with an increasing deterministic intensity converging to infinity at the final horizon. In order to obtain an analytically tractable model, we further assume that the asset prices observed over discrete time come from an unobserved continuous-time stochastic process, which is independent of the sequence of arrival times. It is reasonable for us to think about the continuous time process as an asset price process based on fundamentals independent of time-illiquidity, which would be actually observed if trading occurred at all times.

We investigate an optimal investment problem in the illiquid market as described in the previous section. Let us consider an utility function U defined on $(0, +\infty)$ that is strictly increasing, strictly concave, and $C^1 \in (0, \infty)$, satisfying the Inada conditions: $U'(0^+) = \infty$, $U'(\infty) = 0$. We make the following additional assumptions on the utility function U:

Assumption 5.1 There exist two constants $C > 0$ and $P \in (0,1)$ such that

$$U^+(x) \leq C(1 + X^P) \quad \text{and} \quad (\forall)\, X > 0,$$

where $U^+ = \max(U, 0)$.

Assumption 5.2 There exist two constants $C' > 0$ and $p' < 0$ such that

$$U^-(x) \leq C'(1 + x^{p'}) \quad \text{and} \quad (\forall)\, X > 0,$$

where $U^- = \max(-U, 0)$.

These assumptions include the most popular utility functions, particularly those with constant relative risk aversion $1 - p > 0$, in the following form:

$$U(x) = \frac{(x^p - 1)}{p}, \quad x > 0.$$

Given a chosen positive initial wealth $X_0 > 0$, we consider the following optimal investment problem:

$$V_0 = \sup_{\alpha \in A} E[U(X_T)] = \sup_{X \in \chi} E[U(X_T)] \tag{5.34}$$

We try to provide an analytic solution to the control problem (5.34) by using direct dynamic programming, that is, first solve the dynamic programming equation analytically and then perform a verification argument. Therefore, there is no need to either define the value function at later times or to prove the dynamic programming principle.

Assume that u^* is the solution to the fixed point dynamic programming Equation 5.35. We now state a verification theorem for the fixed point Equation 5.35, which provides the optimal portfolio strategy in the feedback form.

$$\begin{cases} \mathfrak{I}w = w \\ \lim_{t \to T,\, x' \to x} w(t, x') = U(x) \end{cases} \tag{5.35}$$

where

$$\Im w = \sup_{\pi \in [0,1]} \int_t^T \int_{-1}^\infty \lambda(s) e^{-\int_t^s \lambda(u)\,du} w(s, x(1+\pi z)) p(t, s, dz)\,ds \tag{5.36}$$

Denote $V_0 = v^*(0, X_0)$. Then an optimal control $\hat{\alpha} \in A$ is given by

$$\hat{\alpha}_n = \hat{\pi}\left(\tau_n, \hat{X}_{\tau_n}\right) \hat{X}_{\tau_n}, \quad n \ge 0 \tag{5.37}$$

where $\hat{\pi}$ is a measurable function on $[0,T) \times (0,\infty)$ and solution to

$$\hat{\pi}(t, x) \in \arg\max_{\pi \in [0,1]} \int_t^T \int_{(-1,\infty)} \lambda(s) e^{-\int_t^s \lambda(u)\,du} v^*\left(s, X(1+\pi z)\right) P(t, s, dz)\,ds \tag{5.38}$$

and $\left(\hat{X}_{\tau_n}\right)_{n \ge 0}$ is the wealth given by the following recurrence equation

$$\hat{X}_{\tau_{n+1}} = \hat{X}_{\tau_n} + \tilde{\alpha}_n Z_{n+1}, \quad n \ge 0 \tag{5.39}$$

with its initial condition $\hat{X}_0 = X_0$.

An identical verification argument can be performed for an investor to start at time t with initial capital x. In this way, we prove that v^* is actually the value function of the control problem. In addition, it shows that the dynamic programming Equation 5.35 has a unique solution.

5.1.6 Optimal Time of Removing Quarantine Bans: Objectives with Infinite Integrals

Import restrictions on biological materials are used widely as protection against exotic invasive pests and pathogens (Anke and Donald, 2011). While a scientific risk assessment is needed to justify an import ban under the WTO SPS Agreement, economics plays little role in determining the choice of import regime. By using a real option framework, we model the uncertain and irreversible cost from lifting an import ban and derive decision rules about the optimal timing, when ex ante research and development yield positive but uncertain benefits. The insights gained are applied to Australia's current import policy for bananas.

Biological invasions by exotic species can cause significant damage to human health and the natural environment. Since the Quarantine Act 1908, Australia's default position on quarantine has been to ban imports of a biological material unless it is shown to be safe otherwise. This trade policy stance continues to create tension with Australia's trade partners in the World Trade Organization (WTO). Members of the WTO are bound by the rules defined by the Agreement on the Application of Sanitary and Phytosanitary Measures (hereafter, the SPS Agreement), which, in general, requires greater flexibility with respect to quarantine policy than an outright import ban. However, economics is allowed to play only a very limited role in defining a member's quarantine policy and it has been argued in Anderson et al. (2001) that this constraint creates a profound weakness in determining a member's SPS policy.

The problem of pursuing the optimal quarantine policy in the next instant can be modeled as a real options problem in continuous time. In this framework, the welfare consequences of a free trade policy are evaluated specifically against the risks and costs associated with importing exotic invasive pests and pathogens. It is assumed that the importing country is free of these commodity-specific pests and pathogens as long as a trade ban is maintained. The benefit of delaying the removal of import restrictions is linked to learning about the bioeconomic parameters of these invasive organisms and to the development of an adaptation strategy, such as an efficient control measure, which is designed to minimize invasion damages.

First, let us look at the objective function. Let $F(u,C(t,u))$ be the total discounted welfare from the production and consumption of an agricultural product in a single country. It is a function of the prevailing trade regime u and the total cost $C(t,u)$ associated with exotic species. The decision to lift the import ban in the next instant is based on maximizing the present value of the total net benefit

$$F(u,C(t,u)) = \max_u \int_0^\infty e^{-\rho dt} \left[(1+uv)N + (1-u\pi)P - C(t,u) \right] dt \qquad (5.40)$$

where u is a continuous control variable, satisfying $u \in [0,1]$.

A value of $u=0$ represents the status quo of a complete and unconditional import ban, whereas $u=1$ implies free trade, and $0<u<1$ is some intermediate trade policy, where imports are subject to specific regulations. The parameter $\rho>0$ in Equation 5.40 is the real rate of discount and N and P are the consumer and producer surpluses, respectively. Lower domestic prices, following the removal of a trade ban ($u>0$), benefit consumers by ($v>0$) and hurt producers by ($0 \le \pi \le 1$). The total cost $C(t,u)$ associated with exotic crop pests and pathogens varies with time (t) and the adopted trade regime (u).

It is assumed that the importing country conducts research into the development of an effective adaptation strategy while the import ban is in place. With a progressively open trade policy, expenditure on research and development is increasingly redirected toward another type of defensive expenditure, namely, the implementation of adaptation actions, such as monitoring for signs of domestic outbreak and applying control measures. As adaptive actions affect the extent but not the probability of biological invasions, the total cost $C(t,u)$ is the sum of two components

$$C(t,u) = \phi\Omega(t,u) + \Omega(t,u) \tag{5.41}$$

where $\phi\Omega(t,u)$ is the defensive expenditure with ϕ being an exogenously determined proportion of invasion damage $\Omega(t,u)$. Invasion damage is a stochastic dynamic variable, which depends on the adopted quarantine policy and defensive expenditure. Invasions can only occur, in the absence of invasions caused by travelers, with some degree of open trade; and potential invasion damage decreases when more time and financial resources are dedicated to R&D. Over time, the invasion damage $\Omega(t,u)$ is assumed to follow the following mixed geometric Brownian and Poisson process:

$$\frac{d\Omega}{\Omega} = -(1-u)\alpha\phi\,dt + (1-u)^{(1/2)}\phi^{-(1/2)}\sigma\,dz + u\phi^{-1}dq \tag{5.42}$$

where $0 < \alpha < 1$, and owing to the development of effective adaptation measures, σ is the standard variance, which is exogenous, the stochastic variable variance z is a standardized Wiener process satisfying that $dz \sim N(0,dt)$. The last term in Equation 5.42 models the damage from random invasion events as a decreasing function in proportion to R&D expenditure,

$$\frac{dC}{C} = -\left(1-u\right)\alpha\phi\,dt + \left(1-u\right)^{(1/2)}\phi^{-(1/2)}\sigma\,dz + u\phi^{-1}dq \tag{5.43}$$

The chosen trade policy u affects C in several ways. Under the current no-trade regime $(u = 0)$, the only costs are from bioeconomic research into the development of an efficient adaptation strategy to control known exotic pests or pathogens that could become established after trade bans are lifted. Equation 5.46 under a trade ban thus becomes

$$\left.\frac{dC}{C}\right|_{u=0} = -\alpha\phi\,dt + \phi^{-(1/2)}\sigma\,dz \tag{5.44}$$

At the other extreme case of free trade $(u = 1)$, the research expenditure on invasive species is redirected toward the implementation of the adaptation strategy that was developed while $u = 0$. The intuition is that unregulated free trade of a specific agricultural commodity requires confidence within the home country that potential outbreaks of exotic pests and pathogens, while still damaging, will be met by efficient strategic responses and adequate control options. Hence, under free trade, $C(t,u)$ is constant until the next random introduction and establishment of a new exotic pest and pathogen causes it to jump upward. The extent of the jump is the damage from the invasion of exotic species, which is a function of resources that were allocated to R&D expenditure before trade bans were lifted. That is, we have

$$\left.\frac{dC}{C}\right|_{u=1} = \phi^{-1} dq \tag{5.45}$$

Second, the constraint conditions are given as follows:

$$F(\infty) = 0 \tag{5.46}$$

$$F = f \tag{5.47}$$

$$F_C = f_C \tag{5.48}$$

Third, let us look at the solution. By setting $u* = 0$ in Equation 5.45, we obtain the ordinary differential equation

$$\rho F = N + P - C - \alpha\phi C F_C + \frac{1}{2}\phi^{-1}\sigma^2 C^2 F_{CC} \tag{5.49}$$

with the general solution

$$F = A_1 C^{\beta_1} + A_2 C^{\beta_2} \tag{5.50}$$

which yields the expression for $C(T)$. The level of C where lifting the import ban is optimal is given by

$$C(T) = \left(\frac{\beta}{\beta-1}\right)\frac{(\rho+\alpha\phi)(\rho\phi-\lambda)}{\lambda+\alpha\phi^2}\left(\frac{vN-\pi P}{\rho}\right) \tag{5.51}$$

which stands for the total cost associated with invasive species at the time when trade restrictions are lifted.

5.1.7 Risk Premium and Exchange Rates: Objectives with Utility Function

The adoption of a Taylor-type monetary policy rule and an inflation target for emerging market economies that choose a flexible exchange rate regime is often advocated (Choi, 2011). The model in this subsection describes the issue of exchange rate determination when interest-rate feedback rules are implemented in a continuous-time optimizing model of a small open economy facing an imperfect global capital market. It is demonstrated that when a risk premium on external debt affects the monetary policy transmission mechanism, the Taylor principle is not a necessary condition for the determinacy of equilibrium. On the other hand, it is shown that exchange rate dynamics critically depend on whether monetary policy is active or passive. In terms of optimal monetary policy, it is demonstrated that the degree of responsiveness of the nominal interest rate to inflation should be related to the stock of foreign debt. Specifically, it is optimal to implement a more passive monetary policy stance in response to larger levels of the outstanding debt denominated in foreign currencies.

Consider what could potentially occur in the framework of a small open economy in which the transmission mechanism of monetary policies is crucially affected by a risk premium on external debt. When a monetary policy is passive, an upward perturbation in inflation would cause the real interest rate to decrease and private consumption to increase, analogously to the case of a closed economy. However, the increase in consumption tends to stimulate foreign debt accumulation over time, hence leading to an increase in the country-specific risk premium. This makes the interest rate on foreign debt rise. Ceteris paribus, international parity conditions precluding arbitrage opportunities, requires an increase in the domestic real interest rate, which reduces aggregate demand and inflation. As a result, active monetary policies are not necessary to guarantee macroeconomic stability.

Despite the fact that saddle-path stability does not require an aggressive interest rate policy, the study of transitional dynamics we perform demonstrates that exchange rate adjustment in response to exogenous disturbances depends in a critical way on whether the monetary policy is active or passive. An increase in external debt brings about an increase in the country-specific risk premium and hence in the nominal interest rate. That is what faces the small open economy. As a consequence, the risk-adjusted interest rate parity condition requires an increase in the domestic nominal interest rate net of domestic currency depreciation. The domestic real interest rate must also be raised because of the purchasing power parity (PPP) condition. When a monetary policy is active (passive), this rise in the domestic real interest rate may take place if and only if there is an increase (decrease) in the exchange depreciation rate. This explains why exchange rate dynamics are crucially affected by no matter whether the monetary policy stance is active or passive.

Now, let us consider a small open economy that operates in a world of ongoing inflation and flexible exchange rates. The economy is described by a one good-monetary model and consists of four types of agents: consumers, firms, the government, and the central bank. All agents have perfect foresight, and time is continuous.

The domestic economy produces and consumes only one tradable and nonstorable good. Purchasing power parity (PPP) is assumed to hold at all times:

$$P = P^* E \qquad (5.52)$$

where
  $P(P^*)$ is the domestic (foreign) price
  E is the nominal exchange rate, defined as units of domestic currency per unit of foreign currency

In percentage terms, the PPP is given by

$$\pi = \pi^* + e \qquad (5.53)$$

where
  $\pi(\pi^*)$ is the inflation rate of the good in terms of the domestic (foreign) currency
  e is the rate of exchange depreciation of the domestic currency

Domestic residents may hold three assets: the domestic currency, domestic government bonds, and foreign assets. Domestic currency and government bonds are not held by foreigners. Foreign assets are internationally traded and are denominated in the foreign currency. However, the home country has no access to a perfect world capital market, but faces an upward-sloping supply curve of foreign debt, along the lines suggested by Turnovsky (1997). From the standpoint of the borrowing economy, denoting by f the level of real foreign debt and y the domestic output, the nominal interest rate on the foreign debt R^* can then be expressed as follows:

$$R^* = i^* + \sigma(f) \qquad (5.54)$$

where
  i^* is the interest rate prevailing in the world market
  $\sigma(f)$ is the country-specific risk premium

The function $\sigma(\cdot)$ is assumed to be continuous, increasing in f, and strictly positive.
  The mobility pf international capital implies that a risk-adjusted interest parity of the following type holds true:

$$R = R^* + e \qquad (5.55)$$

where R is the nominal rate of interest on bonds issued by the domestic government.
  The infinitely lived representative consumer faces the following lifetime utility function:

$$\int_0^\infty [U(c,l)+V(m)]e^{-\beta t}\,dt \tag{5.56}$$

where

β is the attenuation rate of time preferences

c, l, and m denote the consumption, labor, and real money balances, respectively

Then, the functions $U(\cdot)$ and $V(\cdot)$ satisfy the following conditions:

$$U_c>0,\quad U_l<0,\quad V'>0,\quad U_{cc}<0,\quad U_{ll}<0,\quad U_{cl}<0,\quad V''<0 \tag{5.57}$$

The flow budget constraint in real terms is

$$\dot{m}+\dot{b}+\dot{a}=wl+z-\tau-l+(R-\pi)b+(R^*-\pi^*)a-\pi m, \tag{5.58}$$

$$m(0)=\frac{M_0}{P(0)},\quad b(0)=\frac{B_0}{P(0)},\quad a(0)=\frac{A_0}{P^*} \tag{5.59}$$

where

b denotes government bonds

a denotes foreign assets

w denotes the wage rate

z denotes profits

τ denotes lump-sum taxes

Notice that, by definition, $a=-f$.

The representative agent chooses the optimal plan for c, l, m, b, and a in order to maximize his lifetime utility (5.56) subject to (5.58) given the initial conditions $m(0)$, $b(0)$, and $a(0)$.

Note that consumers take the rate at which the country can borrow from abroad as given in making their decisions. In other words, R^* is intended to be increasing in the aggregate level of foreign debt, which each consumer assumes he is unable to influence. The solution to the consumer's optimization problem yields the following conditions:

$$U_c(c,\ell)-\mu=0,$$

$$U_\ell(c,l)+w\mu=0,$$

$$V'(m)-\mu\pi=-\dot{\mu}+\mu\beta, \tag{5.60}$$

$$\mu(R-\pi)=-\dot{\mu}+\mu\beta,$$

$$\mu(R^*-\pi^*)=-\dot{\mu}+\mu\beta,$$

Together with the flow budget constraint (5.58), the initial conditions and the transversality conditions are

$$\lim_{t \to \infty} \mu m e^{-\beta t} = \lim_{t \to \infty} \mu b e^{-\beta t} = \lim_{t \to \infty} \mu a e^{-\beta t} = 0 \qquad (5.61)$$

where $\mu e^{-\beta t}$ is the discounted Lagrange multiplier associated with the wealth accumulation Equation 5.58. Perfectly competitive firms face a standard neoclassical production function of labor:

$$y = \lambda \phi(l), \quad \phi(\cdot) > 0, \quad \phi''(\cdot) < 0, \qquad (5.62)$$

where

  y denotes the output
  λ denotes a positive technology parameter

Each firm hires labor in order to maximize profits. At the optimum, the marginal productivity of labor is equal to the real wage rate $\lambda \phi'(l) = w$.

The domestic government faces the following flow budget constraint expressed in real terms:

$$\dot{m} + \dot{b} = g - \tau + (R - \pi)b - \pi m, \qquad (5.63)$$

where g is the government spending. The government is assumed to adopt a tax policy consisting in balancing the budget at all times:

$$\tau = g + (R - \pi)b - \pi m \qquad (5.64)$$

The monetary authorities set the nominal interest rate as an increasing function of the inflation rate, as in Benhabib et al. (2001):

$$R = i + \rho(\pi) \qquad (5.65)$$

where

  $\rho(\cdot)$ is continuous and nondecreasing, and there exists at least one $\bar{\pi} > -\beta$ such that $i + \rho(\bar{\pi}) = \beta + \bar{\pi}$
  i is a positive parameter capturing exogenous deviations from the feedback component of the rule

The dynamic equation describing the accumulation of net foreign assets is given by the trade balance plus interest payments

$$\dot{a} = y - c - g + (R^* - \pi^*)a \qquad (5.66)$$

Equation 5.66 can be rewritten in terms of net foreign debt accumulation as follows:

$$\dot{f} = c + g - y + (R^* - \pi^*)f \tag{5.67}$$

5.1.8 Endowment Risk and Monetary Policy: Objectives with Integrals of Utility Functions

Under normal circumstances, economic agents face uninsurable endowment risk when there is limited asset market participation. Because an initial money injection goes first to those who can access the asset market, money is nonneutral and monetary policy redistributes consumption across agents. Thus, monetary policy plays a risk-sharing role that provides crude insurance.

In this section, a limited participation model is constructed to study the risk-sharing role of monetary policy. This model is built on Choi (2011). There are two types of households: traders who participate in the asset market and nontraders who do not. These types are determined exogenously. In each time period, traders receive constant endowments, whereas nontraders face uninsurable endowment shocks. In the asset market, the government injects money through open market operations and traders initially receive the money injection.

In equilibrium, money is nonneutral and monetary policy redistributes consumption between traders and nontraders. The government money injection increases traders' consumption and decreases nontraders' consumption, because traders get the money injection through the asset market whereas nontraders suffer from inflation. If nontraders all receive the same endowment shock, then monetary policy is a perfect risk-sharing tool that can smooth out consumption across traders and nontraders. However, if nontraders receive idiosyncratic endowment shocks, then monetary policy is not enough to perfectly insure nontraders. Although monetary policy does not achieve a Pareto optimal allocation, it can mitigate the dispersion of consumption across nontraders. The optimal growth rate of money can be either positive or negative depending on the endowment distribution. The Friedman rule is not optimal in general.

We all know that monetary policies cannot smooth out the dispersion of consumption across nontraders, because it redistributes consumption as groups. Although any monetary policy does not provide perfect risk-sharing insurance, the government can determine the growth rate of money by maximizing welfare as an alternative:

$$W_t = \max_{\mu_t} \left[\alpha u(c_{r,t}) + (1 - \alpha) \int u(c_{i,t}) dF_t \right] \tag{5.68}$$

where

c represents the perishable consumption, u is the utility function, satisfying

$$c_{r,t}(\mu_t) = \{[y + (\mu_t Y_{t-1})/\alpha] / (1 + \mu_t)\}(Y_t / Y_{t-1}) \tag{5.69}$$

$$c_{i,t}(\mu_t) = [y_{i,t-1}/(1+\mu_t)](Y_t/Y_{t-1}) \tag{5.70}$$

Now, let us look for the constraint conditions.

The welfare reaches a maximum when the growth rate of money is

$$\hat{\mu}_t = \frac{\alpha}{Y_{t-1}}\left[\left(\frac{y_{t-1}^n}{\tilde{y}_{t-1}^n}\right)^{(1/\gamma)} - y\right]. \tag{5.71}$$

where $\tilde{y}_t^n = \int (y_{i,t})^{1-\gamma} dF_t$. Meanwhile, the consumption of traders and nontraders is, for all i,

$$\hat{c}_{r,t} = \frac{Y_t}{A_t}\cdot\left(\frac{(y_{t-1})^{1-\gamma}}{\tilde{y}_{t-1}^n}\right)^{(1/\gamma)} \tag{5.72}$$

$$\hat{c}_{i,t} = \frac{Y_t}{A_t}\cdot\left(\frac{y_{i,t-1}}{y_{t-1}^n}\right) \tag{5.73}$$

where

$$A_t = 1 + \alpha\left\{\left(\frac{y_{t-1}^n}{\tilde{y}_{t-1}^n}\right)^{(1/\gamma)} - 1\right\} \geq 1 \tag{5.74}$$

Clearly, traders and nontraders do not consume equally.

Monetary policies are effective to shuffle individual nontraders' consumption to some extent. The government determines not only the redistributional effects between traders' endowments and nontraders' endowments but also the individual effects on nontraders' consumption.

Next, $\hat{\mu}_t$ can be either positive or negative depending on the sign of $\left(y_{t-1}^n/\tilde{y}_{t-1}^n\right)^{1/\gamma} - y$. The Friedman rule is not optimal in general. If $\left(y_{t-1}^n/\tilde{y}_{t-1}^n\right)^{1/\gamma} > y$, then the government needs to inject money $\hat{\mu}_t > 0$ to redistribute consumption from nontraders to traders, because nontraders receive relatively larger endowments. On the other hand, $\hat{\mu}_t$ becomes negative if $\left(y_{t-1}^n/\tilde{y}_{t-1}^n\right)^{1/\gamma} < y$. The government extracts money to redistribute consumption from traders to nontraders.

Furthermore, when the economy is inefficient, by Jensen's inequality $\hat{\mu}_t \geq \mu_t^*$, the growth rate of money can be at the optimum level $\hat{\mu}_t = \mu_t^*$ and $\gamma = 1$, but it cannot still smooth out the consumption across nontraders with

$$\hat{c}_{i,t}(\mu_t) = \frac{y_{i,t-1}}{y_{t-1}^n}\cdot Y_t. \tag{5.75}$$

5.2 Optimal Asset Allocation: Continuous Objective Functions

After the liberalization of the capital account, national residents can allocate their assets globally. So, their investment profit can be denoted as

$$\max \quad J_c - \pi_{c_0} = \int_0^{+\infty} e^{-rt} [e^{-rt} \bar{R}(x(t)) - C(I_c(t))] dt \tag{5.76}$$

$$s.t. \quad \dot{x}(t) + \delta x(t) = I(t) \tag{5.77}$$

$$x(0) = x_0$$

where
  R is a differentiable function
  C is continuous function

Notice that $\bar{R}(K_c)$ is the revenue generated from K_c, which is the capital stock, and $I(K_c)$ represents the investment cost. Without loss of generality, in this segment, we consider the following simple situation:

$$\bar{R}(x(t)) = x(t)e^{Rt} \tag{5.78}$$

where R is a constant and denotes the rate of return $C(x(t)) = \varepsilon \cdot x(t)$, where ε is a constant and denotes the transaction costs. According to the three conditions of I, $X(t)$ has three different variations (see constraint (5.77)):

$$\because \dot{X}(t) + \delta X(t) = I(t)$$

$$\therefore X = X_0 + e^{-\delta t} \int_0^t e^{\delta \tau} I(\tau) d\tau = \begin{cases} X_1(t) & \text{(case 1)} \\ X_2(t) & \text{(case 2)} \\ X_3(t) & \text{(case 3)} \end{cases}$$

Investors have three different investment strategies corresponding to these different cases.

Strategy 1 When the stock market is in equilibrium, the total investment adds ΔI to the average amount I_0 or subtracts ΔI from it as the market fluctuating, namely,

$$I(t) = I_0 + \Delta I \sin \alpha t$$

where α represents approximate fluctuating cycle, and

$$X(t) = X_0 + \frac{I_0}{\delta}\left(1 - e^{-\delta t}\right) + \Delta I \cdot \frac{\delta \sin \alpha t - \alpha\left(\cos \alpha t - e^{-\delta t}\right)}{\delta^2 + \alpha^2} \qquad (5.79)$$

In the convergence condition of $0 < R < r$, $r + \delta > R$, we have

$$J_c - \pi_{c_0} = -\frac{x_0 + \dfrac{I_0}{\delta}}{R - r}e^{-rt} + \frac{\dfrac{I_0}{\delta} - \dfrac{\alpha \Delta I}{\delta^2 + \alpha^2}}{R - r - \delta}e^{-rt} - \frac{\varepsilon I_0}{r}$$

$$+ \frac{\alpha e^{-rt}\Delta I}{\delta^2 + \alpha^2}\left[\frac{\delta + (R - r)}{\alpha^2 + (R - r)^2}\right] - \varepsilon\Delta I \frac{\alpha}{\alpha^2 + r^2} \qquad (5.80)$$

$$\frac{\partial(J_c - \pi_{c_0})}{\partial \alpha} = -\frac{e^{-rt}\Delta I}{R - r - \delta}\frac{\delta^2 - \alpha^2}{(\delta^2 + \alpha^2)^2} - \varepsilon\Delta I \frac{r^2 - \alpha^2}{(\alpha^2 + r^2)^2}$$

$$+ (\delta + R - r)\Delta I e^{-rt}\frac{\left[\alpha^2 + (R - r)^2\right](\delta^2 - \alpha^2) - 2\alpha^2(\delta^2 + \alpha^2)}{(\delta^2 + \alpha^2)^2\left[\alpha^2 + (R - r)^2\right]^2}$$

For the sake of convenience, let us denote $\delta = r$. Then, it can be concluded from $\dfrac{\partial(J_c - \pi_{c_0})}{\partial \alpha} = 0$ that

$$-\frac{e^{-rt}}{R - 2r}(r^2 - \alpha^2) - \varepsilon(r^2 - \alpha^2) - \varepsilon(r^2 - \alpha^2)$$

$$+ Re^{-rt}\frac{\left[\alpha^2 + (R - r)^2\right](r^2 - \alpha^2) - 2\alpha^2(\delta^2 + \alpha^2)}{\left[\alpha^2 + (R - r)^2\right]^2} = 0$$

After some trial and error attempts, the equation $\alpha = r$ can be detracted.

By putting the aforementioned equation into the objective function, the following maximum value can be obtained:

$$\left(J_c - \pi_{c_0}\right)_{\max} = \frac{I_0 e^{-rt}}{2(R - 2r)r} - \frac{r x_0 + I_0}{(R - r)r}e^{-rt} + \frac{e^{-rt}\Delta I}{2r}\left(\frac{R}{r^2 + (R - r)^2} - \varepsilon e^{rt}\right) - \frac{\varepsilon I_0}{r} \qquad (5.81)$$

It demonstrates that the investment profit is at its maximum when $\alpha = r$ if the stock market is in equilibrium, namely, the approximate market fluctuating cycle is equal to the risk-free rate.

Strategy 2 When the stock market is bearish, the total investment amount is falling, decreased by ΔI, namely,

$$I(t) = I_0 - \Delta I(1 - e^{-\beta t})$$

where β is average selling speed, and we have

$$X(t) = X_0 + \frac{I_0 - \Delta I}{\delta}(1 - e^{-\delta t}) + \frac{\Delta I}{\delta - \beta}(e^{-\beta t} - e^{-\delta t}) \tag{5.82}$$

In the convergence condition of $r > R$, $r + \beta > R$, $r + \delta > R$,

$$J_c - \pi_{c_0} = -\frac{e^{-rt}}{R - r}\left(x_0 + \frac{I_0 - \Delta I}{\delta}\right) + \frac{e^{-rt}}{R - r - \delta}\left(\frac{\Delta I}{\delta - \beta} + \frac{I_0 - \Delta I}{\delta}\right)$$

$$- \frac{\Delta I \cdot e^{-rt}}{(\delta - \beta)(R - r - \beta)} - \varepsilon\left(\frac{I_0 - \Delta I}{r} + \frac{\Delta I}{r + \beta}\right) \tag{5.83}$$

$$\frac{\partial(J_c - \pi_{c_0})}{\partial \beta} = \frac{\Delta I e^{-rt}}{(R - r - \delta)(R - r - \beta)^2} + \frac{\varepsilon \Delta I}{(r + \beta)^2}$$

By letting $\dfrac{\partial(J_c - \pi_{c_0})}{\partial \beta} = 0$, we get

$$\beta = \frac{R\sqrt{e^{rt}\varepsilon(r + \delta - R)}}{1 + \sqrt{e^{rt}\varepsilon(r + \delta - R)}} - r \tag{5.84}$$

It demonstrates that the investment profit is at its maximum if (5.84) holds true when the stock market is bearish, namely, the average selling speed is equal to β in (5.84). Since

$$\frac{\partial\left(J_c - \pi_{c_0}\right)}{\partial \delta} = \frac{e^{-rt}(I_0 + \Delta I)}{\delta^2}\left[\frac{2\delta + r - R}{(R - r - \delta)^2} + \frac{1}{R - r}\right]$$

$$- \frac{e^{-rt}\Delta I}{(\delta - \beta)^2}\left[\frac{1}{R - r - \beta} + \frac{R - r - \beta}{(R - r - \delta)^2}\right]$$

from $\dfrac{\partial\left(J_c - \pi_{c_0}\right)}{\partial \delta} = 0$, we have the following equation:

$$(\delta - \beta)^2(I_0 + \Delta I)\left[\frac{2\delta - (R - r)}{(R - r - \delta)^2} + \frac{1}{R - r}\right] = \delta^2 \Delta I\left[\frac{R - r - \beta}{(R - r - \delta)^2} + \frac{1}{R - r - \beta}\right]$$

By considering the condition of $\beta \ll 1$, $\beta \approx 0$, the aforementioned equation can be converted to

$$(\delta - \beta)^2 (I_0 + \Delta I)[2\delta - (R - r)] = \delta^2 \Delta I[-\beta + (R - r)]$$

For the sake of convenience, let us denote $R - r = \tilde{R}$ and $I_0 + \Delta I = \tilde{I}$. Then the following equation can be derived:

$$\delta^3 + \delta^2 \left(\frac{\Delta I \beta - \tilde{R}\tilde{I} - 4\beta\tilde{I} - \Delta I \tilde{R}}{2\tilde{I}} \right) + \delta \left(\frac{2\beta^2 \tilde{I} + 2\beta\tilde{R}\tilde{I}}{2\tilde{I}} \right) - \frac{\beta^2 \tilde{R}}{2} = 0$$

By introducing $\dfrac{\Delta I \beta - \tilde{R}\tilde{I} - 4\beta\tilde{I} - \Delta I \tilde{R}}{2\tilde{I}} = P_1$, $\dfrac{2\beta^2 \tilde{I} + 2\beta\tilde{R}\tilde{I}}{2\tilde{I}} = P_2$, and $-\dfrac{\beta^2 \tilde{R}}{2} = P_3$, the aforementioned equation can be converted to the following cubic algebraic equation:

$$\delta^3 + P_1\delta^2 + P_2\delta + P_3 = 0,$$

By denoting $\bar{\delta} = (P_1/3) + \delta$, the aforementioned equation can be converted to the standard form of cubic algebraic equation

$$\bar{\delta}^3 + p\bar{\delta} + q = 0,$$

where $p = \left(P_1^2/3 \right) + P_2$ and $q = \left(2P_1^3/27 \right) - (P_1 P_2/3) + P_3$.

Let us introduce a function as follows

$$f(\bar{\delta}) = \bar{\delta}^3 + p\bar{\delta} + q \qquad (5.85)$$

According to the character of the algebraic Equation 5.85, the condition of the positive real root is

$$f(\bar{\delta}) < 0 \quad \text{and} \quad f'(\bar{\delta}) > 0$$

That is,

$$p = \left(P_1^2/3 \right) + P_2 \geq 0 \quad \text{and} \quad q = \left(2P_1^3/27 \right) - \left(P_1 P_2/3 \right) + P_3 < 0. \qquad (5.86)$$

Under the condition (5.86), the positive real root of cubic algebraic equation (5.85) can be denoted as

$$\bar{\delta} = \left[-\frac{q}{2} + \sqrt{\left(\frac{q}{2}\right)^2 + \left(\frac{p}{3}\right)^3} \right]^{1/3} + \left[-\frac{q}{2} - \sqrt{\left(\frac{q}{2}\right)^2 + \left(\frac{p}{3}\right)^3} \right]^{1/3}$$

Then by putting $\delta = \bar{\delta} - (P_1/3)$ into the objective function, we can solve its maximum value as follows:

$$(J_c - \pi_{c0})_{\max} = -\frac{e^{-rt}}{R-r}\left(x_0 + \frac{I_0 - \Delta I}{\delta - \frac{P_1}{3}}\right) + \frac{e^{-rt}}{\left(R-r-\bar{\delta}+\frac{P_1}{3}\right)}\left(\frac{\Delta I}{\delta - \frac{P_1}{3} - \beta} + \frac{I_0 - \Delta I}{\delta - \frac{P_1}{3}}\right)$$

$$-\frac{\Delta I \cdot e^{-rt}}{\left(\delta - \frac{P_1}{3} - \beta\right)(R-r-\beta)} - \varepsilon\left(\frac{I_0 - \Delta I}{r} + \frac{\Delta I}{r+\beta}\right) \tag{5.87}$$

At this point, the investment profit is optimal.

Strategy 3 When the stock market is bullish, the total investment amount is climbing, increased by ΔI, namely, we have

$$I(t) = I_0 + \Delta I(1 - e^{-\beta t})$$

$$X(t) = X_0 + \frac{I_0 + \Delta I}{\delta}(1 - e^{-\delta t}) - \frac{\Delta I}{\delta - \beta}(e^{-\beta t} - e^{-\delta t})$$

In the convergence condition of $r > R$, $r + \beta > R$, $r + \delta > R$, we have

$$J_c - \pi_{c0} = -\frac{e^{-rt}}{R-r}\left(x_0 + \frac{I_0 + \Delta I}{\delta}\right) + \frac{e^{-rt}}{R-r-\delta}\left(-\frac{\Delta I}{\delta - \beta} + \frac{I_0 + \Delta I}{\delta}\right)$$

$$-\frac{\Delta I \cdot e^{-rt}}{(\delta - \beta)(R-r-\beta)} - \varepsilon\left(\frac{I_0 + \Delta I}{r} - \frac{\Delta I}{r+\beta}\right) \tag{5.88}$$

$$\frac{\partial(J_c - \pi_{c0})}{\partial \beta} = \frac{\Delta I e^{-rt}}{(R-r-\delta)(R-r-\beta)^2} - \frac{\varepsilon \Delta I}{(r+\beta)^2} - \frac{2\Delta I e^{-rt}}{(R-r-\delta)(\delta - \beta)^2}$$

By solving β from $\frac{\partial(J_c - \pi_{c0})}{\partial \beta} = 0$, we get in a special case $R - r \approx \delta$ that

$$\beta = R - r \tag{5.89}$$

It demonstrates that the investment profit is at its maximum under the condition (5.89). When the stock market is bullish, namely, the average purchasing speed is equal to β in (5.89), we have

$$\frac{\partial(J_c - \pi_{c_0})}{\partial \delta} = \frac{e^{-rt}(I_0 - \Delta I)}{\delta^2}\left[\frac{2\delta + r - R}{(R - r - \delta)^2} + \frac{1}{R - r}\right]$$

$$+ \frac{e^{-rt}\Delta I}{(\delta - \beta)^2}\left[\frac{1}{R - r - \beta} + \frac{R - r - \beta}{(R - r - \delta)^2}\right]$$

From $\dfrac{\partial(J_c - \pi_{c_0})}{\partial \delta} = 0$, we can derive the following equation:

$$(\delta - \beta)^2(I_0 - \Delta I)\left[\frac{2\delta - (R - r)}{(R - r - \delta)^2} + \frac{1}{R - r}\right] = -\delta^2 \Delta I\left[\frac{R - r - \beta}{(R - r - \delta)^2} + \frac{1}{R - r - \beta}\right]$$

By considering the condition of $\beta \ll 1$ and $\beta \approx 0$, the aforementioned equation can be converted to

$$(\delta - \beta)^2(I_0 - \Delta I)[2\delta - (R - r)] = -\delta^2 \Delta I[-\beta + (R - r)]$$

For the sake of convenience, by denoting $R - r = \tilde{R}$ and $I_0 - \Delta I = \tilde{I}$, then the following equation can be derived:

$$\delta^3 + \lambda_1 \delta^2 + \lambda_2 \delta + \lambda_3 = 0 \tag{5.90}$$

where

$$\lambda_1 = \frac{\Delta I \tilde{R} - \tilde{R}\tilde{I} - 4\beta\tilde{I} - \Delta I\beta}{2\tilde{I}}, \quad \lambda_2 = \frac{2\beta^2\tilde{I} + 2\beta\tilde{R}\tilde{I}}{2\tilde{I}}, \quad \lambda_3 = -\frac{\beta^2\tilde{R}}{2}$$

In the same way as shown in Strategy 2, we can obtain the maximum value as follows:

$$(J_c - \pi_{c_0})_{max} = -\frac{e^{-rt}}{R - r}\left(x_0 + \frac{I_0 - \Delta I}{\delta}\right) + \frac{e^{-rt}}{R - r - \delta}\left(\frac{\Delta I}{\delta - \beta} + \frac{I_0 - \Delta I}{\delta}\right)$$

$$- \frac{\Delta I \cdot e^{-rt}}{(\delta - \beta)(R - r - \beta)} - \varepsilon\left(\frac{I_0 - \Delta I}{r} + \frac{\Delta I}{r + \beta}\right) \tag{5.91}$$

where δ is the positive root of cubic algebraic equation (5.90).

5.3 Verdier Equation: Differential Constraint Conditions

5.3.1 Introduction

Optimization models, as a considerably mature mathematical method, have their widespread applications in natural science, such as physics. In the field of applied economics, it is very common to build optimization models to research various problems about capital account liberation. Different constraint conditions have appeared when an optimization model is constructed to research the opening of capital account. Those relevant constraint conditions are listed in Table 5.1. Besides, Verdier equation also considers recursive constraints (Verdier, 2003).

Verdier (2003) considered the following optimization model:

$$\max_{\{C_t, Z_t, K_t\}} \sum \beta^t \cdot \frac{C_t^{1-\sigma} - 1}{1 - \sigma} \tag{5.92}$$

$$s.t. \, (1+n)(1+g)Z_{t+1} = (1-\delta)Z_t - C_t + \left[aK_t^{\rho} + (1-a)Z_t^{\rho} \right]^{\eta/\rho} \tag{5.93}$$

$$a\eta \left(\frac{d_t}{v_t} \right)^{\rho} \frac{y_t}{d_t} = r + \delta \tag{5.94}$$

This system can be rewritten as a single log-linear law of motion for Z_t. By taking a first-order Taylor log-linear approximation of Equation 5.94, we have

$$(\rho - 1)\widehat{d_t} + \left(\frac{\eta}{\rho} - 1 \right) a\rho \left(\frac{d^*}{v^*} \right)^{\rho} \widehat{d_t} + \left(\frac{\eta}{\rho} - 1 \right)(1-a)\rho \left(\frac{z^*}{v^*} \right)^{\rho} \widehat{z_t} = 0 \tag{5.95}$$

Table 5.1   Types of Optimization Models

Objective Function	Constraint Condition	Research Theme
Expectation with infinite integral	Differentiable form	Optimal portfolio
Infinite integral form	Integral form	Consumption model
Utility function form	Linear constraint condition	Tax decision model
Utility function form	Nonlinear constraint condition	Optimal tax model
Damage function form	Implicit constraint condition	Currency crisis model

where *hat* refers to the variable's log-deviation from its steady state denoted by a symbol *. This expression can be used to rewrite Equations 5.93 and 5.94 into a single log-linear law of motion in z, which has the following familiar solution:

$$\log z_t = \lambda' \log z_0 + (1 - \lambda') \log z^* \tag{5.96}$$

where $1 - \lambda$ denotes the speed of convergence. Since y, k, and d can be written as functions of z, Equation 5.96 implies that any change in net foreign debt takes the form

$$\log d_t - \log d_0 = -(1 - \lambda') \log d_0 + (1 - \lambda') \log d^*. \tag{5.97}$$

What is the function behind this result and what does it imply for the dynamics of debt and capital flows? To understand the intuition, it is useful to suppose $\rho = 0$, that is, the production technology is Cobb–Douglas. For now, this is both innocuous and useful for deriving the qualitative predictions of the model as well as estimating the reduced form. The model retains the same qualitative predictions while the intuition is made clearer. If $\rho = 0$, we have $Y_t = K_t^\alpha Z_t^\eta (\theta_t L_t)^{1-\alpha-\eta}$.

In that case, combining the credit constraint $(k_t = d_t)$, the small-open-economy assumption $(R_{kt} = r + \delta)$, and profit maximization implies that Equation 5.94 becomes

$$d_t = k_t = \frac{\alpha}{r + \delta} y_t \tag{5.98}$$

so that k/y has a constant path to the steady state. The production function can therefore be rewritten as $y_t = Bz_t^\varepsilon$ with $B = (\alpha/(r+\delta))^{\alpha/(1-\alpha)}$ and $\varepsilon = \eta/(1-\alpha)$. Given profit maximization and this collapsed production functions, Equation 5.93 is now becoming

$$(1+n)(1+g)Z_{t+1} = (1-\alpha)y_t - c_t + (1-\delta)Z_t \tag{5.99}$$

With the saving rate being a constant, equilibrium requires that domestic savings must equal the investment in domestic capital minus net factor payments on debt so that

$$(1+n)(1+g)Z_{t+1} = (1-\delta)Z_t + sBZ_t^\varepsilon \tag{5.100}$$

where

$$\varepsilon = \frac{\eta}{1-\alpha}, \quad B = \left(\frac{\alpha}{\gamma+\delta}\right)^{\alpha/(1-\alpha)}$$

Obviously, Equation 5.100 is a first-order nonlinear recursive equation. However, it generates a little question for us: We can only get the general solution of Equation 5.100, because it lacks an initial condition in order for us to find the particular solution.

To analyze it more readily, let us add an initial condition here:

$$Z\big|_{t=0} = Z_0$$

By denoting $a = (1-\delta)/((1+n)(1+g))$ and $b = sB/((1+n)(1+g))$, Equation 5.93 can be transformed into

$$\begin{cases} Z_{t+1} = aZ_t + bZ_t^\varepsilon \\ Z\big|_{t=0} = Z_0 \end{cases} \tag{5.101}$$

5.3.2 Solution of Verdier Equation

We refer Equation 5.94 to as Verdier equation and solve for its solution for some particular conditions

5.3.2.1 In the Case of $\varepsilon = 2$

Obviously, Verdier equation deteriorates into a geometric series of common ratio of $(a+b)$ when $\varepsilon = 1$, namely, $Z_{t+1} = (a+b)Z_t$. This condition can be ignored.

If $a = 0$, Equation set (5.101) can be represented as

$$Z_{t+1} = bZ_t^2 = Z_0^{2^{t+1}} b^{2^{t+1}-1} \tag{5.102}$$

If $a \neq 0$, let us take the following linear substitution for Equation set (5.101):

$$Z_t = mx_t + n$$

then the following can be concluded from Equation set (5.101):

$$mx_{t+1} = bm^2 x_t^2 + m(a + 2bn)x_t + n^2 b + an - n$$

Through a square transformation, it can be simplified into

$$m(x_{t+1} + l) = bm^2(x_t + l)^2 + [nb^2 + (a-1)n + ml - bm^2 l^2] \tag{5.103}$$

where $l = \dfrac{a + 2bn}{2bm}$.

Let

$$\begin{cases} nb^2 + (a-1)n + ml + bm^2l^2 = 0 \\ m = bm^2 \end{cases}$$

From the aforementioned equation set, m and n can be readily determined as follows:

$$\begin{cases} m = \dfrac{1}{b} \\ n = \dfrac{-l(1+l)}{ab - b + b^3} \end{cases}$$

Then Equation 5.103 can be transformed into

$$(x_{t+1} + l) = (x_t + l)^2$$

According to the general term formula of Equation 5.102, the aforementioned recursive form can be solved out as follows:

$$x_t = (x_0 + l)^{2^t} - l$$

By inserting the aforementioned expression into Z_t recursively, we obtain the following formula:

$$Z_t = \frac{1}{b} \left[(x_0 + l)^{2^t} - l \right] - \frac{l(1+l)}{ab - b + b^3} \tag{5.104}$$

Next, we try to seek for the maximum value of the objective function. By considering one of the special conditions: $(1-a)^{\eta/\rho} Z_t^\eta = \delta Z_t$, and $\eta = 1$, namely, $\delta = (1-a)^{1/\rho}$, and by denoting $(1+n)(1+g)b = \tilde{b}$, we have

$$C_t = (1-\delta)Z_t + \left[ak_t^\rho + (1-a)Z_t^\rho \right]^{\eta/\rho} - (1+n)(1+g)Z_{t+1}$$

$$= (1-\delta)Z_t + (1-a)^{\eta/\rho} Z_t^\eta - (1+n)(1+g)bZ_t^2$$

Therefore, $C_t = Z_t - \tilde{b}Z_t^2$ can be established. By putting C_t into the Verdier objective function, we obtain the following infinite series:

$$\sum_{t=0}^{\infty} \beta^t \cdot \frac{C_t^{1-\sigma}-1}{1-\sigma} = \sum_{t=0}^{\infty} \beta^t \cdot \frac{\left(Z_t - \tilde{b}Z_t^2\right)^{1-\sigma}-1}{1-\sigma}$$

In this section, we only consider the special case when $\sigma = 0$. So, the Verdier objective function can be rewritten as

$$\varphi = \sum_{t=0}^{\infty} \beta^t \left[\left(Z_t - \tilde{b}Z_t^2 \right) - 1 \right]$$

$$= \sum_{t=0}^{\infty} \beta^t Z_t - \tilde{b} \sum_{t=0}^{\infty} \beta^t Z_t^2 - \frac{1}{1-\beta} \tag{5.105}$$

Under the condition of $\beta < 1$ and $x_0 + l < 1$, when $t \to +\infty$, $\beta^t Z_t$, $\beta^t Z_t^2$ converge to zero at a significant speed. Hence, the main part of φ consists approximately of the sum of its first three items. For the sake of simplicity, we select l as the root of the following quadratic equation with one unknown:

$$l^2 + l + a - 1 + b^2 = 0$$

In this situation, Z_t takes the simplest form as follows:

$$Z_t = \frac{1}{b}(x_0 + l)^{2^t},$$

By denoting $a = x_0 + l$, we have $Z_t = (1/b)a^{2^t}$. Let us insert this result into the Verdier objective function φ, and consider the sum of the initial three terms of the infinite series:

$$\varphi_3 = \frac{1}{1-\beta} + \frac{a}{b} + \frac{1}{b}[\beta - (1+n)(1+g)]a^2 + \frac{1}{b}[\beta^2 - (1+n)(1+g)\beta]a^4$$

In order to seek for its maximum value, let

$$\frac{\partial \varphi}{\partial a} = \frac{1}{b} + \frac{2}{b}[\beta - (1+n)(1+g)]a + \frac{4}{b}[\beta^2 - (1+n)(1+g)\beta]a^3 = 0$$

that is,

$$a^3 + C_1 a + C_0 = 0 \qquad (5.106)$$

where

$$C_0 = \frac{1}{4\beta[\beta - (1+n)(1+g)]}, \quad C_1 = \frac{1}{2\beta},$$

For the sake of simplicity and clarity, we only discuss the case that the cubic algebraic equation (5.106) has a unique real root. By denoting this real root as $\hat{x}$, we have

$$\hat{x} = \left[-\frac{C_0}{2} + \sqrt{\left(\frac{C_0}{2}\right)^2 + \left(\frac{C_1}{3}\right)^3} \right]^{1/3} + \left[-\frac{C_0}{2} - \sqrt{\left(\frac{C_0}{2}\right)^2 + \left(\frac{C_1}{3}\right)^3} \right]^{1/3}$$

And, the condition for this case can be readily derived as follows:

$$C_0 < 0, \quad C_1 \geq 0, \quad \left(\frac{C_0}{2}\right)^2 + \left(\frac{C_1}{3}\right)^3 \geq 0.$$

Under this condition, the maximum value of (5.105) can be derived as follows:

$$\varphi_{\max} \approx \varphi_3 = \frac{x}{4b}(2m\hat{x}+3) \qquad (5.107)$$

where $m = \beta - (1+n)(1+g)$.

At this point, the liberalization of the capital account reaches its optimal condition under all the various constraints it faces.

5.3.2.2 In the Case of $\varepsilon = -1$

Under this condition, Equation 5.101 can be transformed into the following score iteration form:

$$Z_{t+1} = aZ_t + \frac{b}{Z_t} \qquad (5.108)$$

If a fixed point exists in the iteration Equation 5.108, then we denote it as $\overline{Z} = \lim_{t \to \infty} Z_t$. From Equation 5.108, the fixed point $\overline{Z}$ satisfies the following algebra equation:

$$\lambda = a\lambda + \frac{b}{\lambda}$$

If $a<1$, then Equation 5.108 has two fixed points (see Figure 5.1):

$$\overline{Z}_\pm = \pm\sqrt{\frac{b}{1-a}} \tag{5.109}$$

Let us select $\overline{Z}_+$, which contains some concrete economic signification, and omit $\overline{Z}_-$ from our discussion in this section. So, under the conditions $\eta = \rho = 1$, $\delta + (1+n) \times (1+g)a<1$, and $\dfrac{a\alpha}{r+\delta}B < (1+n)(1+g)b$, from the constraint condition (5.93), we obtain

$$C_t = AZ_t$$

where $A = 2 - \delta - (1+(1+n)(1+g)a)$.

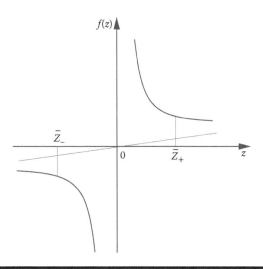

Figure 5.1   Fixed points of Equation 5.108.

Plugging $C_t = AZ_t$ into objective function (5.92), we can readily conclude the following:

$$\phi_{max} = \frac{A^{1-\sigma}\left(\dfrac{b}{1-a}\right)^{1-\sigma/2}}{1-\sigma} \cdot \frac{1}{1-\beta} \qquad (5.110)$$

5.3.2.3 In the Case of a≪1

In this case, we can readily conclude from (5.101) the following:

$$\ln Z_{t+1}\left(1 - a \cdot \frac{Z_t}{Z_{t+1}}\right) = \varepsilon \ln Z_t + \ln b \qquad (5.111)$$

If $|x|<1$, we have the following approximation (see Figure 5.2):

$$\ln(1-x) = -2\ln 2 - \ln x$$

Then we obtain the following first-order approximation:

$$\ln Z_{t+1} - 2\ln 2 - \ln\left(a \cdot \frac{Z_t}{Z_{t+1}}\right) = \varepsilon \ln Z_t + \ln b$$

Namely, $\ln Z_{t+1} - \dfrac{\varepsilon+1}{2}\ln Z_t = \ln 2\sqrt{ab}$.

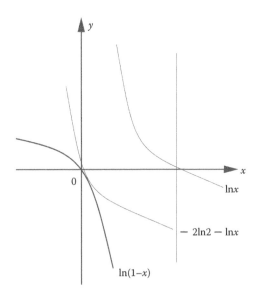

Figure 5.2   An approximation of ln(1−x).

By denoting $x_{t+1} = \ln Z_{t+1}$, we derive the following first-order recursive equation:

$$\begin{cases} x_{t+1} - \dfrac{1+\varepsilon}{2} x_t = \ln 2\sqrt{ab} \\ x\big|_{t=0} = x_0 \end{cases}$$

(5.112)

From elementary algebra, its analytic solution can be obtained after a simple translation transformation as follows:

$$x_t = \left(x_0 + \frac{2\ln 2\sqrt{ab}}{\varepsilon - 1} \right)\left(\frac{1+\varepsilon}{2} \right)^t - \frac{2\ln 2\sqrt{ab}}{\varepsilon - 1}$$

By inserting it back into the variable Z_t, we establish the following formula:

$$Z_t = \frac{\exp\left\{ \left(x_0 + \dfrac{2\ln 2\sqrt{ab}}{\varepsilon - 1} \right)\left(\dfrac{1+\varepsilon}{2} \right)^t \right\}}{(4ab)^{1/(\varepsilon - 1)}}$$

$$= \frac{Z_0^{\left(\frac{1+\varepsilon}{2} \right)^t}}{(4ab)^{\frac{1}{\varepsilon - 1}}} \cdot \exp\left\{ \frac{2\ln 2\sqrt{ab}}{\varepsilon - 1} \cdot \left(\frac{\varepsilon + 1}{2} \right)^t \right\}$$

(5.113)

By denoting

$$A_1 = 2 - a - \delta, \quad A_2 = \frac{\alpha a}{r + \delta} \cdot B, \quad A_3 = -2(1+n)(1+g)\sqrt{ab},$$

$$k_1 = -\left(x_0 + \frac{2\ln 2\sqrt{ab}}{\varepsilon - 1} \right), \quad k_2 = k_1\varepsilon, \quad k_3 = k_1\frac{1+\varepsilon}{2},$$

we rewrite the constraint condition as follows:

$$C_t = A_1 Z_t + A_2 Z_t^{\varepsilon} + A_3 Z_t^{(1+\varepsilon)/2}$$

(5.114)

By inserting (5.114) into the objective function (5.92), we can derive

$$\varphi = A_1 \sum_{t=0}^{\infty} e^{-k_1\left(\frac{1+\varepsilon}{2} \right)^t} \beta^t + A_2 \sum_{t=0}^{\infty} e^{-k_2\left(\frac{1+\varepsilon}{2} \right)^t} \beta^t + A_3 \sum_{t=0}^{\infty} e^{-k_3\left(\frac{1+\varepsilon}{2} \right)^t} \beta^t - \frac{1}{1-\beta}$$

Notice that the exponential function converges to zero very quickly. Therefore, we obtain an approximate maximum value of the objective function as follows:

$$\varphi_{max} \approx \frac{A_1}{e^{k_1}} + \frac{A_2}{e^{k_2}} + \frac{A_3}{e^{k_3}} + \beta \left(\frac{A_1}{e^{\frac{1+\varepsilon}{2}k_1}} + \frac{A_2}{e^{\frac{1+\varepsilon}{2}k_2}} + \frac{A_3}{e^{\frac{1+\varepsilon}{2}k_3}} \right) - \frac{1}{1-\beta} \qquad (5.115)$$

About the general optimal problem, we leave it as an exercise for the reader to solve.

5.4 Asset Pricing Based on Quadric Programming: Discrete Objective Function

5.4.1 Introduction

Harte and Newman (2014) presented the viewpoint of maximum entropy as follows: in order to deduce the distribution under the condition of insufficient information, we should take probability distribution that has the maximum entropy value and satisfies some restrict conditions. And, this strategy is the only agonic choice we can make, any other choice means that we have taken in other qualification or hypothesis, which cannot be made with our information.

Some scholars have been paying attention to asset pricing models, where multiple stationary equilibria are often encountered when one assumes negative-exponential utility with Gaussian uncertainty. Walker and Whiteman (2007) demonstrated that there are exactly two stationary equilibria, which are due solely to the presence of nonlinearity. By applying the theories of linear programming and super-hedging, Musiela and Rutkowsk (1997) gave a measure of pricing the finite (state) security market asset by applying two kinds of securities. Qin and Ying (2004) solved the asset pricing problem when the equivalent martingale measures are not unique in the finite (state) security market, and presented two methods to obtain the unique price of the asset under some specific conditions. By applying the duality principle of linear programming, martingale measure theory, and perfect hedging method, Qin and Ying (2002) considered a kind of calculation method for the seller's arbitrage price and the buyer's arbitrage price of any contingent claim in a finite security market. They studied the following two cases:

1. Allow short selling security, and borrowing and lending cash and
2. Do not allow short selling security, but allow borrowing and lending cash,

where arbitrage prices are obtained by solving corresponding linear programming problems in martingale measure spaces.

Recently, Ayadi and Kryzanowski (2008) used the asset-pricing kernel methodology to examine the sensitivity of various measures of portfolio performance. Alfarano et al. (2008) brought the herding framework into a simple equilibrium asset pricing model and derived closed-form solutions for the time-variation of high moments.

In this section, through applying martingale measure, maximum entropy, and nonlinear programming, we study the asset pricing problem for when the equivalent martingale measures are not unique in a finite (state) security market. And we put forward a method for obtaining the unique price of the asset under some specific conditions, which characterize the case most possibly to appear according to information theory.

5.4.2 Modeling

According to the hypothesis of our finite (state) security market, we therefore consider two moments: the first moment is $t=0$ and the second moment is $t=T=1$ (year). Suppose that there are n kinds of securities, namely, $s(1),\ldots,s(n)$ ($n\geq 1$ is a real number) in the financial market. Generally speaking, any security $s(i)$ appears in one of the two random states at moment $t=1$: one state is premium, the other is depremium, which are denoted as $s^u(i)$ and $s^d(i)(i=1,2,\ldots,n)$, respectively. In light of the combination theory, at moment $t=0$, we may anticipate that there are 2^n random states in the security market at moment $t=1$. Each set of random states in the security market responds to a combination of random states of the n securities, each of which is either premium or depremium at the moment $t=1$. It is obvious that $2^n > n+1$. Assume that the value of asset X is X_j in the jth state, the riskless interest rate is $r>0$, and the value of $s(i)$ at the jth state is $(s_j(i)\ i=1,2,\ldots,n; j=1,2,\ldots,2^n)$. At the moment $t=0$, the value of $s(i)$ is $s_0(i)(i=1,2,\ldots,n)$.

Accounting to the theory of entropy, at the moment $t=1$, the martingale of each state is assumed to be p_j ($j=1,2,\ldots,2^n$). That is, we have

$$s_0(i)(1+r) = \sum_{j=j_1,\ldots,j_{2^n-1}} s_j^u(i)p_j + \sum_{j=j_{2^{n-1}+1},\ldots,j_{2^n}} s_j^d(i)p_j \qquad (5.116)$$

$$p_j \geq 0 \, (j=1,2,\ldots,2^n) \qquad (5.117)$$

$$\sum_{j=1}^{2^n} p_j = 1 \qquad (5.118)$$

where the series $(j_1,\ldots,j_{2n})$ in Equation 5.116 represents a range of the natural number series $(1,\ldots,2^n)$, $(j_1,\ldots,j_{2^{n-1}})$ one of the subranges of the range $(j_1,\ldots,j_{2n})$.

We call the subrange $(j_1,\ldots,j_{2^{n-1}})$ an increasing range for the jth security, provided that the price of the jth security goes up responding to these kinds of 2^{n-1} random states of the security market, and denote the jth security as $s_j^u(i)$ $(j = j_1,\ldots,j_{2^{n-1}})$. And, a subrange $(j_{2^{n-1}+1},\ldots,j_{2^n})$ is called a decreasing range for the jth security, provided that the price of the jth security comes down responding to those kinds of 2^{n-1} random states of security market, and denote the jth security as $s_j^d(i)$ $(j = j_{2^{n-1}+1},\ldots,j_{2^n})$.

As there are 2^n unknown to be calculated for the equations (5.116) and (5.118) in the martingale measure, and the number $n+1$ of equations in (5.116) and (5.118) is smaller than 2^n, Equations 5.116 and 5.118 are made up of indeterminate equations. Hence, the pricing of the assets is not unique. Then, a natural question arises: which price is the most probable among all the possible prices? In another words, which martingale measure is the most probable one?

Note that if the most probable martingale measure is $\bar{p} = (\bar{p}_1, \bar{p}_2,\ldots,\bar{p}_{2^n})^{\mathrm{T}}$, then we can obtain the price of asset X at a kind of measure as follows:

$$c_0(X) = \sum_{j=1}^{2^n} \frac{\bar{p}_j}{1+r} X_j \tag{5.119}$$

which represents the most probable price of asset X. Assume that at the moment of $t=0$, the transcendental probability of each state at the moment of $t=1$ is known to us. We will discuss the asset pricing problem by applying the maximum entropy and nonlinear programming theory. Suppose that the transcendental probability of the security market at the moment of $t=1$ is p_j^0 $(j=1,2,\ldots,2^n)$. Then we introduce the following model:

$$(P_1)\begin{cases} \min \sum_{j=1}^{2^n} \left(p_j - p_j^0\right)^2 \\ \text{s.t.} \quad s_0(i)(1+r) = \sum_{j=j_1,\ldots,j_{2^{n-1}}} s_j^u(i)p_j + \sum_{j=j_{2^{n-1}+1},\ldots,j_{2^n}} s_j^d(i)p_j \\ p_j \geq 0\, (j=1,2,\ldots,2^n) \\ \sum_{j=1}^{2^n} p_j = 1 \end{cases}$$

In fact, according to the principles of maximum entropies, it is ready for us to obtain the following model under the condition that the state probability is known:

$$(P_2) \begin{cases} \min \sum_{j=1}^{2^n} p_j \dfrac{\log p_j}{p_j^0} \\ \text{s.t.} \quad s_0(i)(1+r) = \sum_{j=j_1,\ldots,j_{2^{n-1}}} s_j^u(i)p_j + \sum_{j=j_{2^{n-1}+1},\ldots,j_{2^n}} s_j^d(i)p_j \\ p_j \geq 0 \ (j=1,2,\ldots,2^n) \\ \sum_{j=1}^{2^n} p_j = 1 \end{cases}$$

Theorem 5.1 (P_2) is equivalent to (P_1).

Proof. It is sufficient to show that the object function of (P_2) is equivalent to that of (P_1), because they have the same constrain conditions.

When $p_j = p_j^0$ $(j=1,2,\ldots,2^n)$, it is obvious that $\sum_{j=1}^{2^n} \left(p_j - p_j^0\right)^2$ and $\sum_{j=1}^{2^n} p_j \log p_j / p_j^0$ are zero at the same time. So, by applying Taylor's formula for the series $\sum_{j=1}^{2^n} p_j \log p_j / p_j^0$ in a neighborhood of p_j^0 $(j=1,2,\ldots,2^n)$, we have

$$\sum_{j=1}^{2^n} p_j \log p_j / p_j^0$$

$$= \sum_{j=1}^{2^n} p_j \log[1-(1-p_j/p_j^0)]$$

$$= -\sum_{j=1}^{2^n} (p_j - p_j^0)(1 - p_j/p_j^0) - \sum_{j=1}^{2^n} p_j^0(1 - p_j/p_j^0)$$

$$\quad - \frac{1}{2} \sum_{j=1}^{2^n} \frac{p_j}{(p_j^0)^2} \cdot (p_j - p_j^0)^2 + o((p_j - p_j^0)^2)$$

$$= \sum_{j=1}^{2^n} \frac{1}{2p_j^0}(p_j - p_j^0)^2 + o((p_j - p_j^0)^2)$$

Hence, the object function of (P_2) is equivalent to that of (P_1).

Theorem 5.2 The solution set of (P_1) is convex.

Proof. First, we show that the solution set of (P_1) is nonempty.

Let $m = 2^n$ and Λ be the solution set of (P_1). Let us consider the following matrix

$$S = (s_j(i))_{n \times m} = \begin{pmatrix} s_1(1) & s_2(1) & \cdots & s_{2^n}(1) \\ s_1(2) & s_2(2) & \cdots & s_{2^n}(2) \\ & & \cdots & \\ s_1(n) & s_2(n) & \cdots & s_{2^n}(n) \end{pmatrix}$$

In the security market, it is impossible for two securities to behave exactly the same in all 2^n different states of the security market. That is, it implies that *rank* $S = n$.

Let us select two different elements along the same row in the matrix S, for instance, $s_1(1) \neq s_2(1)$. Then there exists a unique solution $\left(p_1^0, p_2^0 \right)$ to the following linear equations:

$$\begin{cases} s_1(1)p_1 + s_2(1)p_2 = s_0(1)(1 + r) \\ p_1 + p_2 = 1 \end{cases}$$

Therefore, the vector $p^0 = (p_1^0, p_2^0, 0, \ldots, 0)$ is one of the nonzero elements of the solutions set. Hence, we have shown that $\Lambda \neq \phi$.

Second, we show that Λ is convex.

Suppose that $p^{(1)} = \left(p_1^{(1)}, p_2^{(1)}, \ldots, p_{2^n}^{(1)} \right)$ and $p^{(2)} = \left(p_1^{(2)}, p_2^{(2)}, \ldots, p_{2^n}^{(2)} \right)$ are two solutions of (P_1). Take an arbitrary point $p^{(t)}$ on the straight line between points $p^{(1)}$ and $p^{(2)}$. That is, we have

$$p^{(t)} = (1 - t)p^{(1)} + tp^{(2)}, \quad 0 < t < 1$$

By substituting $p^{(t)}$ into Equation 5.116, we get

$$[(1 - t) + t]s_0(i)(1 + r) = \sum_{j = j_1, \ldots, j_{2^n - 1}} s_j^u(i)\left[(1 - t)p_j^{(1)} + tp_j^{(2)} \right]$$

$$+ \sum_{j = j_{2^n - 1} + 1, \ldots, j_{2^n}} s_j^d(i)\left[(1 - t)p_j^{(1)} + tp_j^{(2)} \right]$$

It means that $p^{(t)}$ satisfies Equation 5.116. Obviously, $p^{(t)}$ also satisfies Equations 5.117 and 5.118. Hence, we have shown that $p^{(t)}$ is an element of Λ, implying that Λ is convex.

5.4.3 Solutions of (P₁)

Theorem 5.3 The optimal solution $\bar{p}_j (j = 1,\ldots,2^n)$ of (P_1) satisfies the following equations:

$$\bar{p}_j = -\frac{1}{2}\sum_{i=1}^{n} \lambda_i s_j(i) - \frac{1}{2}\lambda_{n+1} + p_j^0 \tag{5.120}$$

$$s_0(i)(1+r) = \sum_{j=j_1,\ldots,j_{2^{n-1}}} s_j^u(i)\bar{p}_j + \sum_{j=j_{2^{n-1}+1},\ldots,j_{2^n}} s_j^d(i)\bar{p}_j \tag{5.121}$$

$$\sum_{j=1}^{2^n} \bar{p}_j = 1 \tag{5.122}$$

where $\lambda_i (i = 1,2,\ldots,n+1)$ are parameters that need to be confirmed.

Proof. Let us take the Lagrange function

$$L(p,\lambda,\mu) = -\sum_{j=1}^{2^n} \left(p_j - p_j^0 \right)^2$$

$$+ \sum_{i=1}^{n} \lambda_i \left[s_0(i)(1+r) - \sum_{j=j_1,\ldots,j_{2^{n-1}}} s_j^u(i)p_j - \sum_{j=j_{2^{n-1}+1},\ldots,j_{2^n}} s_j^d(i)p_j \right]$$

$$+ \lambda_{n+1}\left(\sum_{j=1}^{2^n} p_j - 1 \right) + \sum_{j=1}^{2^n} \mu_j p_j$$

where $\lambda_i (i = 1,2,\ldots,n+1)$ and $\mu_j \geq 0$ $(j = 1,2,\ldots,2^n)$ are parameters. Any optimal solution of (P_1) must satisfy following conditions:

$$\frac{\partial L}{\partial p_j} = 0 \quad (j = 1,2,\ldots,2^n) \tag{5.123}$$

$$\frac{\partial L}{\partial \lambda_i} = 0 \quad (i = 1,2,\ldots,n+1) \tag{5.124}$$

$$\mu_j \geq 0 \quad (j = 1, 2, ..., 2^n) \tag{5.125}$$

$$\mu_j p_j = 0 \quad (j = 1, 2, ..., 2^n) \tag{5.126}$$

From (5.123), we can derive

$$2\left(\bar{p}_j - p_j^0\right) + \sum_{i=1}^{n} \lambda_i s_j(i) + \lambda_{n+1} + \mu_j = 0$$

It is obvious that $p_j > 0$ ($j = 1, 2, ..., 2^n$). Therefore, from (5.125) and (5.126), we know that $\mu_j = 0$ ($j = 1, 2, ..., 2^n$). Hence, (5.120) holds true. As $\sum_{j=1}^{2^n} \left(p_j - p_j^0\right)^2$ is a concave function about variables p_j ($j = 1, ..., 2^n$), (P_1) is a problem of concave programming, and the results listed in (5.120) through (5.122) are optimal solutions of (P_1).

From Theorems 5.2 and 5.3, we know that we can obtain the martingale measure that is to be calculated by solving these $n + 1$ equations, which contain $n + 1$ parameters λ_i ($i = 1, 2, ..., n + 1$). And the optimal solutions can be obtained in the light of numerical method of nonlinear equations.

5.4.4 Example

Assume that at moment $t = 0$, the prices of securities $s(1)$, $s(2)$ are all 1 and the riskless interest rate is $r = 5\%$, and at moment $t = 0$, we anticipate that there are four market probability states. Respectively, $s(1)$, $s(2)$, and the values of the asset X at different states are given in Table 5.2.

Assume that the transcendental probability of each state at the moment $t = 1$ is $p_0 = (0.20, 0.30, 0.30, 0.20)$. Then, by using the software package MATLAB®, the optimal solution is obtained as follows:

$$\bar{p} = (\bar{p}_1, \bar{p}_2, \bar{p}_3, \bar{p}_4) = (0.32, 0.24, 0.18, 0.20).$$

The value of the asset X at moment $t = 0$ is 0.97.

Table 5.2 Values of Asset X at Different States at Moment $t = 1$

	State 1	State 2	State 3	State 4
$s(1)$	1.20	1.20	0.80	0.80
$s(2)$	1.30	0.70	1.30	0.70
X	1.25	0.90	1.10	0.75

5.4.5 Conclusions

The duality principle of linear programming, together with the martingale measure theory and the perfect hedging method, can help us present a kind of calculation method for the seller's and the buyer's arbitrage prices of any contingent claims in a finite security market. By applying martingale measure theory, maximum entropy theory, and nonlinear programming, we study the asset pricing problem when the equivalent martingale measures are not unique in the finite (state) security market. We then present a method for obtaining the unique price of the asset under some specific conditions, which characterize such a case that would most possibly appear according to information theory. Our main results are outlined in the following paragraphs:

1. The price of an asset is not unique if market equipollence martingale measure is not unique. Even so, for investors, it is beneficial and necessary to make sure which price is the most probable one to appear. Here, we studied this problem by applying martingale measure theory, maximum entropy theory, and nonlinear programming.
2. This section only studied the situation of discrete states. Even so, from the process of reasoning, we learned that our method is also suitable for the situation of continuous states.
3. Theoretically, it seems that we can hedge and price an asset by using all of the market securities. However, in reality, it is very difficult to use all of the market securities to hedge or price an asset because of the constraints of trading costs, terms, and liquidity. The number of future states is possibly very large. So it is of practical and academic significance to study the pricing of probable claim on the condition that the martingale measure is not unique.

5.5 Abnormal Flows of Capital: Discrete Constraint Conditions

5.5.1 Introduction

Most studies of international capital account liberalization and crises link economic performance to either the net inflow of capital or the gross inflow (outflow) defined as the change in foreign (domestic) holdings of domestic (foreign) assets over a time period. Edison and Warnock (2008) investigated the impact of two types of capital account liberalizations on short- and long-horizon capital flows to emerging markets in such a framework that controls for push and pull factors. The first type of liberalization, a reduction in capital controls, is countrywide but uncertain, because its extent and permanence are not known with certainty. The second type, a cross-border listing, is a firm-level liberalization that involves no uncertainty. They found that a deterministic cross-listing results in an immediate but short-lived increase in capital inflows. Thorsten and Daniel (2013) decomposed the net inflow into four rather than two components, and showed that four-way decompositions can be more informative than both the net capital inflow and two-way decompositions of the net inflow.

The cross-border flows of abnormal capital have become hot and difficult issues that have attracted close attention in China's economic operation. Since the beginning of February 2010, SAFE (State Administration of Foreign Exchange) has organized a special operation to deal with and to combat illegal foreign hot money flows in 13 provinces, totaling 3,470,000 cross-border transactions and U.S. $440 billion in the cumulative amount. As of the end of July 2010, 190 cases of violation were confirmed, involving as much money as U.S. $7.35 billion, which accounts for 1.67% of the total amount of investigation. The results of the special operation confirm the fact that beneath cross-border transactions of legal business capital flows, the related illegal activities have indeed posted a great risk to the economic stability of China. This indicates that China needs to further strengthen the monitoring and analysis of abnormal cross-border capital flows, and improve its ability to protect itself against international financial risks. According to the data of RMB credit receipts and payments of financial institutions, as announced by the People's Bank of China in November 2010, the growth of the monetary base for the month of November due to the expansion of funds outstanding for foreign exchanges amounted to ¥319.643 billion. That was the second highest point of the year, only below ¥519.047 billion reached in October.

According to the most recent balance sheet data of the monetary authority from the People's Bank of China, as of the end of December 2013, the foreign currency balance of the central bank amounted to ¥26.427 trillion, indicating a growth of ¥223.342 billion over the end of November and a continuous growth of the sixth month. And, the foreign currency balance of financial institutions amounted to ¥28.630383 trillion, indicating an increase of ¥272.9 billion over the month of November. For the year of 2013, the financial institutions' funds outstanding for foreign exchange increased ¥2.777 trillion, almost unchanged from a year ago, and equal to 5.6 times of that of 2012. Accompanying the growth in the funds outstanding for foreign exchange is the significant 3% appreciation of the RMB against the U.S. dollar in 2013. The data, released by the Bank for International Settlements, shows that the real effective exchange rate of RMB is 118.79 in December 2013, indicating an annual appreciation of 7.89%. All these data imply that the pressure of cross-border capital net inflow increases in 2013, while the capital flows remain volatile.

What is found in practice is that since China still maintains a modest level of capital control, most of the abnormal capital flows are facilitated under the guise of legality of multinational corporations. That makes the abnormal cross-border capital flows of China have the following significant characteristics:

- In terms of the form, the way of flow resembles the kind of movement of *ants*, involving the characteristics of multiple points and penetration. As of this writing, no organized and large-scale inflows were found.
- In terms of the flow direction, after entering China and passing the settlement, most foreign capital inflows play the role of supporting economic development by investing in the real economy, while others go into real estates, the markets of securities, national debts, and RMB deposits market.
- In terms of the purpose of the inflows, it has been mainly for long-term investment. Under the premise of long-term high growth of China's economy, the exchange rate and interest rate of RMB provide an opportunity for long-term arbitrage. That is why abnormal cross-border capitals have not been moving in and out of China quickly for fast profit. Instead, they focus on long-term benefits.

5.5.2 Model

The main channels of abnormal cross-border capital flows under the name of Trade in Goods are the following:

First, loopholes existing in trade credit management have been taken advantage of. The operating mechanism of this channel is that advance payments are not accompanied by the corresponding flows of goods: The relevant companies first make the exchange earnings, and then go through the formalities of exchange settlement on paper for verifying export earnings with the customs' inspection after declaring export of goods. Companies with RMB payment needs but temporarily unable to provide the original verification only need those companies that receive foreign payments to provide proofs.

Second, loopholes in the cash flows of foreign direct investments (FDI) are taken advantage of. Most of the FDI make exchange settlements through authorized banks in the name of capital. The capital inflow of a foreign investment is only required to register. The simple registration procedure provides a convenient channel for foreign capital to enter China for the purpose of chasing after short-term interests.

Third, loopholes of transfer pricing of related transactions have been taken advantage of. The operating mechanism of this channel is that the relevant companies earn more foreign exchange with less payment through the method of reporting higher quoted prices of exports and lower quoted prices of imports. Because the general trade transactions between associated subsidiary corporations are priced by the parent company, the parent company would transfer the interest through adjustments of the prices of the products that are imported and exported.

Fourth, loopholes of the gray area in supervision are taken advantage of. A notable feature of re-exports is that although there are no actual products to make import and export declarations, there are still full payments in foreign exchange funds. That is, flows of capital and goods are completely uncorrelated. The authenticity of the trade background is difficult to determine. So, it is easy for the phenomena of receiving without expending or expending more or expending less to occur. And without going through the online verification of foreign exchange, one can enter into the foreign currency account of the current account and handle the settlement by showing import–export contracts and other information. Currently, the re-export business has become a convenient opportunity of exchange arbitrage for domestic and foreign-affiliated companies. Additionally, since the aforementioned opportunity of arbitrage appears between domestic and foreign affiliates, it possesses the characteristics of high liquidity, large scales of capital, and frequent transactions, which provide the convenience for facilitating abnormal flows of capital.

Our investigative study found that some companies use such operation mechanism to accommodate illegal capital inflows: Under the name of processing business with imported materials, the marketing, pricing, collection methods, and deadlines of the domestic subsidiaries are controlled by the parent company overseas. The subsidiary corporations usually employ the method of deferred payments or lower payments or even no payment when importing, and the method of full receipts or advanced receipts when exporting. They make use of the time difference between receipts and payments to help the foreign capital to stay in China and to move into fixed-term savings or into various investment fields in order to maximize the profit. Because of the specifics of multinational companies, the authenticity of their trade backgrounds in general is difficult to verify. Multinational affiliated companies could accommodate cross-border capital inflows by using methods such as buying low and selling high, fabricated trades, etc., and remit funds through methods such as transfer pricing, returning profit, etc., in order to enjoy the benefit of the RMB appreciation. These companies make use of the time difference between receipts and payments to make as much foreign capital as possible stay in the importing country in order to maximize the return.

Based on the previous discussion, we define the revenue function $H(r)$ that a company helps to accommodate foreign capital stay in the importing country as follows. Suppose that this company has N export proceeds in monitoring period $0 \leq t \leq T$, the points of time are $0 = t_0 < t_1 < t_2 < \ldots < t_N = T$, and the corresponding receipt amount is $c(t_i)$ at each point of time. Meanwhile, there are M importing payments with the points of time being $0 = \tau_0 < \tau_1 < \tau_2 < \ldots < \tau_M = T$, and the corresponding payment amount is $c(\tau_i)$ at each point in time. If this company wants to maximize the amount of foreign capital that stay in the importing country, then the following two principles need to be followed.

Principle 1 (Monotonicity principle) The exchange earnings cycle needs to be as short as possible and the payment cycle as long as possible. That is, the time series $\Delta t_i = t_{i+1} - t_i$ has an upward trend, and $\Delta \tau_i = \tau_{i+1} - \tau_i$ has a downward trend.

We can easily make use of the monotonicity of 5-day average system of these two time series to judge whether the monotonicity principle is established or not.

Principle 2 (Maximization principle) The arbitrage company enjoys the maximum benefits. In particular, it means that the following revenue function reaches the extreme along with constraints.

$$\max_{\substack{t_{j(i)}\in\{t_0,t_1,\cdots,t_N\} \\ \tau_{j(i)}\in\{\tau_0,\tau_1,\cdots,\tau_N\}}} H(r) = \sum_{i=1}^{M} c(\tau_i)e^{r(T-\tau_{j(i)})} - \sum_{i=1}^{N} c(t_i)e^{r(T-t_{j(i)})}$$

$$s.t. \sum_{i=1}^{M} c(\tau_i) - \sum_{i=1}^{N} c(t_i) = c \geq 0 \qquad (5.127)$$

where

r is the fixed income rate of the importing country in the time period of $0 \leq t \leq T$

$j(i)$ stands for a new permutation different from the natural orderings of $12...i...M$ or $12...i...N$

The constant c in the previous optimization model (5.127) is relatively small. The arbitrage company tries to let c as close to zero as possible in order to avoid excesses between receipts and payments so that the arbitrage behavior will not be found by the monitoring authority. The restrictive condition can be understood as legality constraints.

Denote

$$\min_{1\leq i\leq M}\{T-\tau_{j(i)}\} = \Delta\tau_1 \leq \Delta\tau_2 \leq \cdots \leq \Delta\tau_M = \max_{1\leq i\leq M}\{T-\tau_{j(i)}\},$$

$$\min_{1\leq i\leq M}\{c(\tau_i)\} = \Delta c(\tau_1) \leq \Delta c(\tau_2) \leq \cdots \leq \Delta c(\tau_M) = \max_{1\leq i\leq M}\{c(\tau_i)\},$$

$$\min_{1\leq i\leq N}\{T-t_{j(i)}\} = \Delta t_1 \leq \Delta t_2 \leq \cdots \leq \Delta t_N = \max_{1\leq i\leq N}\{T-t_{j(i)}\},$$

$$\min_{1\leq i\leq N}\{c(t_i)\} = \Delta c(t_1) \leq \Delta c(t_2) \leq \cdots \leq \Delta c(t_N) = \max_{1\leq i\leq N}\{c(t_i)\}.$$

According to the optimization theory, we obtain the optimal solution for model (5.127) as follows:

$$H_{max} = \sum_{i=1}^{M} c_i^+ e^{r\Delta\tau_i} - \sum_{i=1}^{N} c_i^- e^{r\Delta t_{N-i}} \qquad (5.128)$$

In light of the approximate formula $e^x \approx 1 + x$ (x is sufficient small), we have

$$H_{max} = \sum_{i=1}^{M} c(\tau_i) - \sum_{i=1}^{N} c(t_i) + \sum_{i=1}^{M} c_i^+ \cdot r\Delta\tau_i - \sum_{i=1}^{N} c_i^- \cdot r\Delta t_{N-i}$$

$$\approx c + \left(\sum_{i=1}^{M} c_i^+ \cdot \Delta\tau_i - \sum_{i=1}^{N} c_i^- \cdot \Delta t_{N-i} \right) r \tag{5.129}$$

5.5.3 Visualization

Recently, there has been a large amount of research in the area of reasoning, cognition, and perception that can be applied to the framework of visual analytics and to the investigation of complex financial problems. Much of the empirical literature deals with early-warning models that rely on conventional statistical methods, such as the single variable signals approach or multivariate logit/probit models. However, the phenomenon of abnormal capital flows represents a complex event driven by nonlinearly related financial factors. These nonlinearities derive, for example, from the fact that abnormal capital flows become more likely as the number of fragilities increases. Potentially due to restrictive assumptions, for example, on linearity or error distributions, conventional statistical techniques may fail in modeling these events. Given the changing nature of the occurrences of these extreme events, stand-alone numerical analyses are unlikely to be able to comprehensively describe these extreme events. As a complement, this challenge motivates the development of innovative tools with clear visual capabilities and intuitive interpretability, enabling an amplification of understanding through the biological perceptions of man.

Data reduction techniques can provide summaries of data by compressing information, while dimensionality reduction (or projection) provides low-dimensional representations of similarity relations in data. The self-organizing map (SOM) (Kohonen, 2001) holds the promise of conquering the difficulty by combining the aims of data and dimensionality reduction. It is capable of providing an easily interpretable nonlinear description of the multidimensional data distribution on a 2D plane without losing sight of individual indicators. The 2D output of the SOM makes it particularly useful for visualizations, or summarizations, of large amounts of information. By 2005, over 7700 pieces of research works had featured the SOM. While extensively applied to topics in engineering and medicine, the literature has been short of thorough testing of the SOM for financial stability surveillance. In the emerging market context, Sarlin and Marghescu (2011) have applied the SOM to indicators of general economic and financial performance. The SOM has not, to the best of our knowledge, been earlier applied to monitoring systemic risk or assessing the global dimensions of financial stability, including global macrofinancial proxies as well as individual advanced and emerging market economies.

Inspired by the vindicated financial fragility view of a creditor asset cycle of Minsky (1982) and Kindleberger (1996), the SOFSM introduces the notion of financial stability cycles. Thus, the SOFSM can be used to monitor macrofinancial vulnerabilities by locating a country in the financial stability cycle: be it either in the precrisis, crisis, postcrisis, or tranquil state. We illustrate how the SOFSM can be used for visualizing samples of the panel dataset, that is, cross-sectional data over time on a country level, as well as different levels of aggregation, such as emerging and advanced economies. Likewise, we illustrate how the SOFSM can be used for visualizing scenario analysis results by showing how positive and negative shocks on domestic and global levels affect the location of the euro area in the financial stability cycle.

Standard Euclidean batch SOM algorithm is used frequently, in which data are processed in batches rather than sequentially and distances are compared using the Euclidean metric. Because of the importance of convergence when using the batch SOM, Kohonen (2001) modified the SOM in the following three steps:

Step 1: Determine two eigenvectors v_1 and v_2 with the largest eigenvalue from the covariance matrix of all data.

Step 2: Let v_1 and v_2 span a 2D linear subspace and fit a rectangular array along it.

Step 3: Identify the initial value of the reference vectors $m_i(0)$ with the array points.

5.5.4 Numerical Example

Suppose that the situation about the earnings cycle ($M = 5$) and the payment cycle ($N = 3$) of a commercial company during a trade year is shown in Table 5.3.

It is straightforward to obtain the revenue series for this commercial company between two arbitrary neighboring time points that we denote $\{\nabla c(t_i)\}_{i=1}^{M+N}$, where

$$\nabla c(t_i) = c(t_{i+1})e^{r(T-t_{i+1})} - c(t_i)e^{r(T-t_i)}.$$

Another revenue series according to optimization model (5.127) is denoted as $\{\nabla \hat{c}(t_i)\}_{i=1}^{M+N}$.

Table 5.3 Earnings and Payment Situation of a Commercial Company

τ_i	0	0.1	0.3	0.4	0.5	0.7	0.9	1
$c(\tau_i)$	−225	610	−83	95	−243	−130	−320	295

Note: $c(t_i) > 0$ means payment and $c(t_i) < 0$ means earnings.

According to the SOM and Kohonen's modified approach in the Section 5.4.3, we select 2D vectors in the aforementioned two revenue series. After some simple calculation, we obtain the following two covariance matrix:

$$\Sigma = \begin{pmatrix} 7519.1572 & -3193.4520 \\ -3193.4520 & 6506.6572 \end{pmatrix}, \quad \hat{\Sigma} = \begin{pmatrix} 7337.8932 & -2420.2290 \\ -2420.2290 & 7200.1132 \end{pmatrix}$$

The largest and the smallest eigenvectors according to these two covariance matrix are shown as follows:

$$v_1 = (-0.7605 \quad 0.6494)^T, \quad v_2 = (-0.6494 \quad -0.7605)^T$$

$$\hat{v}_1 = (-0.7171 \quad 0.6970)^T, \quad \hat{v}_2 = (-0.6970 \quad -0.7171)^T.$$

After normalization, all these vectors can be represented in the 2D plane as in Figure 5.3.

After finding the eigenvectors corresponding to the largest eigenvalue, the region nearby the eigenvectors can be defined as suspicious region on the basis of this eigenvalue. We can properly divide the region between the largest eigenvalue and the smallest eigenvalue in the unit circle and dye them with different colors so that we will be able to judge whether a cash flow of the company is abnormal or not during the period T by using a visualization technology.

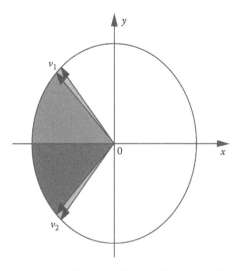

Figure 5.3   Eigenvectors in the unit circle.

5.5.5 *Conclusions*

An abnormal flow of cross-border capital has a significant negative impact on the receiving country's economy. When the direction, scale, and structure of capital flows deviate from the needs of the development of the real economy, the capital flows are prone to have negative effects on the economic development and financial system of capital inflow country, leading to increased currency appreciation pressure, deteriorating terms of trade, and influencing macroeconomic stability. Especially the *churning* of short-term capitals tend to seriously affect the economic and financial security in emerging economies, which would lead to rapid expansion of financial assets, and increase the fragility of the banking system and the volatility of the stock market, triggering a systemic risk or even resulting in regional or global financial crises. Currently, international speculative capital or illegal funds usually flow into the territory of an emerging economy through the channel of current account. These funds often avoid the regulatory policy through misrepresenting import and export prices, public welfare donations, direct investments, foreign investment profit remittances, re-export business, and so on, in order to achieve the goal of inflowing under a legal cloak. After making quick profits, speculative capitals will outflow in the same way or with the help of underground banks, which encourage and facilitate illegal foreign exchange transactions, thereby disturbing the operating order of the domestic foreign exchange market. This section summarizes the main channels of abnormal cross-border capital flows under the trade in goods and two principles, that is, the monotonicity principle and the maximization principle, which are followed by such international trade companies that are motivated to engage in abnormal flows of cross-border capital. This section also establishes an optimization model that those international trade companies that participate in abnormal flows of cross-border capital employ to obtain additional revenue through introducing the concept of revenue functions. Then we obtain the optimal solution of the model by using the optimization method, and provide a theoretical basis for detecting an abnormal flow of cross-border capital in virtue of visualization technology.

Chapter 6

From Underground Economics to Financial Contagion: Regressions for Capital Account Liberation

6.1 General Methods of Regression Analysis

In this chapter, we investigate the finite-sample properties of least-squares applied to a random sample in the linear regression model. In particular, we calculate the finite-sample mean and covariance matrix and propose standard errors for the coefficient estimates.

6.1.1 Sample Mean

To start with the simplest setting, let us consider the intercept-only model

$$y_i = \mu + e_i$$

$$E(e_i) = 0$$

which is equivalent to the regression model with $k = 1$ and $x_i = 1$. In the intercept model, $\mu = E(y_i)$ is the mean of y_i. And, the least-squares estimator $\hat{\mu} = \bar{y}$ is equal to the sample mean.

We now calculate the mean and variance of the estimator $\bar{y}$. Since the sample mean is a linear function of the observations, its expectation is simple to calculate as follows:

$$E(\bar{y}) = E\left(\frac{1}{n}\sum_{i=1}^{n} y_i\right) = \frac{1}{n}\sum_{i=1}^{n} Ey_i$$

This shows that the expected value of the least-squares estimator (the sample mean) is equal to the projection coefficient (the population mean). An estimator with the property that its expectation is equal to the parameter it is estimating is called unbiased.

Definition 6.1 *An estimator $\hat{\theta}$ for θ is unbiased if $E\hat{\theta} = \theta$.*

We next calculate the variance of the estimator $\bar{y}$. By making the substitution $y_i = \mu + e_i$, we have

$$\bar{y} - \mu = \frac{1}{n}\sum_{i=1}^{n} e_i$$

and

$$\mathrm{var}(\bar{y}) = E(\bar{y} - \mu)^2 = \frac{1}{n^2}\sum_{i=1}^{n}\sum_{j=1}^{n} E(e_i e_j) = \frac{1}{n}\sigma^2$$

The second to the last equality is due to the facts that $E(e_i e_j) = \sigma^2$ for $i = j$ yet $E(e_i e_j) = 0$ for $i \neq j$ because of independence.

We have shown that $\mathrm{var}(\bar{y}) = (1/n)\sigma^2$. This is the familiar formula for the variance of the sample mean.

6.1.2 Linear Regression Model

Let us now consider the linear regression model. Throughout the remainder of this chapter, we maintain the following.

Assumption 6.1 (Linear Regression Model) The available observations (y_i, x_i) come from a random sample and satisfy the linear regression equation

$$y_i = x_i'\beta + e_i \tag{6.1}$$

$$E(e_i \mid x_i) = 0 \qquad\qquad (6.2)$$

The variables have finite second moments

$$E y_i^2 < \infty$$

$$E \|x_i\|^2 < \infty,$$

and an invertible design matrix

$$Q_{xx} = E(x_i x_i') > 0.$$

We will consider both the general case of heteroskedastic regression, where the conditional variance

$$E\left(e_i^2 \mid x_i\right) = \sigma^2(x_i) = \sigma_i^2$$

is unrestricted, and the specialized case of homoskedastic regression, where the conditional variance is constant. In the latter case, we add the following assumption.

Assumption 6.2 (Homoskedastic Linear Regression Model) In addition to Assumption 6.1, we assume

$$E\left(e_i^2 \mid x_i\right) = \sigma^2(x_i) = \sigma_i^2 \qquad\qquad (6.3)$$

is independent of x_i.

6.1.3 Mean of Least-Squares Estimator

In this section, we show that the ordinary-least squares (OLS) estimator is unbiased in the linear regression model. The relevant calculation can be done by using either summation notation or matrix notation. And we will use both.

First let us take summation notation. Observe that under (6.1) and (6.2), we have

$$E(y_i \mid X) = E(y_i \mid x_i) = x_i'\beta \qquad\qquad (6.4)$$

The first equality states that the conditional expectation of y_i given $\{x_1,\dots,x_n\}$ only depends on x_i, since the observations are independent across i. The second equality is the assumption of a linear conditional mean.

We have the linearity of expectations, (6.4), and properties of the matrix inverse as follows:

$$E(\hat{\beta}|X) = E\left[\left(\sum_{i=1}^{n} x_i x_i'\right)^{-1} \left(\sum_{i=1}^{n} x_i y_i\right) \middle| X\right]$$

$$= \left(\sum_{i=1}^{n} x_i x_i'\right)^{-1} E\left[\left(\sum_{i=1}^{n} x_i y_i\right) \middle| X\right]$$

$$= \left(\sum_{i=1}^{n} x_i x_i'\right)^{-1} \sum_{i=1}^{n} E(x_i y_i \mid X)$$

$$= \left(\sum_{i=1}^{n} x_i x_i'\right)^{-1} \sum_{i=1}^{n} x_i E(y_i \mid X)$$

$$= \left(\sum_{i=1}^{n} x_i x_i'\right)^{-1} \sum_{i=1}^{n} x_i x_i' \beta$$

$$= \beta$$

Now let us show the same result by using the matrix notation. Equation 6.4 implies

$$E(y \mid X) = \left(\begin{array}{c} \vdots \\ E(y_i|X) \\ \vdots \end{array}\right) = \left(\begin{array}{c} \vdots \\ x_i'\beta \\ \vdots \end{array}\right) = X\beta \tag{6.5}$$

Similarly, we have

$$E(e \mid X) = \left(\begin{array}{c} \vdots \\ E(e_i \mid X) \\ \vdots \end{array}\right) = \left(\begin{array}{c} \vdots \\ E(e_i \mid x_i) \\ \vdots \end{array}\right) = 0 \tag{6.6}$$

Similarly, the linearity of expectations, (6.5), and the properties of the matrix inverse imply

$$E(\hat{\beta} \mid X) = E((X'X)^{-1} X'y \mid X) = (X'X)^{-1} X'X\beta = \beta$$

At the risk of belaboring the derivation, another way to calculate the same result is as follows:

$$\hat{\beta} = (X'X)^{-1}(X'(X\beta + e)) = \beta + (X'X)^{-1} X'e \tag{6.7}$$

This is a useful linear decomposition of the estimator $\hat{\beta}$ into the true parameter β and the stochastic component $(X'X)^{-1}X'e$. Once again, we can calculate that

$$E(\hat{\beta}-\beta \mid X) = E((X'X)^{-1}X'e \mid X)$$
$$= (X'X)^{-1}X'E(e \mid X)$$
$$= 0$$

Regardless of the method, we have shown that $E(\hat{\beta} \mid X) = \beta$. By applying the law of iterated expectations, we find that

$$E(\hat{\beta}) = E(E(\hat{\beta} \mid X)) = \beta.$$

So, we have shown the following result.

Theorem 6.1 (Mean of Least-Squares Estimator) In the linear regression model (Assumption 6.1), the following hold true:

$$E(\hat{\beta} \mid X) = \beta \tag{6.8}$$

and

$$E(\hat{\beta}) = \beta \tag{6.9}$$

Equation 6.9 says that the estimator $\hat{\beta}$ is unbiased for β, meaning that the distribution of $\hat{\beta}$ is centered at β. Equation 6.8 says that the estimator is conditionally unbiased, which is a stronger result than the previous one. It says that $\hat{\beta}$ is unbiased for any realization of the regression matrix X.

6.1.4 Variance of Least-Squares Estimator

In this section, we calculate the conditional variance of the OLS estimator. To this end, for any $r \times 1$ random vector Z, let us define the $r \times r$ covariance matrix

$$\text{var}(Z) = E(Z - EZ)(Z - EZ)'$$
$$= EZZ' - (EZ)(EZ)'$$

and for any pair (Z,X) let us define the conditional covariance matrix

$$V_{\hat{\beta}} \overset{def}{=} \text{var}(\hat{\beta} \mid X)$$

of the regression coefficient estimates. Let us now derive its form.

The conditional covariance matrix of the $n \times 1$ regression error e is the $n \times n$ matrix

$$D = E(ee' \mid X)$$

The ith diagonal element of D is

$$E\left(e_i^2 \mid X\right) = E\left(e_i^2 \mid x_i\right) = \sigma_i^2$$

while the ijth off-diagonal element of D is

$$E(e_i e_j \mid X) = E(e_i \mid x_i)E(e_j \mid x_j) = 0$$

where the first equality uses independence of the observations and the second is (6.2). Thus, D is a diagonal matrix with the ith diagonal element σ_i^2:

$$D = diag\left(\sigma_1^2, \ldots, \sigma_n^2\right) = \begin{pmatrix} \sigma_1^2 & 0 & \cdots & 0 \\ 0 & \sigma_2^2 & \cdots & 0 \\ \vdots & \vdots & \ddots & \vdots \\ 0 & 0 & \cdots & \sigma_n^2 \end{pmatrix} \tag{6.10}$$

In the special case of the linear homoskedastic regression model (6.3), we have

$$E\left(e_i^2 \mid x_i\right) = \sigma_i^2 = \sigma^2$$

and the simplification

$$D = I_n \sigma^2$$

In general, however, D need not necessarily take this simplified form.

For any $n \times r$ matrix $A = A(X)$, define

$$var(A'y \mid X) = var(A'e \mid X) = A'DA \tag{6.11}$$

In particular, we can write $\hat{\beta} = A'y$ when $A = X(X'X)^{-1}$, and thus we have

$$V_{\hat{\beta}} = var(\hat{\beta} \mid X) = A'DA = (X'X)^{-1} X'DX(X'X)^{-1}$$

It is useful to note that the following weighted version of $X'X$:

$$X'DX = \sum_{i=1}^{n} x_i x_i' \sigma_i^2,$$

In the special case of the linear homoskedastic regression model, $D = I_n \sigma^2$ so that $X'DX = X'X\sigma^2$, and the variance matrix is simplified to

$$V_{\hat{\beta}} = (X'X)^{-1} \sigma^2$$

Theorem 6.2 (Variance of Least-Squares Estimator) In the linear regression model (Assumption 6.1), the following holds true:

$$V_{\hat{\beta}} = \text{var}(\hat{\beta} \,|\, X) = (X'X)^{-1} (X'DX)(X'X)^{-1} \qquad (6.12)$$

And, in the homoskedastic linear regression model (Assumption 6.2), the following holds true:

$$V_{\hat{\beta}} = (X'X)^{-1} \sigma^2.$$

6.1.5 Gauss–Markov Theorem

Now let us consider the class of estimators of β each of which is a linear function of the vector y, and thus can be written as

$$\tilde{\beta} = A'y$$

where A is an $n \times k$ function of X. As noted before, the least-squares estimator is the special case obtained by setting $A = X(X'X)^{-1}$. As for the question, what is the best choice of A, the Gauss–Markov theorem, which we now present, says that the least-squares estimator is the best choice among linear unbiased estimators when the errors are homoskedastic in the sense that the least-squares estimator has the smallest variance among all unbiased linear estimators.

To see this end, from $E(y|X) = X\beta$, it follows that for any linear estimator $\tilde{\beta} = A'y$, we have

$$E(\tilde{\beta} \,|\, X) = A'E(y\,|\,X) = A'X\beta,$$

so $\tilde{\beta}$ is unbiased if (and only if) $A'X = I_k$. Furthermore, we can conclude from (6.11) that

$$\mathrm{var}(\tilde{\beta} \mid X) = \mathrm{var}(A'y \mid X) = A'DA = A'A\sigma^2$$

where the last equality is obtained by using the homoskedasticity assumption $D = I_n\sigma^2$. The *best* unbiased linear estimator is obtained by finding a matrix A_0 satisfying $A_0'X = I_k$ such that $A_0'A_0$ is minimized in the positive definite sense in that for any other matrix A satisfying $A'X = I_k$, the expression $A'A - A_0'A_0$ is positive semidefinite.

Theorem 6.3 (Gauss–Markov) Firstly, in the homoskedastic linear regression model, the best (minimum-variance) unbiased linear estimator is the least-squares estimator

$$\tilde{\beta} = (X'X)^{-1}X'y$$

And, secondly, in the linear regression model, the best unbiased linear estimator is

$$\tilde{\beta} = (X'D^{-1}X)^{-1}X'D^{-1}y \tag{6.13}$$

We give a proof of the first part of the following theorem, and leave the proof of the second part for readers as an exercise.

Proof. Let A be an arbitrary $n \times k$ function of X such that $A'X = I_k$. The variance of the least-squares estimator is $(X'X)^{-1}\sigma^2$ and that of $A'y$ is $A'A\sigma^2$. It is sufficient to show that the difference $A'A - (X'X)^{-1}$ is positive semidefinite. To this end, let us set $C = A - X(X'X)^{-1}$. Since $X'C = 0$, we do the following calculation:

$$A'A - (X'X)^{-1} = (C + X(X'X)^{-1})'(C + X(X'X)^{-1}) - (X'X)^{-1}$$

$$= C'C + C'X(X'X)^{-1} + (X'X)^{-1}X'C + (X'X)^{-1}X'X(X'X)^{-1} - (X'X)^{-1}$$

$$= C'C$$

The matrix $C'C$ is indeed positive semidefinite as required.

The first part of the Gauss–Markov theorem is a limited efficiency justification for the least-squares estimator. The justification is limited because the class of models is restricted to homoskedastic linear regression and the class of potential estimators is restricted to linear unbiased estimators. This latter restriction is particularly unsatisfactory as the theorem leaves open the possibility that a nonlinear or biased estimator could have lower mean squared error than the least-squares estimator.

The second part of the theorem shows that in the (heteroskedastic) linear regression model, within the class of linear unbiased estimators, the best estimator is not least-squares but is (6.13). This is called the generalized least-squares (GLS) estimator. The GLS estimator is infeasible as the matrix D is unknown. So, this result does not suggest a practical alternative to least squares.

6.1.6 Residuals

In this section, we discuss in the context of the linear regression mode properties of the residuals $\hat{e}_i = y_i - x_i'\hat{\beta}$ and prediction errors $\tilde{e}_i = y_i - x_i'\hat{\beta}_{(-i)}$.

Notice that we can write the residuals in the vector notation as

$$\hat{e} = Me$$

where $M = I_n - X(X'X)^{-1}X'$ is the orthogonal projection matrix. By using the properties of conditional expectation, we have

$$E(\hat{e}|X) = E(Me|X) = ME(e|X) = 0$$

and

$$\text{var}(\hat{e}|X) = \text{var}(Me|X) = M\,\text{var}(e|X)M = MDM \qquad (6.14)$$

where D is defined in (6.10).

We can simplify this expression under the assumption of conditional homoskedasticity

$$E\left(\hat{e}_i^2\,\middle|\,x_i\right) = \sigma^2$$

In this case, (6.14) can be simplified to

$$\text{var}(\hat{e}|X) = M\sigma^2 \qquad (6.15)$$

In particular, for a single observation i, we can find the (conditional) variance of $\hat{e}_i$ by taking the ith diagonal element of (6.14). Since the ith diagonal element of M is $1-h_{ii}$, we obtain

$$\text{var}(\hat{e}_i|X) = E\left(\hat{e}_i^2\,\middle|\,X\right) = (1-h_{ii})\sigma^2 \qquad (6.16)$$

As this variance is a function of h_{ii} and hence x_i, the residuals $\hat{e}_i$ are heteroskedastic even if the errors e_i are homoskedastic.

Similarly, prediction errors $\tilde{e}_i = (1 - h_{ii})^{-1} \hat{e}_i$ can be written in the vector notation as $\tilde{e} = M * \hat{e}$, where $M*$ is a diagonal matrix with the ith diagonal element being $(1 - h_{ii})^{-1}$. Thus, $\tilde{e} = M * Me$. We can calculate that

$$E(\tilde{e} \mid X) = M * ME(e \mid X) = 0$$

and

$$\text{var}(\tilde{e} \mid X) = M * M \,\text{var}(e \mid X)MM* = M * MDMM *$$

which can be simplified under homoskedasticity to

$$\text{var}(\tilde{e} \mid X) = M * MMM * \sigma^2$$

$$= M * MM * \sigma^2$$

The variance of the ith prediction error is then

$$\text{var}(\tilde{e}_i \mid X) = E\left(\tilde{e}_i^2 \mid X\right)$$

$$= (1 - h_{ii})^{-1}(1 - h_{ii})(1 - h_{ii})^{-1}\sigma^2$$

$$= (1 - h_{ii})^{-1}\sigma^2.$$

A residual with constant conditional variance can be obtained by rescaling. The standardized residuals are

$$\bar{e}_i = (1 - h_{ii})^{-1/2}\hat{e}_i \tag{6.17}$$

and in the vector notation

$$\bar{e} = (\bar{e}_1, \ldots, \bar{e}_n)' = M^{*1/2} Me.$$

From our above calculations, under homoskedasticity, we have

$$\text{var}(\bar{e} \mid X) = M^{*1/2} MM^{*1/2} \sigma^2$$

and

$$\text{var}(\bar{e}_i \mid X) = E\left(\bar{e}_i^2 \mid X\right) = \sigma^2 \tag{6.18}$$

and thus, these standardized residuals have the same bias and variance as the original errors when the latter are homoskedastic.

6.1.7 Estimation of Error Variance

The error variance $\sigma^2 = Ee_i^2$ can be a parameter of interest, even in a heteroskedastic regression or a projection model. This variance σ^2 measures the variation in the *unexplained* part of the regression. Its method of moment estimator (MME) is the sample average of the squared residuals:

$$\hat{\sigma}^2 = \frac{1}{n}\sum_{i=1}^{n}\hat{e}_i^2$$

In the linear regression model, we can calculate the mean of $\hat{\sigma}^2$. In light of the properties of projection matrices and the trace operator, we can observe that

$$\hat{\sigma}^2 = \frac{1}{n}\hat{e}'\hat{e} = \frac{1}{n}e'MMe = \frac{1}{n}e'Me = \frac{1}{n}tr(e'Me) = \frac{1}{n}tr(Mee')$$

so that we have

$$E(\hat{\sigma}^2 \mid X) = \frac{1}{n}tr(E(Mee' \mid X)) = \frac{1}{n}tr(MD) \tag{6.19}$$

By adding the assumption of conditional homoskedasticity $E\left(e_i^2 \mid x_i\right) = \sigma^2$, we have $D = I_n\sigma^2$ so that (6.19) is simplified to

$$E(\hat{\sigma}^2 \mid X) = \frac{1}{n}tr(M\sigma^2)$$

$$= \sigma^2\left(\frac{n-k}{n}\right),$$

This calculation shows that $\hat{\sigma}^2$ is biased toward zero. The order of the bias depends on k/n, the ratio of the number of estimated coefficients over the sample size.

Another way to see this is to use (6.16). Note that

$$E(\hat{\sigma}^2 \mid X) = \frac{1}{n}\sum_{i=1}^{n}E\left(e_i^2 \mid X\right) = \frac{1}{n}\sum_{i=1}^{n}(1 - h_{ii})\sigma^2 = \left(\frac{n-k}{n}\right)\sigma^2 \tag{6.20}$$

Since the bias takes a scale form, a classic method to obtain an unbiased estimator is by rescaling the estimator. So let us define

$$s^2 = \frac{1}{n-k}\sum_{i=1}^{n}\hat{e}_i^2. \tag{6.21}$$

From the aforementioned calculation, we have

$$E(s^2 \mid X) = \sigma^2 \qquad (6.22)$$

Hence,

$$E(s^2) = \sigma^2$$

and the estimator s^2 is unbiased for σ^2. Consequently, s^2 is known as the *bias-corrected estimator* for σ^2, and in empirical practice, s^2 is the most widely used estimator for σ^2.

Interestingly, this is not the only method to construct an unbiased estimator for σ^2. An estimator constructed with the standardized residuals $\bar{e}_i$ from (6.17) is

$$\bar{\sigma}^2 = \frac{1}{n}\sum_{i=1}^{n}\bar{e}_i^2 = \frac{1}{n}\sum_{i=1}^{n}(1-h_{ii})^{-1}\hat{e}_i^2. \qquad (6.23)$$

It is straightforward to show that

$$E(\bar{\sigma}^2 \mid X) = \sigma^2 \qquad (6.24)$$

and thus, $\bar{\sigma}^2$ is unbiased for σ^2 (in the homoskedastic linear regression model).

When k/n is small, the estimators $\hat{\sigma}^2$, s^2, and $\bar{\sigma}^2$ are likely to be close. However, if not then s^2 and $\bar{\sigma}^2$ are generally preferred to $\hat{\sigma}^2$. Consequently, it is best to use one of the bias-corrected variance estimators in applications.

6.1.8 Mean-Square Forecast Error

A major purpose of estimated regressions is to predict out-of-sample values. Consider an out-of-sample observation (y_{n+1}, x_{n+1}) where x_{n+1} will be observed but not y_{n+1}. Given the coefficient estimate $\hat{\beta}$, the standard point estimate of $E(y_{n+1} \mid x_{n+1}) = x'_{n+1}\beta$ is $\tilde{y}_{n+1} = x'_{n+1}\hat{\beta}$. The forecast error is the difference between the actual value y_{n+1} and the point forecast, $\hat{e}_{n+1} = y_{n+1} - \tilde{y}_{n+1}$. The mean-squared forecast error (*MSFE*) is

$$MSFE_n = E\tilde{e}_{n+1}^2.$$

In the linear regression model, $\tilde{e}_{n+1} = e_{n+1} - x'_{n+1}(\hat{\beta}-\beta)$, so we have

$$MSFE_n = Ee_{n+1}^2 - 2E(e_{n+1}x'_{n+1}(\hat{\beta}-\beta)) + E(x'_{n+1}(\hat{\beta}-\beta)(\hat{\beta}-\beta)'x_{n+1}) \qquad (6.25)$$

The first term in (6.25) is σ^2. The second term in (6.25) is zero since $e_{n+1}x'_{n+1}$ is independent of $\hat{\beta} - \beta$ and both are mean zero. The third term in (6.25) is

$$tr(E(x_{n+1}x'_{n+1})E(\hat{\beta}-\beta)(\hat{\beta}-\beta)') = tr(E(x_{n+1}x'_{n+1})E(\hat{\beta}-\beta)(\hat{\beta}-\beta)')$$

$$= tr(E(x_{n+1}x'_{n+1})EV_{\hat{\beta}})$$

$$= Etr((x_{n+1}x'_{n+1})V_{\hat{\beta}})$$

$$= E(x'_{n+1}V_{\hat{\beta}}x_{n+1}) \qquad (6.26)$$

where we use the fact that x_{n+1} is independent of $\hat{\beta}$ and the fact that $V_{\hat{\beta}} = E((\hat{\beta}-\beta)(\hat{\beta}-\beta)' \mid X)$.

Thus, we obtain

$$MSFE_n = \sigma^2 + E(x'_{n+1}V_{\hat{\beta}}x_{n+1}).$$

Under conditional homoskedasticity, this equation is simplified to

$$MSFE_n = \sigma^2(1 + E(x'_{n+1}(X'X)^{-1}x_{n+1})).$$

If $\tilde{e}_i = y_i - x'_i\hat{\beta}_{(-i)} = \hat{e}_i(1-h_{ii})^{-1}$, then a simple estimator for the *MSFE* can be calculated as follows:

$$E\tilde{\sigma}^2 = E\tilde{e}_i^2$$

$$= E(e_i - x'_i(\hat{\beta}_{(-i)}-\beta))^2$$

$$= \sigma^2 + E(x'_i(\hat{\beta}_{(-i)}-\beta)(\hat{\beta}_{(-i)}-\beta)'x_i)$$

By the same calculations as in (6.26), we find

$$E\tilde{\sigma}^2 = \sigma^2 + E\left(x'_iV_{\hat{\beta}_{(-i)}}x_i\right) = MSFE_{n-1}.$$

This is the *MSFE* based on a sample of size $n-1$ rather than size n. The difference arises because the in-sample prediction errors $\tilde{e}_i$ for $i \leq n$ are calculated using an effective sample size of $n-1$, while the out-of-sample prediction error $\tilde{e}_{n+1}$ is calculated from a sample with the full n observations. Unless n is very small, we should expect $MSFE_{n-1}$ (the *MSFE* based on $n-1$ observations) to be close to $MSFE_n$ (the *MSFE* based on n observations). Thus, $\tilde{\sigma}^2$ is a reasonable estimator for $MSFE_n$.

Theorem 6.4 (*MSFE*) In the linear regression model (Assumption 6.1), the following holds true

$$MSFE_n = E\tilde{e}_{n+1}^2 = \sigma^2 + E(x_{n+1}'V_{\hat{\beta}}x_{n+1})$$

where $V_{\hat{\beta}} = \text{var}(\hat{\beta} \mid X)$.

6.1.9 Covariance Matrix Estimation under Homoskedasticity

For the purpose of drawing inferences, we need an estimate of the covariance matrix $V_{\hat{\beta}}$ of the least-squares estimator. To this end, in this section, we consider the homoskedastic regression model (Assumption 6.2).

Under homoskedasticity, the covariance matrix takes the following relatively simple form

$$V_{\hat{\beta}} = (X'X)^{-1}\sigma^2$$

which is known up to the unknown scale σ^2. In Section 6.1.8, we discussed three estimators of σ^2. The most commonly used choice is s^2, leading to the classic covariance matrix estimator

$$\hat{V}_{\hat{\beta}}^0 = (X'X)^{-1}s^2. \tag{6.27}$$

Since s^2 is conditionally unbiased for σ^2, it is easy to calculate that $\hat{V}_{\hat{\beta}}^0$ is conditionally unbiased for $V_{\hat{\beta}}$ under the assumption of homoskedasticity:

$$E\left(\hat{V}_{\hat{\beta}}^0 \mid X\right) = (X'X)^{-1}E(s^2 \mid X)$$

$$= (X'X)^{-1}\sigma^2$$

$$= V_{\hat{\beta}}.$$

This estimator was the dominant covariance matrix estimator in applied econometrics for many years, and is still the default method in most regression packages.

If the estimator (6.27) is used while the regression error is heteroskedastic, it is possible for $\hat{V}_{\hat{\beta}}^0$ to be quite biased for the correct covariance matrix $V_{\hat{\beta}} = (X'X)^{-1}(X'DX)(X'X)^{-1}$. For example, suppose $k = 1$ and $\sigma_i^2 = x_i^2$ with $Ex_i = 0$. Then the ratio of the true variance of the least-squares estimator to the expectation of the variance estimator is

$$\frac{V_{\hat{\beta}}}{E\left(\hat{V}_{\hat{\beta}}^0 \mid X\right)} = \frac{\sum_{i=1}^n x_i^4}{\sigma^2 \sum_{i=1}^n x_i^2} \approx \frac{Ex_i^4}{\left(Ex_i^2\right)^2} = k$$

(Notice that we have used the fact that $\sigma_i^2 = x_i^2$ implies $\sigma^2 = E\sigma_i^2 = Ex_i^2$.) The constant k is the standardized fourth moment (or kurtosis) of the regressor x_i, and can be any number greater than one. For example, if $x_i \sim N(0, \sigma^2)$, then $k = 3$; so the true variance $V_{\hat{\beta}}$ is three times larger than the expected homoskedastic estimator $\hat{V}_{\hat{\beta}}^0$. But k can be much larger. Suppose, for example, that $x_i \sim \chi_1^2 - 1$. In this case $k = 15$ so that the true variance $V_{\hat{\beta}}$ is 15 times larger than the expected homoskedastic estimator $\hat{V}_{\hat{\beta}}^0$. While this is an extreme and constructed example, the point is that the classical covariance matrix estimator (6.27) may be quite biased when the homoskedasticity assumption fails.

6.1.10 Covariance Matrix Estimation under Heteroskedasticity

In the previous section, we showed that the classical covariance matrix estimator can be highly biased if homoskedasticity fails. In this section, we try to show how to construct covariance matrix estimators, which do not require the assumption of homoskedasticity.

Recall that the general form for the covariance matrix is

$$V_{\hat{\beta}} = (X'X)^{-1}(X'DX)(X'X)^{-1}.$$

This depends on the unknown matrix D, which we can write as

$$D = diag\left(\sigma_1^2, \ldots, \sigma_n^2\right)$$

$$= E(ee' \mid X)$$

$$= E(D_0 \mid X)$$

where $D_0 = diag\left(e_1^2, \ldots, e_n^2\right)$. Thus, D_0 is a conditionally unbiased estimator for D. If the squared errors e_i^2 were observable, we could construct the unbiased estimator as follows:

$$\hat{V}_{\hat{\beta}}^{ideal} = (X'X)^{-1}(X'D_0X)(X'X)^{-1}$$

$$= (X'X)^{-1}\left(\sum_{i=1}^n x_i x_i' e_i^2\right)(X'X)^{-1}.$$

Indeed, we have

$$E\left(\hat{V}_{\hat{\beta}}^{ideal} \mid X\right) = (X'X)^{-1}\left(\sum_{i=1}^{n} x_i x_i' E\left(e_i^2 \mid X\right)\right)(X'X)^{-1}$$

$$= (X'X)^{-1}\left(\sum_{i=1}^{n} x_i x_i' \sigma_i^2\right)(X'X)^{-1}$$

$$= (X'X)^{-1}\left(X'DX\right)(X'X)^{-1}$$

$$= V_{\hat{\beta}}$$

verifying that $\hat{V}_{\hat{\beta}}^{ideal}$ is unbiased for $V_{\hat{\beta}}$.

Since the errors e_i^2 are unobserved, $\hat{V}_{\hat{\beta}}^{ideal}$ is not a feasible estimator. However, we can replace the errors e_i with the least-squares residuals $\hat{e}_i$. By making this substitution, we obtain the following estimator:

$$\hat{V}_{\hat{\beta}}^{W} = (X'X)^{-1}\left(\sum_{i=1}^{n} x_i x_i' \hat{e}_i^2\right)(X'X)^{-1}. \tag{6.28}$$

We know, however, that $\hat{e}_i^2$ is biased toward zero. To estimate the variance σ^2, the unbiased estimator s^2 scales the moment estimator $\hat{\sigma}^2$ by $n/(n-k)$. By making the same adjustment, we obtain the following estimator

$$\hat{V}_{\hat{\beta}} = \left(\frac{n}{n-k}\right)(X'X)^{-1}\left(\sum_{i=1}^{n} x_i x_i' \hat{e}_i^2\right)(X'X)^{-1}. \tag{6.29}$$

Alternatively, we could use the prediction errors $\tilde{e}_i$ or the standardized residuals $\bar{e}_i$ to yield the estimators

$$\tilde{V}_{\hat{\beta}} = (X'X)^{-1}\left(\sum_{i=1}^{n} x_i x_i' \tilde{e}_i^2\right)(X'X)^{-1}$$

$$= (X'X)^{-1}\left(\sum_{i=1}^{n} (1-h_{ii})^{-2} x_i x_i' \hat{e}_i^2\right)(X'X)^{-1} \tag{6.30}$$

and

$$\bar{V}_{\hat{\beta}} = (X'X)^{-1}\left(\sum_{i=1}^{n} x_i x_i' \tilde{e}_i^2\right)(X'X)^{-1}$$

$$= (X'X)^{-1}\left(\sum_{i=1}^{n} (1-h_{ii})^{-1} x_i x_i' \hat{e}_i^2\right)(X'X)^{-1} \tag{6.31}$$

The four estimators $\hat{V}_{\hat{\beta}}^{W}$, $\hat{V}_{\hat{\beta}}$, $\tilde{V}_{\hat{\beta}}$, and $\bar{V}_{\hat{\beta}}$ are collectively called robust, heteroskedasticity—consistent, or heteroskedasticity—robust covariance matrix estimators. The estimator $\hat{V}_{\hat{\beta}}$ was first developed by Eicker (1963) and introduced to econometrics by White (1980). That is why it is sometimes called the Eicker–White or White covariance matrix estimator. The scaled estimator $\hat{V}_{\hat{\beta}}$ is the default robust covariance matrix estimator implemented in Stata. The estimator $\tilde{V}_{\hat{\beta}}$ was introduced by Andrews (1991) based on the principle of leave-one-out cross-validation. The estimator $\bar{V}_{\hat{\beta}}$ was first introduced by Horn et al. (1975).

Since $(1-h_{ii})^{-2}>(1-h_{ii})^{-1}>1$, it is straightforward to show that

$$\hat{V}_{\hat{\beta}}^{W} < \bar{V}_{\hat{\beta}} < \tilde{V}_{\hat{\beta}} \tag{6.32}$$

The inequality $A<B$ when applied to matrices means that matrix $B-A$ is positive definite.

In general, the bias of the covariance matrix estimators is quite complicated, but they become greatly simplified under the assumption of homoskedasticity (6.3). For example, using (6.16), we can calculate that $\tilde{V}_{\hat{\beta}}$ is biased away from zero, specifically

$$E(\tilde{V}_{\hat{\beta}}| X) \geq (X'X)^{-1}\sigma^2 \tag{6.33}$$

while the estimator $\bar{V}_{\hat{\beta}}$ is unbiased

$$E(\bar{V}_{\hat{\beta}}| X) = (X'X)^{-1}\sigma^2 \tag{6.34}$$

It might seem rather odd to compare the bias of heteroskedasticity-robust estimators under the assumption of homoskedasticity, but it does give us a baseline for comparison.

We have introduced five covariance matrix estimators, $\hat{V}_{\hat{\beta}}^{0}$, $\hat{V}_{\hat{\beta}}^{W}$, $\hat{V}_{\hat{\beta}}$, $\tilde{V}_{\hat{\beta}}$, and $\bar{V}_{\hat{\beta}}$. Which should we use? The classical estimator $\hat{V}_{\hat{\beta}}^{0}$ is typically a poor choice, as it is only valid under the unlikely homoskedasticity restriction. For this reason, it is not typically used in contemporary econometric research. Unfortunately, standard regression packages set their default choice as $\hat{V}_{\hat{\beta}}^0$, so users must intentionally select a robust covariance matrix estimator.

Of the four robust estimators, $\hat{V}_{\hat{\beta}}^W$ and $\hat{V}_{\hat{\beta}}$ are the most commonly used. And in particular $\hat{V}_{\hat{\beta}}$ is the default robust covariance matrix option in Stata. However, $\tilde{V}_{\hat{\beta}}$ and $\bar{V}_{\hat{\beta}}$ are preferred based on their improved bias. As $\tilde{V}_{\hat{\beta}}$ and $\bar{V}_{\hat{\beta}}$ are simple to implement, this should not be a barrier.

6.1.11 Measures of Fit

As we described in the previous chapter, a commonly reported measure of regression fit is the regression R^2 defined as

$$R^2 = 1 - \frac{\sum_{i=1}^{n} \hat{e}_i^2}{\sum_{i=1}^{n} (y_i - \bar{y})^2} = 1 - \frac{\hat{\sigma}^2}{\hat{\sigma}_y^2} \tag{6.35}$$

where $\hat{\sigma}_y^2 = n^{-1} \sum_{i=1}^{n} (y_i - \bar{y})^2$. This regression R^2 can be viewed as an estimator of the population parameter

$$\rho^2 = \frac{\text{var}(x_i'\beta)}{\text{var}(y_i)} = 1 - \frac{\sigma^2}{\sigma_y^2}$$

However, $\hat{\sigma}^2$ and $\hat{\sigma}_y^2$ are biased estimators. Therefore, Theil (1961) proposed to replace these by the unbiased versions s^2 and $\tilde{\sigma}_y^2 = (n-1)^{-1} \sum_{i=1}^{n} (y_i - \bar{y})^2$, yielding the following so-called adjusted R-squared:

$$\bar{R}^2 = 1 - \frac{s^2}{\tilde{\sigma}_y^2} = 1 - \frac{(n-1) \sum_{i=1}^{n} \hat{e}_i^2}{(n-k) \sum_{i=1}^{n} (y_i - \bar{y})^2} \tag{6.36}$$

While $\bar{R}^2$ is an improvement on R^2, a much better improvement is

$$\tilde{R}^2 = 1 - \frac{\sum_{i=1}^{n} \tilde{e}_i^2}{(n-k) \sum_{i=1}^{n} (y_i - \bar{y})^2} = 1 - \frac{\tilde{\sigma}^2}{\hat{\sigma}_y^2}$$

where $\tilde{e}_i$ are the prediction errors and $\tilde{\sigma}^2$ is the MSFE. As described in Section 6.1.9, $\tilde{\sigma}^2$ is a good estimator of the out-of-sample MSFE, so $\tilde{R}^2$ is a good estimator of the percentage of the forecast variance that is explained by the regression forecast. In this sense, $\tilde{R}^2$ is a good measure of fit.

One problem with R^2, which is partially corrected by $\bar{R}^2$ and fully corrected by $\tilde{R}^2$, is that R^2 necessarily increases when regressions are added to a regression model. This occurs because R^2 is a negative function of the sum of squared residuals, which cannot increase when a regression is added. In contrast, both $\bar{R}^2$ and $\tilde{R}^2$ are nonmonotonic in the number of regressions. And $\tilde{R}^2$ can even be negative, which occurs when an estimated model predicts worse than a constant-only model.

In the statistical literature the MSFE, $\tilde{\sigma}^2$ is known as the leave-one-out cross-validation criterion and is popular for model comparison and selection, especially in high-dimensional (nonparametric) contexts. It is equivalent to using $\tilde{R}^2$ or $\tilde{\sigma}^2$ to compare and select models. Models with high $\tilde{R}^2$ (or low $\tilde{\sigma}^2$) are better models in terms of expected out-of-sample squared error. In contrast, R^2 cannot be used for model selection, as it necessarily increases when regressions are added to a regression model. And $\bar{R}^2$ is also an inappropriate choice for model selection (it tends to select models with too many parameters), though a justification of this assertion requires a study of the theory of model selection. Unfortunately, $\bar{R}^2$ is routinely used by some economists, possibly as a hold-over from previous generations.

In summary, it is recommended to calculate and report $\tilde{R}^2$ and/or $\tilde{\sigma}^2$ in regression analysis, and omit R^2 and $\bar{R}^2$.

6.2 Who Controls the Future? Presidential Election and Economic Policy in America

Many economics problems can be solved by using OLS, such as relationship of economic policy and electoral outcomes, financial development and the underground economy, and so on. A few relevant models are shown in this chapter.

In general, economic trick is one of the pejorative terms its detractors use to describe the art and science of econometrics. No doubt, these terms are well deserved in many instances. Some of the problems stem from the econometrics' connection with statistics, which had its origin in the analysis of experimental data. In a typical experiment, the analyst can hold the levels of variables not of interest constant, alter the levels of treatment variables, and measure both the treatment variables and the outcome with high accuracy. With some confidence, the statistician can assert that changes in the treatment variable cause changes in the outcome and then can quantify the relationship.

Analysts began to apply statistical tools appropriate to this experimental setting to economic and business data that were clearly not the outcome of any experiment. The question of cause and effect became murky. Statisticians relied on economic theory to guide them; they had few other choices. So was born econometrics: the use of statistical analysis, combined with economic theory, to analyze economic data.

One of the pioneers of econometric forecasting was Charles Sarle. His essay describing a single equation model to forecast the price of hogs won the Babson prize in 1925 and was published in a special supplement to the American Economic Review. The Babson prize was awarded for the best essay submitted by a student, as judged by a committee of eminent economists. At $650, the prize could have bought young Charles his first car. Sarle was several decades ahead of his time. He used lagged explanatory variables, so their values were known at the time of forecast; he performed both within-sample and out-of-sample forecasts. Although his work was published in the leading economic journal, it was then largely ignored. Such is the fate of many pioneers.

Econometricians are a diverse and large group of quantitative analysts. For the last 60 years or so, the group has focused on one key problem: what nonexperimental data violate many of the statisticians' standard assumptions?

The principal tool of the econometrician is regression analysis, using several causal variables. Other methods of causal modeling exist, but they are not discussed in this chapter. Compared with univariate modeling, multivariate analysis opens up many additional choices for the investigator: the set of variables to include in the analysis; the structure, that is, the number of equations relating the set of variables to each other and the causal variables to use in each equation, if more than one equation is included; and the functional form of the equation: whether it is linear or nonlinear in parameters.

In this section, we discuss a problem about the relationship between economic policy and electoral outcomes in America.

6.2.1 Background

Let us recall the circumstances when the 2009 elections took place, which began with an evolution of taxes and quantitative restrictions on exports from the AFC and then moved on to the impact on production.

In March 2008, when Resolution 125 was passed by the National Ministry of Economy—later rejected by the Congress—a conflict between the agromanufacturing sector and the national government reached a crisis point. The conflict had been developing for some time, provoked by the high and increasing tax pressure on the sector's production, and by the introduction of different policies of market intervention—mainly quantitative restrictions on exports.

The development in time was as follows. The taxes on exports that were practically nonexistent during the 1990s were reintroduced by Resolution 11 in 2002 by the National Ministry of Economy in the context of a serious of socioeconomic, fiscal, financial, and political crises. The tax rate fluctuated between 5% and 20%, depending on the product, and was modified over time, reaching values somewhere between 5% and 27.5% by the middle of 2007. At that moment, the ad-valorem tax rates implicit in the quantitative restrictions were around 14% for wheat and 10% for meat. A new rise in tax rates occurred in November 2007 after the elections,

Table 6.1   Taxes on Exports of Agromanufacturing Sector as % of Total Revenues and GDP

Year	2001	2002	2003	2004	2005	2006	2007	2008
Total revenues (%)	0.1	5.8	7.8	6.4	5.7	5.2	5.2	6.6
GDP (%)	0.01	0.93	1.49	1.40	1.29	1.19	1.28	1.70

Source: Porto, A. and Lodola, A., *Journal of Applied Economics*, 14(2), 333, 2013.

which increased those corresponding to soybean and sunflower from 27.5% to 35%, those on wheat from 20% to 28%, and those on corn from 20% to 25%. In Resolution 125, which manifestly led to the conflict, a significant rise was established in the tax rates on soybean and sunflower and a small drop in those on wheat and corn.

As shown in Table 6.1, the rates dropped from 2003 (7.8% of the total national revenue and 1.49% of the gross domestic product of the country) to 2006–2007 (5.2% and 1.28%, respectively), rising significantly in 2008 (6.6% and 1.7%, respectively).

Public policies had a strong impact on the AFC. The sector had a remarkable expansion starting from the nineties. "The growth of Argentine agriculture was further stimulated in the nineties as a result of the stability of domestic prices and the liberalization of the economy, which promoted important advances in technology and in production organization… Human resources' high quality, massive incorporation of technology (zero tilling, latest generation machinery, incorporation of transgenic soya and of chemical fertilizers, increasing use of agrochemicals, incipient use of auxiliary irrigation in the Pampa region), as well as the use of new forms of management have sustained this profound change in the organization and growth of Argentine agriculture" (Lucio, 2008). Such growth extended to a great part of the economy as a result of the strong backward linkages—derived demand for inputs, and forward linkages—demand for goods generated by the income from the AFC. Taxes and quantitative restrictions on exports affected the incentives of agricultural and livestock producers. The results of the 2008/2009 campaign clearly show this fact, aside from the influence of the drought. Derived demand for inputs—among others, fertilizers, combined harvesters, tractors, and seeders—dropped significantly since the end of 2007 (Alejandro, 2009).

In this context the 2009 election took place.

6.2.2 Model

The case of concern here is the connection between the election results and the fiscal policies in the 134 municipalities of the province of Buenos Aires. The results of the 2007 congressional elections for the partial renewal of deputies at the

National House of Representatives were compared to those of the 2009 elections. The municipalities differ greatly in their productive structures so that a policy that is detrimental (beneficial) to a certain sector of activity may be expected to have a greater negative (positive) impact on the election results for the governing party in those municipalities where that sector is quantitatively important. The fiscal theory has been developed gradually with the long-term vision of a benevolent government that makes decisions (e.g., on tax rates) with passive adaptation on the part of producers and consumers. The theory consists of such models that incorporate the economic reactions of these agents by modifying their tax behaviors as a response to the variables set by the government. At a further stage, politicians and bureaucrats came on scene with their own interests, as well as consumers and producers reacting economically (modifying their tax base) and politically (voting). This section focuses on this last group by observing how economic policy decisions may influence electoral results and how the government may appeal to fiscal instruments to compensate for presumably negative reactions on the part of voters. The policy under consideration in this section is the one related to the agrifood chain (AFC), which has been affected by high and increasing tax rates over exports and by quantitative restrictions, which deteriorated the prices paid to the producers (selling quotas, prohibitions on exporting, etc.).

Porto and Lodola (2013) firstly developed a simple model of the connection between election results and economic and fiscal variables. Then, they presented the context of the conflict between the agromanufacturing sector and the national government to the elections prior to the 2009. They also described productive structure of the municipalities of the province of Buenos Aires and constructed an index to measure the importance of the sector for the empirical analysis. In the last section, the authors presented their econometric results. These authors find that the relative importance of the AFC had a positive impact on the party in power in 2007 and a negative one in 2009. Therefore, they suggest that voters take into consideration the effects of public policies at the time of voting and that they change voting decisions according to whether the policies are for or against their interests. The government, foreseeing a negative reaction on the part of voters, appealed to increases in the conditioned transfers to the municipalities, which had a positive impact on the voters for the ruling party, but could not compensate for the negative impact of the policy on the AFC.

The connection between election results and economic and fiscal variables has been the subject of studies, both theoretical and empirical. Theoretically, there have emerged several questions of interest, the answer to which depends on how political markets work. One of them is whether citizens decide their votes according to the evolution of fiscal and economic variables (for which they must be informed); a further question is whether, when voting, they distinguish which governmental level is responsible for which policy (for instance, not to punish a mayor or governor for the wrong development of some variable, which is mainly a responsibility of the national government). An additional line of research involves inquiring not only whether voters take into consideration the governments' performance, but also whether governments make decisions knowing that election results depend on them. In this case, one is inquiring into the possible interdependence between voters' decisions and those taken by politicians.

The 2009 Argentine election for the partial renewal of the legislative bodies at the three levels of government, framed within a prolonged conflict between the government and the agromanufacturing sector, represents a case in point. Specifically, in this section, we focus on the connection between the votes obtained at the level of municipalities of the province of Buenos Aires by the political party in power at the national and provincial governments, and the relative importance of the AFC. Votes correspond to the elections for the partial renewal of the representatives of the province of Buenos Aires in the National House of Representatives. The results are compared to those of the 2007 election, which was prior to the conflict.

In the literature, two dependent variables have been used, namely, the percentage of votes obtained in year t by the political party that is in power because of having won the previous elections or, alternatively, the probability of change of the ruling party in year t elections. The explanatory variables refer to economic and fiscal conditions, apart from certain control variables. For instance, it is believed that the better the fiscal performance is, the greater the percentage of votes will be or the lesser the probability of change of political party in power. Fiscal performance is measured with such variables as the variation in total public expenditure, in capital expenditure, in tax pressure, in fiscal deficit, etc. In the same way, it is believed that the better the global development of economy has been, the greater the percentage of votes will be or the lesser the probability of change. In this case, the relevant variables are changes in the gross domestic product, in the inflation rate, in unemployment, etc.

Let us use Sam model as the starting point, which connects the percentage of votes for the ruling party (Z) with the economic-fiscal performance (P) in the period during which it has been in power (Sam, 1992). It is assumed that a citizen has a normal probability of voting for the party that is in power (K) and that he/she deviates from such a probability according to the impact (actual or estimated) of the ruling party's policies on the variation in his/her utility (U). For the entire group of voters, a linear relationship is assumed between the percentage of votes and the impact of the government's performance on the voters' utilities,

$$Z_t = K + mU_t \tag{6.37}$$

It is assumed that P affects U according to the function

$$U_t = (1-w)P_t + wU_{t-1}, \quad 0 \le w \le 1 \tag{6.38}$$

which conveys the idea that voters remember their evaluation of the performance in the previous elections (U_{t-1}) and they give some weighting (w) to it when calculating U_t. If $w = 0$, it means that the voter calculates the impact on his/her income by referring only to information about the fiscal and economic performance as from the last elections. The greater w, the greater the weighting of the evaluation of the political party's performance carried out in the previous elections. Substituting (6.38) into (6.37) produces the following:

$$Z_t = (1-w)K + m(1-w)P_t + wZ_{t-1} \qquad (6.39)$$

where m and w are constants.

In this case study, estimations of the percentage of votes are presented following the ideas in Sam's model with some modifications according to the availability of information. The main variable included in P_t is the importance of the AFC in the municipality measured with the percentage of the total gross product of the municipality that is generated in this chain. This variable is representative of the actual or potential damage (benefit) to the sector and to the municipality resulting from the national policies supported by the provincial government. The variations in the local per capita expenditures, in the per capita transfers received from the provincial and the national governments, and in the per capita local own revenues are added as further performance variables. Two dummy variables are added, which refer to the political party in power at the municipality. There are also controls in relation to the social and demographic characteristics of the municipalities.

An important related body of scholarly literature (both theoretical and empirical) investigates the connection between trade and trade policies and income distribution. The starting point is the Wolfgang–Paul theorem that shows how commercial flows are determined by comparative advantages (Wolfgang and Paul, 1941). In turn, these comparative advantages are determined by the factor endowment in each country. Some of this work investigates the impact of trade on the prices of capital and skilled and unskilled labor in the context of different models. One important contribution is from Sebastian, who considered economies, which produce three goods, one of them being a nontradable good (Edwards, 2010). Other studies have investigated the reactions of individuals or groups that benefit from trade and trade policies, or lose from them. These studies have also researched how these people urge governments to adopt a determined policy (through, e.g., different forms of *pressure*, contributions to political campaigns, and/or votes).

6.2.3 Data

The geographical area chosen for the case study is the province of Buenos Aires, which presents stark demographic and social contrasts and a great productive heterogeneity among its municipalities. Although the manufacturing sector is the one with the greatest participation in the gross geographic product (GGP) of the province, in 100 of its 134 municipalities, the most important economic activity is agriculture and cattle farming.

Leaving aside the steady growth of services in the productive structure (whose contribution to the GGP rose from 32% in 1964 to 54% in 2008), common to all geographical areas, the contribution of the branches of agriculture and cattle farming fell from 31% of the goods production in 1964 to 10% in 1993, rising back to 16% in 2008. All throughout the period, there was a change in the composition of the agriculture and cattle farming sector's GGP in favor of agriculture, which in 1964 contributed with less than half of the sector's production, rising to 74% by 2008. This profound change reflects the multiproduct nature (agriculture and cattle farming) of exploitations in the province, which make it possible to alternate agriculture with cattle farming, and vary within agriculture, according to their relative profitability.

These changes were reflected geographically. The group of *inner* municipalities came to produce 41% of the provincial GGP in 2008, compared to 36% in 1993.

In order to evaluate the weight of agriculture and cattle farming activities in generating the GGP in each municipality, it is not sufficient to restrict ourselves to primary production. It is more adequate to work with the concept of chain. Due to the availability of information, the study is based on a subgroup of activities that constitute what in this section is called AFC. The AFC involves, in the first place, the direct contribution to the GGP from the agromanufacturing food production (agricultural crops, stockbreeding, agricultural and stock farming services, fishing, and food and drink production); it also involves the main backward linkages (derived demand for inputs) and forward linkages (demand coming from the income and expenditure of the factors involved in the activity).

Figure 6.1 shows the estimated importance of the AFC in the GGP of each group of municipalities. Although the average contribution of the chain to the gross product of the province is 18%, at one extreme for the small inner municipalities, it amounts to 48% of the GGP (there exist municipalities where the contribution

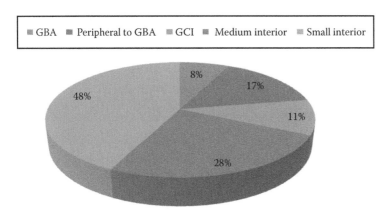

Figure 6.1 Contribution of agrofood chain by groups of municipalities in 2008. (From Porto, A. and Lodola, A., *J. Appl. Econ.*, 14(2), 333, 2013.)

exceeds 70%), while for the Great Centers of the Interior, such proportion is around 11%. In the municipalities of the Great Buenos Aires (GBA), the contribution of the AFC is 8%.

Comparing the contribution of the AFC to the GGP with the average expenditure on food and drinks in each municipality, they are classified as net producers (NPs) or net consumers (NCs) of goods originated in the AFC. In this way, 33 municipalities of the province, where 67% of its population lives, can be classified as NCs, the remaining 101 being NPs.

Figure 6.1, which gives us a contrast among the municipal productive structures, suggests that the responses (acceptance or rejection) to certain national–provincial policies are not homogeneous in the provincial territory. In the following subsection, the results of the 2007 and 2009 elections, framed within the farmers–government conflict, are presented.

6.2.4 Regression Results

It can be showed by OLS the simple relationship, at the level of each municipality, between the relative importance of the AFC in the GGP and the votes obtained by the official party (fpv: Frente para la Victoria) in the 2007 and 2009 elections. The first is positive and the second negative, suggesting a change in the sector's voting behavior. This change relates the importance of the sector with the difference of votes for the Frente para la Victoria (FPV) in each municipality: the higher the difference, the higher the importance of the AFC.

These relations could be spurious because of the lack of control variables. In order to take them into account, this section only presents the results of the econometric estimates for the 2007 and 2009 elections following the lines of the model in Section 6.2.2. The cross-section estimating equation is

$$fpv_j = \alpha_0 + \alpha_1 cadagro_j + cv_j + e_j \tag{6.40}$$

where

cv_j represents the control variable, $j = 1, \ldots, 134$ (municipalities)

e_j is an error term that is assumed to be independent across municipalities

The variable to be explained is the percentage of votes obtained by the list of national deputies of the governing party (*fpv*). The main explanatory variable, representative of the impact of the economic policy (*P*), is the share of the AFC in the gross product of each municipality (*cadagro*). The control variables included are the percentage of the population having unsatisfied basic needs in the municipality (*nbi*) and variables, which represent the demographic characteristics and fiscal performance. The variable *nbi* is justified in that the party in power (Frente para la Victoria), as a major sector of the Justicialismo (Peronist Party), traditionally has a solid electoral support coming from the sectors of relatively low socioeconomic level; one can expect a positive sign for this variable—the greater the percentage of *nbi*, the greater is the percentage of votes. The variable *vargk* measures the change in capital expenditures in each municipality since the prior election. Two dummy variables are included to control political determinants. One of them (*int-fpv*) is assigned value 1 if the mayor belongs to the Frente para la Victoria and the other (*int-radK*) is assigned value 1 if the mayor is a Radical but allied with the government. In this way, the influence of the party in power on the municipality is controlled. Both *varae* and *varpre* refer to the changes in economic activity and prices in each municipality since the prior election (between 2005 and 2007 in this case). Estimates are based on cross-section specifications for each election separately using OLS regression model with correction of heteroskedasticity.

The results for the 2007 elections are presented in columns (1) and (2) of Table 6.2.

The signs of *cadagro* and *nbi* are positive: In the municipalities where the importance of the AFC and/or the percentage of population having unsatisfied basic needs are greater, the percentage of votes for the Frente para la Victoria (*fpv*) is greater as well. The fact that the mayor is a member of the Frente para la Victoria (*int_fpv*=1) or a K Radical allied with the government (*int_radK*=1) also has a positive impact on the municipality. Fiscal variables, in particular *vargk*, and those representatives of demographic variables, were not significant. The constant, representative of the *normal* percentage of votes for the governing party, is positive and significant. The coefficients for *varae* and *varpre* are not significant.

The coincidence between the *nbi* and *cadagro* voters is truly remarkable when taking into account the history of biased policies against *cadagro* coming from the political party in power (Jorge et al., 2008; Lucio, 2008). Maybe the memory of the nineties (absent of any taxes on exports) along with the high and increasing international prices, the high yields per hectare, high level of exchange rate, and the belief that this and other restrictions could be a temporal consequence of the 2001–2002 crisis. But the evolution of the collection shows that the official idea was to maintain and increase taxation on the sector.

The results of the 2009 elections (columns 3 and 4) reveal a significant change in voters' behavior. The population with *nbi* still has a positive impact on the percentage of votes for the Frente para la Victoria (*fpv*), but AFC has a negative sign. The 2007 social coalition (positive votes from *cadagro* and *nbi*) broke down in 2009. In the same way, there is a break in the political alliance: the mayors of the *fpv* have a positive impact, but the dummy variable of the K Radicals (*int_radK*) changed its sign. The coefficients for *varae* and *varpre* are again not significant.

The constant, which represents the normal percentage of votes for the party in government, decreases between 2007 and 2009. This drop reflects a fall in votes common to all the municipalities, which can be associated with the policy of the governing party with different sectors and the reemergence of inflation (Figures 6.2 and 6.3).

Table 6.2 Regression: Percentage of Votes Obtained by the Frente para la Victoria (*fpv*)

	2007		2009	
Variables	(1)	(2)	(3)	(4)
cadagro	0.141***	0.140***	−0.0929**	−0.0908
	(0.0273)	(0.0273)	(0.0438)	(0.0432)
nbi	0.586***	0.590***	0.536***	0.503***
	(0.117)	(0.117)	(0.113)	(0.112)
fpv(t − 1)	0.169***	0.164***	0.404***	0.400***
	(0.0590)	(0.0591)	(0.0751)	(0.0740)
int_fpv	0.0682***	0.0696***	0.0213***	0.0219*
	(0.0129)	(0.0130)	(0.0124)	(0.0122)
int_radK	0.0570***	0.0602***	−0.0323*	−0.0261
	(0.0205)	(0.0207)	(0.0186)	(0.0185)
varae	−0.0479	−0.0464	0.0504	0.0593
	(0.0903)	(0.0903)	(0.129)	(0.127)
varpre	0.0586	0.0605	0.0300	0.0286
	(0.0391)	(0.0391)	(0.0881)	(0.0869)
Constant	0.243***	0.245***	0.0436	0.0422
	(0.0329)	(0.0330)	(0.0299)	(0.0295)
vargk		−0.00135		0.00227**
		(0.00123)		(0.00104)
Observation	134	134	134	134
R-squared	0.446	0.451	0.632	0.645

Note: Standard errors in parentheses, where ***$p < 0.01$, **$p < 0.05$, and *$p < 0.1$.

The variations in capital expenditures, which were not significant in the 2007 elections, had a positive impact on the 2009 results—such expenditures were one of the tools used by the national government in its policy for obtaining more votes. Although this variable is an item in the municipal budgets, in the period under study the national government transferred, directly to the mayors, considerable

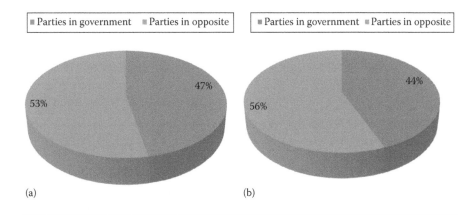

Figure 6.2 Differences at the level of municipalities in 2007 elections. (a) NC municipalities and (b) NP municipalities.

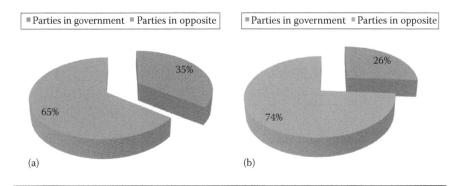

Figure 6.3 Differences at the level of municipalities in 2009 elections. (a) NC municipalities and (b) NP municipalities.

funds which were directed, fundamentally, to housing (Federal Housing Plan, Plan Federal de Viviendas). The quantitative importance of these transfers is reflected on the capital expenditures of the municipalities, which increased from barely over 104 pesos per capita in 2007 (15% of the aggregated municipal expenditure) to $238 per capita (22%) in 2009.

The coefficients for the *fpv*($t - 1$)—votes for the *fpv* in the previous elections (2005 and 2007, respectively)—are positive and significant: voters remembered the evaluation on performance made at the moment of the previous elections and gave it a positive weighting at the time of voting. The coefficient for this variable is greater in the 2009 elections than in the 2007 elections, which can be a consequence of a better evaluation at the moment of the previous elections—in fact, the general socioeconomic situation was better in 2007 than in 2005.

In order to check the robustness of these results, the difference in differences model was estimated using the equation

$$fpv_{jt} = \beta_0 + \delta_0 D_t + \beta_1 cadagro_j + \delta_1 D_t cadagro_j + cv_{jt} + e_{jt} \qquad (6.41)$$

where

D_t = dummy variable such that year 2009 = 1, year 2007 = 0
j = municipalities
t = years

The results in columns (1) and (2) of Table 6.3 are consistent with the previous estimates. The coefficient for $D_t cadagro_j$ that measures the change in *fpv* due to the policies toward the sector in 2009 is significantly negative. In the municipalities where *cadagro* is higher, the loss of votes of the official party is also higher. The R^2 is 0.81.

An additional estimation was carried out in columns (3) and (4) of Table 6.3 considering control (net-consumer municipalities) and treatment groups (NP municipalities). The estimated equation is

$$fpv_{jt} = \beta_0 + \delta_0 D_t + \beta_1 dB_j + \delta_1 D_t dB_j + cv_{jt} + e_{jt} \qquad (6.42)$$

where

D_t = dummy variable with year 2009 = 1, year 2007 = 0
dB = 1 for net-producer municipalities, 0 otherwise
t = years

The coefficient for $D_t dB_j$ in columns (3) and (4) of Table 6.3, which measure the effect of the policy, is significantly negative and corroborates the previous findings. In NP municipalities, the change in the voting behavior is negative in comparison with the NC municipalities.

An interesting result along the same lines as presented in this study can be found in Jorge et al. (2008). These authors estimated the relationship between the number of route blockades in March 2008—one of the moments of greatest tension in the farmers/government conflict—and the importance of the agriculture and cattle farming sector in 457 towns (where the importance is measured in terms of the number and extent of exploitations, and the percentage of the surface sown with soybean). The coefficients turned out to be positive and significant: the greater the importance of the sector, the greater the number of pickets. In that study, the estimation relates a form of political participation—the route blockade or picket—to the importance of the affected sector which wishes to have an influence, by those means, on political results. In this study, the estimation is presented for the result of another form of political participation—votes in the ballot box. Certainly, both

Table 6.3 Percentage of Votes Obtained by the Frente para la Victoria (*fpv*), 2007 and 2009

Variables	(1)	(2)	(3)	(4)
Dt	−0.133***	−0.134***	−0.133***	−0.134***
	(0.0153)	(0.0156)	(0.0151)	(0.0153)
nbi	0.600***	0.594***	0.570***	0.561***
	(0.0879)	(0.0879)	(0.0888)	(0.0880)
int_fpv	0.0489***	0.0486***	0.0498***	0.0493***
	(0.00934)	(0.00935)	(0.00991)	(0.00995)
int_radK	0.0469*	0.0457*	0.0341	0.0328
	(0.0243)	(0.0241)	(0.0265)	(0.0264)
Dt int_radK	−0.0600**	−0.0574*	−0.0388	−0.0358
	(0.0302)	(0.0302)	(0.0317)	(0.0318)
varae	0.00425	· 0.00435	0.0337	0.0335
	(0.0638)	(0.0634)	(0.0584)	(0.0582)
varpre	0.0442	0.0429	0.0633*	0.0613*
	(0.0323)	(0.0324)	(0.0326)	(0.0324)
fpv(t − 1)	0.241***	0.242***	0.221***	0.222***
	(0.0577)	(0.0577)	(0.0592)	(0.0592)
cadagro	0.142***	0.142***		
	(0.0291)	(0.0293)		
Dt cadagro	−0.212***	−0.212***		
	(0.0363)	(0.0364)		
dB			0.0570***	0.0564***
			(0.0129)	(0.0131)
dB			−0.0930***	−0.0932***
			(0.0158)	(0.0159)

(Continued)

Table 6.3 (*Continued*) Percentage of Votes Obtained by the Frente para la Victoria (*fpv*), 2007 and 2009

Variables	(1)	(2)	(3)	(4)
vargk		0.000631		0.000693
		(0.00105)		(0.00113)
Constant	0.217***	0.217***	0.236***	0.236***
	(0.0263)	(0.0263)	(0.0249)	(0.0249)
Observation	268	268	268	268
R-squared	0.812	0.813	0.803	0.803

Note: Standard errors in parentheses, where ***$p < 0.01$, **$p < 0.05$, *$p < 0.1$.

forms of political participation, as actions tending to modify political results, can be complementary.

Studied are the results of the 2007 and 2009 elections for the partial renewal of deputies of the province of Buenos Aires at the National House of Representatives in the 134 municipalities, and their relationship with their productive characteristics. The context of the 2009 elections was the conflict that arose in 2008 between the agromanufacturing sector and the national government supported by the provincial government, due to the rise in taxes on exports (fundamentally grains and oleaginous products) and to market interventions through quantitative restrictions.

Municipalities differ greatly in their productive structures: at one extreme, there exist municipalities in which the contribution of the agromanufacturing chain is over 70% of the total gross product, while at the other extreme, it is about 5% or less. One can assume that a policy that is detrimental to a certain sector will have a greater negative (positive) impact on the election results in those municipalities where the sector is quantitatively important. Results from presenting estimations for the 2007 and 2009 elections suggest that voters take into account the effect of public policies at the time of voting and that they change according to whether they benefit or are damaged. The relative importance of the AFC had a positive impact on the votes obtained by the political party in power in 2007 and a negative one in 2009. The 2007 social coalition (positive votes from the AFC and the population having unsatisfied basic needs) broke down in 2009. In the same way, there was a breakdown in the political alliance with the radical mayors who supported the national and provincial governments.

The change in votes from the *cadagro* between 2007 and 2009 is intriguing and brings up the questions of whether voters are well informed or not and whether they are short-sighted and only take into account recent policies. The votes from *cadagro* supported the party in power in 2007, when taxes and quantitative restrictions on the sector's production already existed. One hypothesis is that the combination of high and increasing international prices, high exchange rate and high yields, may have moved them to accept those policies—or else to believe in their temporal character. The *cadagro* voters do not seem to have recalled the history of biases against the sector on the part of the party in power in other periods further away in time. The reversion of the favorable environment in 2007, combined with the rise in taxes and quantitative restrictions, may explain the change in 2009. In this change, the activity of the sector's entrepreneur unions must have played an important part, carrying out an active plan of action, spreading information and putting the problem in the media. These actions worked to face the *rational ignorance* problem in the political market assuming that the absence of voluntary cooperation is lower in the context of repeated games. The high costs of the measures against the sector encouraged voters to have a rational behavior by assuming the cost of the plans of action and of obtaining information.

The *normal* percentage of votes for the ruling party, which was captured by the constant in the regressions, dropped between 2007 and 2009, reflecting an effect in common among all the municipalities. Partly it was associated with the *economic vote* due to the change in the economic situation, from positive to negative, between lower growth and reemergence of inflation. Political factors, such as the discretionary behavior and authoritarianism perceived by the population, also played a part.

Recently, the national government (by way of compensation and seeking to obtain votes) transferred funds directly to the municipalities in order to finance capital expenditure, fundamentally housing. The rise in capital expenditures was particularly important in the year prior to the 2009 elections and had a positive impact on the percentage of votes for the ruling party—contrary to the 2007 elections where the variations in capital expenditures did not turn out to be significant. It is difficult to ascertain to which level of government the responsibility for such expenditure was attributed, since the executant was at the municipal level and the financer was the national one. In the important official campaign, the responsibility was shared but, probably, closeness with voters favored the municipal governments. The differences between votes for national deputies and municipal councilors point in that direction. Although information is not available for all of the municipalities, the councilors obtained a total of 25% additional votes when compared to the deputies.

6.3 Gone with the Wind: Cigarette Taxes in the State

6.3.1 Background

Many countries impose excise taxes on alcohol, cigarettes, gasoline, and environment-related goods. Excise taxes are relatively small but represent nontrivial sources of revenue. On average across the OECD, they account for almost 11% of government revenues (Organization for Economic Cooperation Development,

2012, p. 63). However, it is widely agreed that the revenues are not the main explanation for which goods are taxed. As Hines (2007) argued: "Instead, excise taxes are intended to discourage consumption of the specific taxed goods, thereby preventing some potential consumers from contributing to pollution, traffic congestion, injury, and poor health."

In neoclassical welfare economics substantial excise taxes on certain goods, including cigarettes, can be justified as efficient Pigouvian taxes that internalize external costs. Work in behavioral economics suggests that much higher excise taxes may sometimes be justified to correct the *internalities* consumers impose on their future selves by unhealthy time-inconsistent decisions, possibly including their decisions to smoke (Ted and Matthew, 2006). The potential for corrective cigarette excise taxes might be limited because smokers can avoid excise taxes by making purchases from nearby tax jurisdictions with lower tax rates. DeCicca et al. (2013a) tried to conduct an applied welfare economics analysis of cigarette tax avoidance. That was the first study to develop an extension of the standard formula for the optimal Pigouvian corrective tax to incorporate tax avoidance. To provide a key parameter for our formula, DeCicca et al. (2013) estimated a structural endogenous switching regression model of border-crossing and cigarette prices. The consequent empirical results in illustrative calculations help us find that for many states, after taking into account of tax avoidance, the optimal tax is at least 20% smaller than the standard Pigouvian tax that simply internalizes external costs. The empirical estimate that tax avoidance strongly responds to the price differential is the main reason for this result. Regardless of how large smoking's externalities or internalities are, tax avoidance reduces the effectiveness of state excise taxes as a corrective policy tool. If tax avoidance and evasion directly generate external costs, such as traffic fatalities or illegality costs, the optimal state excise tax on cigarettes is even lower.

Another line of research in health economics attempts to control for legal consumer tax avoidance and illegal smuggling of cigarettes, but these studies lack direct measures of tax avoidance and mostly focus on developing unbiased estimates of the price elasticity of demand. DeCicca et al. (2013) estimated that the elasticity of border-crossing with respect to the home-state price of cigarettes is 3.1. This implies that border-crossing accounts for almost one quarter of the response of home-state purchases to changes in the home-state price. Although their direct measure of cross-border purchases by consumers has important advantages, an advantage of the indirect approach is that it might better capture organized cigarette smuggling over longer distances, which might help explain at least part of the difference in estimates. Therefore, their applied welfare analysis provides a systemic framework for thinking about current cigarette tax policy debates. Since 2000, 48 states and the District of Columbia have enacted over 100 cigarette tax hikes (DeCicca et al., 2013). Cigarette tax rates currently range from a low of $0.17 per pack in Missouri to a high of $4.35 per pack in New York. Some localities also tax cigarettes, the most notable being New York City's $1.50 per pack tax since 2002, and Chicago and Cook County's combined \$2.68 per pack tax since 2006. Policy makers have realized the potential for tax avoidance created by these large differences in tax rates between sometimes very nearby jurisdictions. In an approach targeted to discourage border-crossing, in 2012, Arkansas established low-tax zones on its side of its borders with lower tax states (DeCicca et al., 2013). In an approach targeted to encourage border-crossing, in 2011, New Hampshire reduced its cigarette tax by \$0.10 per pack to encourage residents of other states to purchase cigarettes in New Hampshire (Love, 2013). While these changes in tax policies seem to have been mainly driven by revenue concerns, our normative analysis here examines the impact of taxes and tax avoidance on social welfare more broadly.

6.3.2 Data

DeCicca et al. (2013a) used data of the individual level on cross-border cigarette purchases in their empirical study. The data are from the 2003 and 2006–2007 cycles of the Tobacco Use Supplements to the U.S. Current Population Survey (TUS–CPS). Each TUS–CPS cycle provides a large nationally representative sample and subsamples that are representative at the state level. In addition to the standard questions about smoking, in the cycles, the TUS–CPS was used to ask smokers whether their last purchase of cigarettes was in a state other than their state of residence or over the Internet or by other means. The *last purchase* can be considered to be a random draw from the distribution of each smoker's purchases. The responses should provide an accurate snapshot of consumer behavior, even though for a specific smoker the last purchase might not be typical of his or her purchases. Smokers might not take the question literally and instead based their responses on their typical or modal purchase location. It is difficult to judge the magnitude or direction of the resultant measurement error. Compared to literal responses, responses about typical purchases might even contain less of the random noise created by nontypical purchases. In any case, because most smokers make fairly frequent cigarette purchases, self-reported data on their most recent purchases seem likely to be reasonably accurate.

DeCicca et al. (2013a) used geographic information on the respondents' location to merge data on excise taxes in their home states and bordering states, as well as to measure their distance to the state border. In order to calculate distance to the state border for each respondent, it is necessary to restrict the sample to residents of 234 Metropolitan Statistical Areas (MSAs) identified in the TUS–CPS. These authors used Google Maps to calculate the driving distance from the geographic center of each respondent's MSA of residence to the closest lower tax border state. Their sample of analysis consists of 29,377 smokers who lived in an MSA and provided valid responses to the questions about border-crossing and cigarette price paid. They matched cigarette excise tax rates from Orzechowski and Walker (2008) to respondents, based on their MSA, the closest border state, and their interview months. When MSAs span state lines, tax rates and assigned distance to the closest

Table 6.4 Descriptive Statistics

Variable	Mean for Noncrossers	Mean for Border-Crossers
Price paid	3.54 (1.13)	2.91 (1.18)
Home state tax	0.95 (0.64)	1.23 (0.65)
Distance to low-tax state border < 100k	1.20 (1.05)	0.52 (0.76)
Age 15–29 (omitted category)	0.22	0.12
Age 30–39	0.22	0.17
Age 40–49	0.25	0.28
Age 50–59	0.18	0.22
Age 60+	0.13	0.21
Female	0.51	0.51
White (omitted category)	0.76	0.82
Black	0.11	0.10
Hispanic	0.09	0.04
Other races	0.04	0.04
Less than high school	0.17	0.13
High school (omitted category)	0.38	0.37
Some college	0.30	0.29
College or higher	0.15	0.20
Family income < 25k	0.32	0.26
Family income 25k–40k	0.22	0.20
Family income 40k–75k	0.29	0.32
Family income 75k+	0.18	0.22
Household size	2.66	2.35
Married	0.42	0.44
Employed (omitted category)	0.68	0.66

(Continued)

Table 6.4 (*Continued*)   Descriptive Statistics

Variable	Mean for Noncrossers	Mean for Border-Crossers
Unemployed	0.07	0.05
Retired	0.08	0.14
Not in the labor force	0.17	0.15
Northeast (omitted category)	0.20	0.32
Midwest	0.28	0.33
South	0.30	0.25
West	0.22	0.10
February 2003 (omitted category)	0.17	0.17
June 2003	0.21	0.20
November 2003	0.19	0.20
May 2006	0.16	0.15
August 2006	0.12	0.13
January 2007	0.16	0.14
N	27,878	1499

Note: The numbers in parenthesis are the standard deviations of the continuous variables.

lower tax border state based on the respondents' state of residence were matched. Respondents in the Chicago and New York City MSAs were also assigned the applicable local cigarette taxes.

It is not necessary to add state and local general sales taxes, because most states' sales taxes are in the range of 4%–7% of the purchase price (DeCicca et al., 2013). Average cigarette prices are around \$3.00–\$3.50 (Table 6.4), so the data without state sales tax will usually distort the comparison of home- and border-state taxes on cigarette purchases by less than \$0.11 (3% of \$3.50). Table 6.4 provides descriptive statistics from the TUS–CPS for the variables used in the following empirical models.

About 5% of TUS–CPS smokers reported that their last cigarette purchase was made across a state border, and less than 1% reported that their last purchase was over the Internet or other means. Therefore, we focus also on cross-border purchases in the empirical work below. Table 6.4 cross-tabulates the means for the other variables by border-crossing status. The average border-crosser lives about 70 miles closer to a lower tax state and pays about $0.60 less for a pack of cigarettes. The difference between the median distances to the border by border-crossing status is also about 70 miles (20 miles versus 93 miles). In addition to the data on border-crossing, in the endogenous switching regression model, DeCicca et al. (2013b) used novel data on self-reported cigarette prices. The TUS–CPS asked smokers to report how much they paid for their last pack or carton of cigarettes after using discounts or coupons. To provide some evidence on the accuracy of the self-reported prices, DeCicca et al. (2013b) compared them to two other sources of price data. The TUS–CPS state-average prices are highly correlated ($r = 0.94$) with the state-average prices reported by Orzechowski and Walker (2008). The TUS–CPS MSA-average prices are also highly correlated ($r = 0.86$) with MSA-average prices in Nielsen supermarket scanner data. Within states and MSAs, the coefficients of variation for cigarette prices are in a fairly narrow range from around 20% to 40%. This degree of price dispersion is generally comparable to that seen for other goods (Baye et al., 2006). Overall, several lines of evidence suggest that the TUS-CPS provides reasonably high-quality data on border-crossing and self-reported cigarette prices.

The numbers in parenthesis are the standard deviations of the continuous variables.

6.3.3 Linear Regression Model

In this section, we estimate a structural endogenous switching regression model of border-crossing and cigarette prices in two regimes: the home price (P^H) and the border price (P^B). Depending on whether individual i crosses the border ($B_i = 0$ or 1), the price paid by the individual switches between the two regimes is as follows:

$$P_i^H = \beta_1 + \beta_2 T_i^H + \beta_3 X_i + \varepsilon_i^H \quad \text{if } B_i = 0 \tag{6.43}$$

$$P_i^B = \gamma_1 + \gamma_2 T_i^B + \gamma_3 X_i + \varepsilon_i^B \quad \text{if } B_i = 1 \tag{6.44}$$

The coefficients β_2 and γ_2 show the rates at which taxes are passed through to prices, because the tax rates are the main drivers of the home-price and the border-price. The price an individual pays for cigarettes also varies for other reasons, for example, between discount brands and premium brands. To capture these influences, Equations 6.43 and 6.44 also include a vector X of socioeconomic variables. We have no priori prediction about the associated parameter vectors β_3 and γ_3.

The individual is assumed to make cross-border purchases of cigarettes in order to increase his or her utility. Assume that the individual receives utility from a composite consumption good g, disutility from miles of distance m traveled, and utility from cigarette consumption c. The latent utility difference behind the observed border-crossing decision is given by

$$\Delta u = u(g^*, m^*, c^*) - u(g^{**}, m^{**}, c^{**}) \tag{6.45}$$

where

g^*, m^*, and c^* are the optimal choices given border-crossing

g^{**}, m^{**}, and c^{**} are the optimal choices given no border-crossing

To develop an empirical version of (6.45), let us consider a first-order Taylor series approximation of the utility function. After simple calculations, we obtain

$$\Delta u = u_g c(P^B - P^H) + u_m(\text{miles to the border}) + r \tag{6.46}$$

where u_g and u_m are partial derivatives of the utility function and r the remainder term for the Taylor series approximation. Because g is the composite consumption good, the difference between the optimal choice of g with and without border-crossing is simply the potential savings from purchasing less expensive cigarettes across the border, that is, $g = g^* - g^{**} = c(P^B - P^H)$. The difference between the optimal choice of m with and without border-crossing is the distance to the border, that is, $m = m^* - m^{**} = $ miles to the border.

Equation 6.46 motivates the following structural equation for the latent utility difference behind observed border-crossing:

$$\Delta u = \delta_0 + \delta_1(P_i^B - P_i^H) + \delta_2(\text{miles to the border})_i + \delta_3 W_i + \xi_i \tag{6.47}$$

Equation 6.47 underlies the empirical model that shows the probability of border-crossing as a function of the price differential $P^B - P^H$ and other explanatory variables. The Taylor series remainder term r in Equation 6.46 is captured in Equation 6.47 by the constant term and the vector of exogenous variables W. The empirical model reported in the following text includes quadratic terms and interaction terms corresponding to a second-order Taylor series approximation. The higher order terms are suppressed in Equation 6.47 for expositional ease. The use of a Taylor series approximation is justified on the grounds that border-crossing to purchase lower price cigarettes results in small changes relative to the typical consumer's total purchases of all goods and total travel for all purposes.

To estimate the endogenous switching model, we use the standard assumption that the error terms ε_i^H, ε_i^B, and ξ_i have a trivariate normal distribution with nonzero covariance. First, we jointly estimate the price Equations 6.43 and 6.44 and a reduced-form version of the border-crossing equation that does not include the endogenous price variables on the right-hand side. Next, we use the estimated price equations to predict P^H and P^B for each smoker in the sample. Finally, we estimate the structural border-crossing equation as a function of the predicted price differential $P^B - P^H$ and the other explanatory variables.

The endogenous switching model has several advantages. First, it solves the problem (Chiou and Muehlegger, 2008) faced in their study of border-crossing: Cigarette prices are only observed for the location of purchase chosen by the consumer. The switching model provides predicted values of both P^H and P^B for all consumers corrected for the endogenous selection of purchase location. Second, the parameters of the structural border-crossing equation can be more tightly linked to a model of consumer behavior. It is also needed to estimate an equation of reduced form that can show the probability of border-crossing as a function of home-state and border-state tax rates and other explanatory variables. However, at this point, DeCicca et al. (2013a) had to rely on the results of the switching model because the interpretation is cleaner. Substituting Equations 6.41 and 6.42 into Equation 6.47 shows that the reduced-form coefficients on the tax variables combine two structural parameters: δ_1, which captures the relationship of interest reflecting consumers' incentives to cross-borders to obtain lower cigarette prices, and β_2 or γ_2, which capture the pass-through of home- or border-state taxes to consumer prices. These authors found systemic differences in the pass-through rates β_2 and γ_2 of home- versus border-state taxes, which makes it problematic to draw inferences about δ_1 from the reduced-form estimates.

The structural endogenous switching model is identified by differences in the vectors of explanatory variables in Equations 6.47, 6.41, and 6.42. Specifically, the exclusion restrictions are that only distance and squared distance enter the border-crossing Equation 6.47, and the home- and border-state tax variables only enter the home- and border-state price Equations 6.41 and 6.42, respectively. The exclusion restrictions are supported by the argument that there is enough competition in retail cigarette markets so that within each state, the price is driven down to approximately the retailers' marginal cost plus the state tax rate. This implies that the consumer's distance to the border does not directly enter the price Equations 6.43 and 6.44. It also implies that retailers in the home state cannot change their prices in response to border-state taxes, and vice versa.

Anecdotal evidence suggests that the retail cigarette market is instead equilibrated by changes in the volume of cigarette sales and the entry and exit of cigarette retailers. For example, Fleenor (2008) related that after the 1995 Michigan cigarette tax increase, "One Michigan convenience store located approximately four miles from the Indiana border lost 98% of its cigarette carton sales. ..." Patrick (1998, 2008) and Amir (2007) related more recent anecdotes along these lines, including a fivefold increase in cigarette sales in Sunland Park, New Mexico, after a cigarette tax hike across the border in El Paso, TX, USA. These large swings in the volume of sales are consistent with price-picking behavior within a state where retailers near borders cannot change their prices in response to border-state taxes.

To further explore the issue empirically in the TUS–CPS data, it is necessary to examine the geographic patterns of cigarette prices paid by consumers who purchased their cigarettes in their home state. Prices do not vary systematically with distance to the border of states with either lower or higher cigarette taxes. Regarding the strength of our identification strategy, in OLS models of the reduced-form versions of Equations 6.47, 6.41, and 6.42, the F-statistics on the identifying variables are 25.8, 388.5, and 9.2, respectively. It is easy for everyone to be unaware of specific tests of weak identification in the endogenous switching model. The values for the F-statistics in two of the reduced-forms exceed the common rule of thumb for linear instrumental variables models that the F-statistic on the excluded IVs should be greater than 10 (Stock et al., 2002). The F-statistic for the border-state tax variable suggests that this might be marginally weak, perhaps due to the much smaller sample size of border crossers whom we can observe paying the border price.

Results in Table 6.5 report estimates of the structural endogenous switching model. The probability of border-crossing is estimated to decrease with distance from the border and to increase with the differential between home- and border-state cigarette prices. The marginal effect of distance is around −0.05. We use the estimated marginal effect of the price differential to calculate that the cross-price elasticity $\eta^B = 3.1$ (with a boot-strapped standard error of 0.497, so it is statistically significantly different from zero at above the 99% confidence level). This is a *cross* price elasticity in two senses. First, η^B shows the elasticity of the probability of border-crossing with respect to the home-state price of cigarettes. Recall that the baseline level of cross-border purchases is 5%. Secondly, an elasticity estimate implies that a 10% increase in the home-state price, holding the border-state price constant, increases the probability of border-crossing by 1.55 percentage points (31% of 5%).

Turning to other results in Table 6.5, the ratio of the marginal effect of distance to the marginal effect of the price differential provides an estimate of the shadow price of distance traveled. The shadow price calculation also requires an estimate of the quantity of cigarettes purchased (c in Equation 6.46). If we assume the average border crosser purchases one carton of 10 packs of cigarettes per trip, and that distance traveled on a round trip is twice the distance to the border, the implied shadow price of distance is $0.06 per mile. DeCicca et al. (2013a) used this estimate below to quantify the benefits of replacing avoidable state taxes with a harder-to-avoid federal tax. Their estimate of the shadow price of distance traveled is lower than standard estimates of travel costs, but this is reasonable if consumers travel to jointly produce cross-border cigarette purchases and other activities. Suppose that consumers make travel decisions by comparing the total travel cost per mile to the total benefits, where the value of the total benefits is equal to the sum of the values placed on the various services that jointly flow from the travel. Their approach isolates one component of the total value.

Although DeCicca et al. (2013b) did not provide a complete discussion of all the results in Table 6.6, we note that the results also show that some socioeconomic factors, such as income, have similar effects on the probability of border-crossing and on the price paid for cigarettes in either of the regimes.

Table 6.5 Structural Endogenous Switching Model

Variables	Probability of Border-Crossing	Home-State Price Paid	Border-State Price Paid
Distance to border	−0.044*** (0.012)		
Distance²	0.007*** (0.002)		
Price difference	0.095*** (0.017)		
Price difference × distance	−0.078*** (0.018)		
Price difference × distance²	0.017*** (0.004)		
Home-state tax		0.866*** (0.043)	
Border-state tax			0.399*** (0.136)
Age 30–39	0.009*(0.004)	−0.146*** (0.020)	−0.134 (0.107)
Age 40–49	0.014*** (0.004)	−0.323*** (0.018)	−0.445*** (0.095)
Age 50–59	0.012** (0.005)	−0.409*** (0.024)	−0.633*** (0.113)
Age 60+	0.027*** (0.008)	−0.497*** (0.032)	−0.741*** (0.134)
Female	0.001 (0.003)	−0.003 (0.013)	0.065 (0.060)
Black	−0.005(0.008)	0.272*** (0.027)	0.516*** (0.077)
Hispanic	−0.004(0.006)	0.230*** (0.027)	0.421*** (0.177)
Other races	0.004 (0.008)	0.057 (0.038)	0.105 (0.167)
Less than high school	−0.005 (0.004)	−0.011 (0.014)	−0.036 (0.097)
Some college	0.005 (0.003)	0.034*** (0.012)	0.097 (0.068)
College or higher	0.015** (0.006)	0.192*** (0.020)	0.281*** (0.082)
Family income 25k–40k	0.006* (0.004)	0.063*** (0.017)	0.081 (0.085)
Family income 40k–75k	0.012*** (0.004)	0.116*** (0.017)	0.081(0.085)
Family income 75k+	0.020*** (0.005)	0.228*** (0.024)	0.392*** (0.084)

(Continued)

Table 6.5 (*Continued*) Structural Endogenous Switching Model

Variables	Probability of Border-Crossing	Home-State Price Paid	Border-State Price Paid
Household size	−0.007*** (0.001)	−0.004 (0.005)	−0.067*** (0.024)
Married	0.003 (0.003)	−0.075*** (0.015)	−0.039 (0.080)
Unemployed	−0.000 (0.005)	−0.043** (0.019)	0.079 (0.120)
Retired	0.016** (0.007)	−0.117*** (0.028)	−0.040 (0.108)
Not in the labor force	0.002 (0.004)	−0.113*** (0.017)	−0.138** (0.069)
Marginal effect of distance (a)	−0.052		
Marginal effect of price difference (a)	0.044		

Note: Robust standard errors (clustered at MSA level) in parentheses. All models also include a constant term and a set of dummies for MSA size, region, and survey month. Sample size for all models is 29,377, where $*p < 0.1$, $**p < 0.05$, and $***p < 0.01$.

6.3.4 Conclusions

The previously described empirical models of cigarette tax avoidance show that consumers respond to the incentives created by excise tax differentials across states. DeCicca et al. (2013a) has shown that taking tax avoidance into account reduces the optimal Pigouvian tax rate on goods that generate negative externalities or internalities (Table 6.6). By combining an existing estimate that the external cost of smoking is $1.36 per pack with our empirical parameter estimates, their illustrative calculations suggest that some states may already impose cigarette excise taxes that are higher than optimal, and there is a strong policy trend toward even more states hiking cigarette taxes. In another exercise, these authors suggested that there would be substantial benefits if the United States moved toward replacing avoidable state cigarette taxes with a higher harder-to-avoid federal cigarette tax. One direction for future research is to explore whether insights from behavioral economics might shed more light on cigarette tax avoidance. For example, Ahmed et al. (2007) suggested that smokers might purchase their cigarettes by the pack instead of the carton as a commitment device to limit their smoking. An interesting question for future research is whether, as a commitment device to limit their smoking, some smokers also avoid opportunities to avoid cigarette taxes. Several policy trends suggest additional directions for future research. First, some policy advocates argue that using higher taxes to discourage unhealthy consumption

Table 6.6   Estimates of Optimal Tax by State

State	Tax in 2003	Optimal Tax in 2003	Tax in 2006/2007	Optimal Tax in 2006/2007
Alabama	0.17	1.25	0.43	1.15
Alaska	1.00	1.36	1.73	1.36
Arizona	1.18	1.32	1.73	1.32
Arkansas	0.51	1.15	0.59	1.19
California	0.87	1.33	0.87	1.34
Colorado	0.20	1.23	0.84	1.26
Connecticut	1.38	1.36	1.51	1.36
Delaware	0.34	1.36	0.55	1.36
District of Columbia	1.00	0.66	1.00	0.81
Florida	0.34	1.20	0.34	1.36
Georgia	0.20	1.20	0.37	1.22
Hawaii	1.23	1.36	1.53	1.36
Idaho	0.47	1.31	0.57	1.36
Illinois	0.98	1.03	0.98	0.96
Indiana	0.56	1.18	0.56	1.23
Iowa	0.36	1.23	0.36	1.23
Kansas	0.79	1.00	0.79	1.11
Kentucky	0.03	1.34	0.30	1.28
Louisiana	0.36	1.19	0.36	1.17
Maine	1.00	1.11	2.00	1.14
Maryland	1.00	0.87	1.00	0.95
Massachusetts	1.51	1.09	1.51	1.14
Michigan	1.25	1.17	2.00	1.27
Minnesota	0.48	1.25	0.82	1.17
Mississippi	0.18	1.25	0.18	1.36

(Continued)

Table 6.6 (*Continued*)   Estimates of Optimal Tax by State

State	Tax in 2003	Optimal Tax in 2003	Tax in 2006/2007	Optimal Tax in 2006/2007
Missouri	0.17	1.24	0.17	1.36
Montana	0.53	1.36	1.70	1.36
Nebraska	0.64	1.03	0.64	0.95
Nevada	0.50	1.17	0.80	1.27
New Hampshire	0.52	1.36	0.80	1.36
New Jersey	1.68	1.24	2.52	1.40
New Mexico	0.44	1.27	0.91	1.32
New York	1.50	1.30	1.50	1.36
North Carolina	0.05	1.13	0.33	1.13
North Dakota	0.44	1.19	0.44	1.36
Ohio	0.55	1.20	1.25	1.13
Oklahoma	0.23	1.29	1.03	1.26
Oregon	1.28	1.32	1.18	1.33
Pennsylvania	1.00	1.10	1.35	1.19
Rhode Island	1.45	1.35	2.46	1.36
South Carolina	0.07	1.15	0.07	1.36
South Dakota	0.46	1.07	0.86	0.98
Tennessee	0.20	1.05	0.20	1.25
Texas	0.41	1.23	0.74	1.26
Utah	0.70	1.17	0.70	1.23
Vermont	1.02	1.15	1.59	1.32
Virginia	0.03	1.36	0.30	1.30
Washington	1.43	1.34	2.03	1.31
West Virginia	0.42	1.08	0.55	1.10
Wisconsin	0.77	1.31	0.77	1.31
Wyoming	0.28	1.36	0.60	1.36

might be a lesson for obesity control to be learned from tobacco control (Baye et al., 2006; Frank, 2011). A number of states are considering new taxes on soft drinks and other sugar-sweetened beverages. Given our results and other recent research on cigarette tax avoidance, another lesson to be learned is that higher state-level excise taxes on soft drink taxes might result in substantial tax avoidance. If states begin to impose new taxes to combat obesity, it will create a rich new set of policy experiments for research on excise tax avoidance in different contexts with different goods and different consumers. A second policy trend is that the U.S. Food and Drug Administration's regulatory authority over the tobacco industry means that nontax tobacco control initiatives are likely. For example, the FDA is considering a complete ban or new restrictions on the sales and promotion of menthol cigarettes. Future research on regulation avoidance could address one of the FDA's concerns: "If menthol cigarettes could no longer be legally sold, is there evidence that illicit trade in menthol cigarettes would become a significant problem?" (Food and Drug Administration, 2013).

6.4 Undercurrents: The Underground Economy and Financial Development

6.4.1 Background

Recent estimates indicate that the underground economy represents 10%–15% of GDP in developed countries and 30%–40% in developing countries. In some countries, such as Panama and Bolivia, almost 70% of GDP is hidden (Dreher and Schneider, 2010). Apart from ethical and political concerns, a large share of underground economy is a serious issue for governments and policy makers since it distorts investments, exacerbates income inequality, and hampers economic growth (Schneider, 2010a,b). Many factors explain the emergence and size of informal economic activities. A high level of taxation, a cumbersome legislation, high payroll taxes, and labor costs are only some of the many factors that may push firms into such informality. Among these factors, the availability of credit and its cost have received little attention. In this section, we introduce choices on how to operate underground (and to what extent) and how this interacts with financial development. As in Ellul et al. (2012), the starting point of our analysis is that the ability to reveal and signal revenues reduces information frictions and the cost of credit. When firms or individuals operate underground, their ability to signal revenues and assets is lower, and the cost of credit higher. As financial markets develop, more efficient intermediaries enter the market, and the cost of credit falls, increasing the opportunity cost of continuing to operate underground. In short, financial market development is negatively correlated with the size of the underground economy.

To clarify our arguments, we propose a simple theoretical model in which agents choose between a low-return technology and a more advanced and rewarding technology. Investing in the low-return technology does not require a loan, while the high-return technology requires external funding. We posit that firms can reduce the cost of credit by pledging more collateral (Jappelli et al., 2005). Since contracts are not completely enforceable, part of the pledged resources can be lost in the case of a dispute, for example, because of judicial costs and inefficiencies. Pledging more collateral, however, is costly because firms must disclose their revenues and assets to financial intermediaries and also to tax officials. Hence, agents choose how much to invest in the two technologies by trading off the reduced financial cost of supplying more collateral against the benefit of hiding revenues and operating with the low-return technology. The choice between the two technologies therefore is also a choice between the underground and the official economy. Financial development reduces the cost of credit and the incentives to operate underground, while making it more profitable to reveal the revenues from high-tech projects.

In this underground economic model, we add two important insights to the existing works. First, technological choice to operate in the underground economy is driven by technological reasons, and the model implies that high-tech firms operate mainly in the formal economy, while low-tech firms operate mainly underground. Second, agents can operate simultaneously in both sectors, because they choose the optimal levels of income and assets to disclose to the tax authorities.

In this section, we challenge the model's predictions with empirical evidence, exploiting the variability in local financial development across Italian regions. The data are drawn from the Bank of Italy's Survey of Households Income and Wealth (SHIW) to build an index of the underground economy based on individual-level data. The index measures the level of work irregularity among Italian workers from 1989 to 2006, and ranges from 0 (activity is only in the formal sector) to 1 (activity is completely hidden). We regress this index on an indicator of financial development and other individual and regional variables. The results show that the underground economy is strongly negatively correlated with financial development. More importantly, in the empirical approach, we control for the endogeneity of financial development using the indicator proposed by Guiso et al. (2004).

Because of the heavy burden on the economy, many studies have examined causes and consequences of underground activities. It is not easy to provide in-depth and exhaustive explanations for why firms and individuals evade taxes or operate irregularly and underground. High levels of taxation, cumbersome legislation, and a tight regulatory system, often considered to be the main determinants of underground activities, provide only partial explanations. Other factors play a role in shaping the underground sector, and among them, the role of institutions is likely to be the most relevant. Indeed, institutional failure such as poor contract enforcement, judicial inefficiency, complex and arbitrary regulation reduce the incentive for firms and individuals to reveal their revenues.

In a recent work, Schneider (2010) found that the underground economy is rooted in a combination of factors such as a large burden of taxation and social security payments, stringent labor market regulation, poor quality of state institutions, and poor tax morale. The institutional setting can significantly affect the choice of informality because the efficiency of public institutions and the quality of public goods provision are important determinants of the opportunity cost to operate underground. For this reason, lack of democratic participation, low level of tax morale, and institutional distrust are all factors, which affect positively the size of underground economy; for more details, see Dreher and Schneider (2010). These factors play a major role, and improvements in the quality of institutions might work much better in reducing the size of the underground economy than other measures of deterrence; for more details, see Schneider (2000, 2010).

Among the many institutions, which have been linked to the underground economy, the degree of financial development has received relatively little attention. Yet informality is associated with a higher cost of credit, which is an important component of the overall opportunity cost to operate underground. To the extent that financial development reduces the cost of credit, it increases the opportunity cost of informality. Some papers explore such relation. In Blackburn et al. (2012), entrepreneurs need external resources for investment and can reduce the level of information costs and the financial outlays by supplying more collateral. Supplying more collateral, however, involves a higher tax burden. Given the financial costs, entrepreneurs choose whether or not to evade taxes and to operate underground. Ellul et al. (2012) suggested that when firms choose accounting transparency, they trade off the benefits of access to more abundant and cheaper capital against the cost of a higher tax burden, and studied this trade-off in a model with distortionary taxes and endogenous rationing of external finance.

On the empirical front, some investigations study the relationship between the underground economy and financial constraints. By using cross-country data, Bose et al. (2012) found that bank development is negatively associated with the size of the underground economy. Ellul et al. (2012) used microeconomic data from Worlds Cope and from the World Bank Enterprise Survey and found that investment and access to finance are positively correlated with accounting transparency and negatively with tax pressure. They also found that transparency is negatively correlated with tax pressure, particularly in sectors where firms are less dependent on external finance, and that financial development encourages greater transparency by firms that are more dependent on external finance. However, existing empirical studies do not address the issue of endogeneity and the potential reverse causality argument that a large underground economy limits the growth of financial intermediaries.

6.4.2 Linear Model

Let us consider an economy with a large number of banks, which lend to a continuum of risk neutral entrepreneurs, denoted by i. Banks have a positive and exogenous cost of issuing a unit of loan, $\bar{R} + \delta = R$, which is the sum of the cost of raising funds, $\bar{R}$, and an intermediation cost δ. Each entrepreneur is endowed with an illiquid asset, A_i, which is uniformly distributed in the interval $[0, \overline{A}]$. The asset (or part of it) can be used as loan collateral. We denote the fraction of A_i disclosed to the bank and employed as collateral as γ_i with $\gamma_i \in [0,1]$. Hence, banks observe $\gamma_i A_i$ but not γ_i or A_i separately. The fraction of the asset that is hidden, $(1-\gamma_i)A_i$, is not observed by any other agent or hence by government. Each entrepreneur can undertake two types of investment, high-tech and low-tech projects (H_T and L_T, respectively). The H_T projects are risky, require a loan, and operate under a technology with constant returns to scale. The L_T projects do not require a loan but operate with a technology with less rewarding and decreasing returns.

The assumption that firms use simultaneously different technologies with different returns is not a new one. Learning costs, financial costs, and other constraints may hamper the adoption of new technologies even when they are readily available. Since Mansfield's seminal contribution (Mansfield, 1963), economists have attempted to study not only the dynamics of interfirm rates of diffusion (technology diffusion between firms) but also intrafirm rates of adoption of new technologies (technology diffusion within a firm). In the presence of frictions and constraints, firms may use for a long period of time different technologies and tend to substitute old for new ones slowly. Nonmonotonicity in the dynamics of adoption of new technologies implies that within the same industry, and in a particular firm, more advanced and mature technology may coexist. Actually, the dynamics of adoption itself may affect the returns of technologies (Arthur, 1989). High rates of adoption lead to innovation and further improvements. The more the technologies are adopted, the more knowledge is gained from their use and the more they are improved upon, a process that Rosenberg (1982) describes as "learning by using."

Competition between technologies can enhance this process, which is the reason why more dynamic and more competitive sectors tend to involve a prevalence of high-return technologies. The opposite applies to mature and stagnant technologies where lack of innovation and increasing costs lead to decreasing returns. Following these arguments, we assume that L_T projects operate in the underground economy, while H_T projects operate in the official economy. The match among L_T and H_T projects as well as the formality of the economy accords also with the idea that operations in the underground economy rely on self-financing and more traditional projects. Firms engaged in the official economy, in contrast, rely more heavily on external finance and implement more technologically advanced projects.

In the remainder of this section, we assume that the L_T project of our interest does not require a loan and that it can be carried out using the illiquid asset A_i to purchase low-tech capital K_{LT}. If entrepreneurs undertake an L_T project, they operate with a decreasing return to scale technology, according to the following production function:

$$Q_{LT} = \Phi K_{LT} \tag{6.48}$$

All L_T projects are completely hidden to both lenders and government. Entrepreneurs invest in these projects the share of the illiquid asset, which is not pledged as collateral. Hence, if $\gamma_i A_i$ is the fraction of the asset disclosed to the bank in order to obtain a loan to finance an H_T project, the capital invested in the L_T project is $K_{LT} = (1-\gamma_i)A_i$.

All H_T projects operate under constant returns to scale. They require a loan L_i and deliver $Q_{HT} = QL_i$ units of output with probability p and 0 unit of output with probability $(1-p)$. Each H_T project has a positive net present value $pQL_i > \tilde{R}L_i$. There is no information asymmetry between borrowers and lenders, and banks can always observe whether projects succeed or fail. However, as in Jappelli et al. (2005), we assume that only part of the proceeds of the investment can be pledged against the loan. In particular, we assume that in case of success, lenders can recover at most a fraction θ of output (QL_i) and a fraction ϕ of the collateral, with $\theta \in [0,1]$ and $\phi \in [0,1]$. The remaining fraction of output $(1-\theta)$ and collateral $(1-\phi)$ can be interpreted as the amount of resources required by the judicial system for its functioning. One can think of this loss as the cost of premature liquidation of the investment or, alternatively, as the cost of judicial efficiency. Thus, in the case that the project succeeds, lenders obtain $\theta QL_i + \phi\gamma_i A_i$ units of output, while in the case of failure, they obtain $\phi\gamma_i A_i$.

We denote by $R_i \geq \tilde{R}$ the agreed repayment per unit of loan. This repayment is set after borrowers supply the collateral $\gamma_i A_i$. In a competitive credit market, banks' expected profits are zero, and hence

$$\tilde{R}L_i = pR_i L_i + (1-p)\min[R_i L_i, \phi\gamma_i A_i] \tag{6.49}$$

Depending on the amount of collateral, the zero profit condition (6.49) determines three possible cases. A first case (Case A) arises if the collateral is sufficient to repay the lender if the project should fail, that is, $\phi\gamma_i A_i \geq R_i L_i$. From Equation 6.49, it is clear that the required interest rate is equal to the lowest possible rate. That is, the bank's cost of supplying the loan is

$$R_i = \tilde{R} \tag{6.50}$$

Only borrowers with large endowments can access this contract. Recalling that $\gamma_i \in [0,1]$ and that the condition $\phi\gamma_i A_i \geq R_i L_i$ must be satisfied to access this contract, the collateral required is $A_i \geq A_{max}$, with $A_{max} \equiv \tilde{R}L_i/\phi$.

A second case (Case B) arises if the collateral would be insufficient to repay the lender and the project were to fail ($\phi\gamma_i A_i < R_i L_i$). By using Equation 6.49, it is straightforward to show that the required interest rate is now

$$R_i = \frac{\tilde{R}}{p} - \frac{1-p}{p}\phi\frac{\gamma_i A_i}{L_i} \tag{6.51}$$

In this case, the interest rate is a decreasing function of the pledged collateral and greater than that in Case A.

The third case (Case C) arises if the amount of the collateral is insufficient to repay the lender and the project is to succeed. This occurs if the collateral is insufficient to cover the bank's cost of funding. Let us denote by A_{min} the level of the endowment A_i, below which the expected return on the project does not cover the cost of funding:

$$A_{min} = \tilde{R}L_i \varphi - \frac{p\theta QL_i}{\varphi} \qquad (6.52)$$

In this case, potential borrowers with endowments $A_i < A_{min}$ are excluded from credit (while borrowers with $A_i \geq A_{min}$ can access the financial contract as in Case B).

For the sake of simplicity, we rule out Case A and focus on a situation in which $\bar{A} < A_{max}$. That is, no borrower has enough collateral to finance an H_T project at the interest rate $\tilde{R}$. Thus, we assume that regardless of the disclosed collateral $\gamma_i \in [0,1]$, all borrowers are financially constrained.

The problem of financially constrained borrowers is to choose the optimal level of the initial asset to disclose to the bank $(\gamma_i A_i)$. This choice involves a trade-off. The higher the level of the pledged collateral $\gamma_i A_i$, the lower will be the cost of the loan (see Equation 6.51) and, in turn, the higher the return on the H_T project. However, by disclosing the asset, borrowers face two costs: a direct cost due to higher taxation, and a higher opportunity cost due to the income loss in operating the L_T project on a smaller scale.

By using the method of optimization, it can be proved that the optimal choice of collateral, correspondingly, the extent to which borrowers invest in the L_T project and hide their income, depends on the relative returns from the two projects.

Now, let us look at how financial market development (a reduction in the cost of credit) affects the relative return and the size of the underground economy. We focus on the effects of improvements in judicial efficiency, tax reforms, and changes in the technology of underground activities. It is well known that financial development is a multifaceted phenomenon, and it involves typically the emergence of new and thicker capital markets, the introduction of new financial instruments, and greater competition between intermediaries. Yet, in general, it is possible to argue that financial development entails a lower cost of raising funds. In our model, we consider financial development as corresponding to a smaller intermediation cost δ and a lower cost of finance $\bar{R} + \delta = \tilde{R}$.

In this framework, each disclosure entails a trade-off. Disclosing collateral reduces the cost of accessing external funding, but increases the tax burden. Furthermore, once the collateral is disclosed, it cannot be used in the L_T sector, which reduces revenues from L_T projects. Financial development reduces the size of the underground economy only if it relaxes the credit constraints, inducing more agents to borrow. Therefore, agents with very low-endowment $(A_i < A_{min})$ are not affected by financial development. In fact, any change in the cost of credit does not affect their investment decision. Instead, entrepreneurs whose assets are above A_{min} are able to access the credit market, set $\gamma_i < 1$, and run both projects. Moreover, for these entrepreneurs, the choice of collateral is a monotonic and increasing function of their endowment, that is, the higher the A_i, the higher the γ_i. Since it is assumed that A_i is distributed uniformly over the interval $[0, \bar{A}]$, the area below the disclosure function measures total disclosed assets.

Let us now look at how financial development affects underground activities. We know from Equation 6.52 that a reduction in $\tilde{R}$ reduces the threshold level of collateral A_{min}, which allows borrowers to access credit. It is easy to understand that reduction in $\tilde{R}$ reduces A_{min} to its new value A'_{min}. Borrowers with $A'_{min} \leq A_i < A_{min}$, who previously were credit-constrained and operated only in the underground economy, now disclose part of their asset, obtain a loan, and run an H_T project. We summarize this conclusion in the following result.

Theorem 6.5 If other variables are held equal and constant, then financial development increases the opportunity cost of tax evasion, lowers underground activity, reduces credit rationing, and stimulates investment in new technologies.

Note that within our setting, financial development also implies technological improvement. That is, more firms operate H_T projects. This is in line with the empirical evidence showing that a reduction in the size of the underground economy is associated with more efficient use of resources and allocation of investments (Farrel, 2004).

As with the cost of financial intermediation, any other factor that affects the relative returns of the two projects also affects the choice of collateral and the choice of γ_i. This implies that an increase in the tax rate t reduces the expected return from H_T projects and the optimal γ_i. For the same reasons, an increase in the productivity of L_T projects (an increase in ϕ) raises the profitability of the project and reduces γ_i. In graphical terms, as t increases and the expected return of H_T projects falls, the disclosure function $\gamma(A_i)$ shifts downward. This implies that each entrepreneur will disclose a lower share of assets as collateral. Notice that in our model, taxation does not affect credit rationing because the tax rate does not enter Equation 6.52 and therefore the value of A_{min}.

The model also suggests that changes in judicial efficiency may affect γ_i. To see this, recall that we interpret the terms $(1 - \theta)QHT$ and $(1 - \phi)A_i$ as the amount of resources lost in the case of a legal dispute, and that an increase in θ or ϕ signals a more efficient judicial system. These two parameters affect γ_i in two ways:

1. By reducing credit rationing
2. By changing the relative return between H_T and L_T projects

The first channel operates, because a better judicial system (an increase in θ or ϕ) reduces the threshold A_{min} (see Equation 6.52) and the region of credit rationing.

So, the amount of disclosed assets increases accordingly, and the underground economy shrinks. An increase in ϕ also raises the return on H_T projects relative to L_T projects. This increases also the incentive to disclose additional assets and to invest in the H_T technology. We summarize the results of this paragraph in the following theorem.

Theorem 6.6 If other variables are held equal and constant, then an improvement in judicial efficiency reduces the size of the underground economy, and it also amplifies the impact of financial development on the size of underground economy.

Empirical evidence shows that the size of the underground economy differs considerably across sectors (Batra et al., 2003; Farrel, 2004). For instance, in the construction industry, underground activities are widespread, while the chemicals and drugs sectors are comprised mostly of formal enterprises. One of the reasons for this is due to labor market regulation, but most of the difference depends on the technologies involved in these sectors. On the other hand, a higher return of H_T projects (Q) does not affect the investment in H_T projects directly, but reduces credit rationing by lowering A_{min} (see Equation 6.61) and reducing the size of the underground economy. Therefore, the size of the underground economy in each sector depends on the relative returns of investment projects and the degree of credit rationing. More dynamic and competitive sectors (e.g., the financial sector or the chemicals industry) tend to have higher returns (Q) from their H_T projects. Firms in these sectors tend to have lower rates of underground activities because they are less likely to be credit-constrained. These sectors are more competitive, more technologically advanced, and are likely to exhibit a lower technological gap between H_T and L_T technologies. The opposite happens in less dynamic sectors (e.g., construction or retail), where new technologies are introduced at slower rates and firms can survive despite the implementation of mature technologies. We summarize the discussion in this paragraph in the following proposition.

Theorem 6.7 The size of the underground economy is dependent on the technological gap between L_T and H_T projects. More mature and less dynamic sectors tend to display higher rates of underground activities. If other variables are held equal and constant, then the impact of financial development on the underground economy is larger in these sectors.

6.4.3 Data

To test the main implications of the model, let us use the Bank of Italy's SHIW, which allows us to construct an index of underground activities based on microeconomic information. SHIW is a biannual cross-section of about 8,000 households and 24,000 individuals, and provides detailed information on demographic variables, income, consumption, and wealth. Survey data are available from 1977, but the main variable of interest for this subsection is available only in 1995, 1998, 2000, 2002, and 2004. The final sample includes 11,781 observations. The SHIW is a representative sample of the Italian resident population. The sample design is similar to the Labor Force Survey conducted by ISTAT (the Italian national statistics agency). Data are collected through personal interviews. Questions concerning the whole household are addressed to the family head or the person most knowledgeable about the family finances; questions about individual incomes are answered by individual household members wherever possible. The unit of observation is the family, which is defined to include all persons residing in the same dwelling who are related by blood, marriage, or adoption. Individuals selected as *partners or other common-law relationships* are also treated as families.

Obviously, tax evasion and underground activities are difficult to detect and measure. Individuals and firms who evade taxes or operate irregularly tend to hide their income from the government, and hence, are unlikely to release information on their hidden activities. This makes it difficult to obtain direct data on underground activities and is the reason why economists have tried different indirect measurement methods, such as the currency demand approach, the gap between effective and potential electricity consumption, or the multiple indicators approach. These methods are based on macroeconomic estimates of the size of the underground economy and have at least two limitations:

1. They are subject to large measurement errors.
2. By construction, the resulting indicators of underground activities are strongly correlated with other macroeconomic variables.

How to solve these measurement problems? One of the possible solutions is to construct an index of underground economy using microeconomic data. Another approach is to construct an alternative measure of underground economy. Following a standard approach of the literature, we proxy underground activities by calculating the fraction of income received in cash. The idea is that informal activities give rise to cash transactions. As our first indicator, this variable is based on the following question available in the SHIW: last year, did you receive part of your income in cash? And, in which fraction? As other proxies, this indicator has some limitations. One limitation is that it might be associated with different payment technologies, which may themselves be related to the level of local financial development. For this reason we use this variable only as a robustness check. As already mentioned, the SHIW provides also an indicator of local financial development. This indicator measures the probability that households have access to credit, that is, that they are not credit-constrained (Guiso et al., 2004). The SHIW asks households to report whether, in the 12 months before the interview, they have been denied credit or did not apply for credit because they thought they would be turned down. Based on this information, and controlling for other relevant variables, Guiso et al. (2004) estimated the probability that a potential borrower is turned down for credit or discouraged from borrowing, controlling for a wide range of individual and regional variables. The regional dummies obtained from the regression model are then normalized to be equal to zero in the region with the maximum value of the coefficient of the regional dummy (Calabria is the least financially developed region, while March and Liguria are the most developed) and therefore vary between 0 and 1 (the highest value is 0.58). Having collected indicators of irregular activities, financial development, and judicial efficiency, we can test some of the implications of the linear model by estimating equations of the form

$$U_{irs} = \alpha X_{irs} + \alpha_2 FD_r + \alpha_3 JUD_r + \alpha_4 Z_r + \mu_s + \varepsilon_{irs} \tag{6.53}$$

where

U_{irs} is an indicator of irregular activities for individual i in region r and sector s
X is a set of socioeconomic indicators
FD is the index of financial development
JUD is a measure of judicial inefficiency
Z is a set of regional indicators
μ_s is sector fixed effects
ε is an error term

Theorem 6.5 suggests that $\alpha_1 < 0$, and Theorem 6.6 suggests that $\alpha_2 > 0$ (because longer length of trails is associated with less judicial efficiency and an increase of underground activities).

It is necessary to analyze the relationship between financial development and the underground economy to address the issue of potential reverse causality and endogeneity of financial development FD in Equation 6.53. In particular, an increase in underground activities (e.g., due to an increase in general taxation) reduces the demand for credit, hampering financial market growth. Similarly, low GDP growth might reduce the demand for loans and financial development, while at the same time increasing underground activities. This implies that simply observing that low financial development is associated with a high level of underground activities does not necessarily mean that low financial development actually causes more underground activities.

It is necessary to analyze the relationship between financial development and the underground economy to address the issue of potential reverse causality and endogeneity of financial development FD in Equation 6.53. In particular, an increase in underground activities (e.g., due to an increase in general taxation) reduces the demand for credit, hampering financial market growth. Similarly, low GDP growth might reduce the demand for loans and financial development, while at the same time increasing underground activities. This implies that simply observing that low financial development is associated with a high level of underground activities does not necessarily mean that low financial development actually causes more underground activities.

In aforementioned microeconomic data, we address the endogeneity of financial development relying on an instrument proposed by Guiso et al. (2004), which is correlated with financial development, but is not affected by the degree of underground economy. The instrument is based on the characteristics of the 1936 Banking Law, which over time has constrained the growth of the Italian banking system and is an exogenous determinant of the trajectories of local financial development. Following a period of frequent banking crises, in 1936, Italian legislators attempted to stabilize the financial system by strictly limiting in each region the number of banks and bank branches. In achieving this aim, the law has worked very well, as witnessed by the fact that the number of new branches in Italy after 1936 has expanded very little. Yet, in some regions and for some local credit institutions (such as savings banks and cooperative banks), the 1936 Banking Law has been less constraining. Therefore, the 1936 law explains a large part of the variability in local financial development even after 60 years. These authors tested this hypothesis by estimating the correlation between the index of regional financial development and the characteristics of the banking system before the 1936 law. They found that 1936 bank branches, local branches, saving banks, and cooperative banks (each in per capita terms) explain 72% of the regional variation in credit supply in the 1990s. In our empirical estimates, we use the same instrument that is uncorrelated to underground economy to control for endogeneity in financial development.

One of regressions is shown in the following text, which is significant at the 1% level, where there are 11,781 observations and R-square is 012:

$$U_{irs} = 1.606 - \underset{(0.012)}{0.034x_1} - \underset{(0.003)}{0.022x_2} - \underset{(0.007)}{0.02x_3} - \underset{(0.008)}{0.079x_4} - \underset{(0.009)}{0.026x_5}$$
$$\underset{(0.121)}{\phantom{U_{irs}=1.606}}$$

$$- \underset{(0.037)}{0.223FD_r} + \underset{(0.026)}{0.083JUD_r} \tag{6.54}$$

where x_1 = male, x_2 = age, x_3 = years of education, x_4 = log disposable income, and x_5 = married.

6.4.4 Conclusions

The existence of a large underground economy represents a relevant burden on society. The underground economy can slow the investment rate, reduce the adoption of new technologies, and limit the ability of governments to raise sufficient resources to pay for public goods and for infrastructure. Eventually, it can affect the allocation of real resources and thwart economic growth. A high level of taxation, cumbersome and inefficient bureaucracy, and poor legal protection are among the factors that have been identified as the major causes of tax evasion and a large underground economy. In this section, we focus on financial development, a factor that has received less attention in recent years from economists.

The main idea is that when individuals and firms hide all or part of their income, they pay less tax, but they also face a higher cost of credit. Therefore, the choice of operating in the underground economy involves a trade-off. By reducing the cost of credit or by granting credit to previously credit-constrained agents, financial development affects the trade-off, increasing the incentive to operate in the formal economy. We capture these ideas in a simple model in which agents choose to disclose their collateral in order to obtain credit for investment in a high-return project. The alternative is to operate in the informal sector in a low-return project using only internal funds. The choice to go underground therefore is also a choice between different technologies. The model predicts that financial development (a reduction in the cost of credit) induces firms to disclose more assets and to invest in a high-tech project, and that this effect is stronger in mature sectors. Furthermore, an improvement in judicial efficiency reduces the cost of credit and the size of the underground economy. In the second part of the section, we test the main implications of the model by using Italian microeconomic data. We build an index of job irregularity by using the 1995–2004 Bank of Italy SHIW, and regress this index over an indicator of local financial development, judicial inefficiency, and other individual and regional variables. The results show that the underground economy is strongly negatively correlated with financial development, even when we control for financial development endogeneity. We find also that more competitive and innovative sectors display a lower level of underground activity and that financial development has a stronger impact in mature sectors (such as construction, retail, and tourism). The effect of judicial inefficiency is in line with the model's predictions, but the coefficient is not statistically different from zero if we control for other regional variables. Our study implies that successful programs to reduce the extent of the underground economy should take into account the structure of credit markets, and implies also that financial market development has important spillover effects. By reducing the incentives to operate in the underground economy, financial market development can stimulate the adoption of new technologies, reduce the size of the underground economy, and increase the levels of tax collection.

6.5 Who Cares about My Health? The Baumol Model

6.5.1 Background

As it is well known, health care expenditures continue to grow throughout the world. For example, health care spending as a percent of GDP in the United States stood at slightly over 5% in 1960; today, it stands at nearly 18%. What is less well known are the various reasons that have been offered for rising health care expenditures. Among the more popular explanations, Newhouse (1992) and others have argued that new medical technologies, growing national income, and an aging population likely account for most of the growth in health care costs throughout the world. Largely overlooked by many analysts is an explanation proposed by Baumol nearly 40 years ago. In his now classic article, Baumol pointed out that industrialized economies contain both progressive (e.g., manufacturing) and nonprogressive (e.g., services) industries (Baumol, 1967). The progressive industries are characterized as being capital intensive in nature and subject to many process innovations over time. Such characteristics help to raise labor productivity in the progressive industries. While the wage rate tends to rise in the progressive industries over time, unit costs remain fairly flat because of proportional increases in labor productivity. In contrast, Baumol viewed nonprogressive industries, such as medical care and education, as being more labor intensive in practice and relatively void of productivity-enhancing innovations that substitute for labor or enable labor to produce more output in a given amount of time. Consequently, labor productivity tends to be relatively stagnant in these industries. Baumol pointed out that wage rates in the nonprogressive industries tend to increase with higher wage rates in the progressive sector. The higher wage rates are necessary for nonprogressive industries to attract more workers over time. However, the higher wage rates in the nonprogressive industries are not necessarily matched with parallel increases in productivity because of the high labor intensity and lack of dynamic efficiencies. Thus, given the rise in wage rates but relatively constant productivity due to stagnant technologies, unit costs are driven up in nonprogressive industries over time. Relative prices also increase in the nonprogressive industries because of the reduced supply of output resulting from higher wage rates. Moreover, because the market demands for services, such as medical care and education, tend to be price inelastic, the rise in price also causes expenditures, the product of price and quantity, to increase on a continual basis.

Additionally, in conjunction with the relative price increase, expenditures may rise if the government subsidizes price in the nonprogressive sector or if the demand increases because of market forces. Despite the logical nature of Baumol's model, relatively little research has subjected his theory to econometric testing, particularly with respect to the medical care sector. The few exceptions are Colombier (2010) and Hartwig (2011), which examined if Baumol's model of unbalanced growth can account for differences in the growth of health care costs among a number of countries belonging to the Organization for Economic Cooperation and Development (OECD). While these few studies found evidence to support Baumol's model, it remains unclear how much Baumol's cost disease theory accounts for the rising health care spending throughout the United States. Understanding whether or not the Baumol phenomenon accounts for a significant portion of the rising health care costs throughout the United States is important from a public policy perspective. For instance, the Patient Protection and Affordable Care Act (PPACA) of 2010 is introduced to *bend the cost curve* through a variety of policies. If unbalanced growth accounts for a significant share of the growth of health care spending, then some of these policy proposals may be ill advised and/or unproductive. This section fills this knowledge gap by modeling and empirically examining whether or not Baumol's cost disease of the service sector theory at least partially accounts for the growth of health care costs in the United States. More specifically, Baumol's cost disease theory is empirically tested using a panel data set of U.S. states over the time period from 1980 to 2009.

6.5.2 Nonlinear Model

Although powerful in its predictions, Baumol's model is relatively simple both conceptually and mathematically. First, Baumol assumed that only one type of input, labor L, generates output in both the nonprogressive Y^{NP} and progressive Y^P sectors. He also assumed that the amount of output generated by labor at time t can be written as

$$Y^{NP}(t) = aL_{NP}(t) \tag{6.55}$$

$$Y^P(t) = bL_P(t)e^{rt} \tag{6.56}$$

Note that Equation 6.55 implies that the output is proportional to labor in the stagnant sector, whereas output in the progressive sector continues to grow at a constant growth rate of r over time. From these two production functions, the corresponding marginal productivities of labor dY/dL at time t can be derived, respectively. Note that the marginal productivity of labor in the stagnant sector remains constant, whereas the marginal productivity of labor in the progressive sector rises over time.

Baumol further supposed that wage rates in the two sectors are equal at each point in time and that they rise over time in conjunction with productivity improvements in the progressive sector. Thus, the wage rate in the two sectors at each point in time can be expressed as

$$W(t) = W_0 e^{rt} \tag{6.57}$$

From our single input production functions, it follows that the unit cost (C) can be written as the ratio of wage rates to marginal productivities in the two sectors as follows:

$$C^{NP}(t) = \frac{W_0 e^{rt}}{a} \tag{6.58}$$

$$C^P(t) = \frac{W_0 e^{rt}}{be^{rt}} = \frac{W_0}{b} \tag{6.59}$$

One implication of Baumol's model is that the unit cost in the progressive sector remains constant over time (Equation 6.59). Another implication is that the unit cost continually rises over time in the stagnant sector based upon the rate of productivity growth in the progressive sector (Equation 6.58). This latter result is referred to as Baumol's cost disease of the service sector.

According to Hartwig (2008), Baumol's theory could be tested by examining if changes in the unit cost in the stagnant sector are directly proportional to the excess of wage rate growth less labor productivity growth in the overall economy, or symbolically:

$$\Delta \log(C^{NP}) = \lambda[\Delta \log(W) - \Delta \log(Y)] \tag{6.60}$$

The left-hand side variable represents the growth of the unit cost in the nonprogressive sector. The expression in brackets, named the Baumol variable by Hartwig, is central to the analysis. The variables W and Y measure the economy-wide wage and output per worker, so the expression in brackets captures the difference between overall wage rate growth and productivity gains. As just discussed, Baumol's theory suggests that laborers in both sectors experience similar wage rate adjustments, but the stagnant sector is entirely responsible for any productivity shortfalls. Hence, overall wage rate increases in excess of productivity improvements drive the unit cost in the stagnant sector. Thus, an estimated positive value for λ provides evidence for the Baumol cost disease according to Hartwig.

Colombier (2010) showed mathematically that Hartwig's analysis, as depicted in Equation 6.60, represents a special case when all labor is allocated to the nonprogressive sector. He showed mathematically that a general formulation of the model can be expressed as

$$\Delta \log(C^{NP}) = \frac{\beta_1[\Delta \log(W) - \Delta \log(Y)]}{L_{NP}/L_T} \tag{6.61}$$

The implication of Colombier's model is that Hartwig's λ approaches 1 asymptotically as the nonprogressive sector's share of total labor (LT) approaches 100% over time. In Equation 6.61, a positive value for β_1 provides support for Baumol's cost disease. For empirical purposes, note that Equation 6.61 can be rewritten as

$$\frac{L_{NP}}{L_T} \cdot \Delta \log(C^{NP}) = \beta_1[\Delta \log(W) - \Delta \log(Y)] \tag{6.62}$$

Similar to Hartwig (2008) and Colombier (2010), this section focuses on the effect of the Baumol variable on the growth of health care costs. Most agree that the health care sector fits the characteristics of a nonprogressive industry due to its high labor intensity and general lack of substitutability between labor and capital (Baumol, 1967; Hartwig, 2008, 2011; Colombier, 2010). Our study differs from what Hartwig and Colombier did because it uses a balanced panel data set of all states within the United States over the time period from 1980 to 2009 to estimate Equation 6.62.

To isolate the impact of the Baumol variable, it is important to control for other factors affecting the growth of health care costs. As mentioned in the introduction, many analysts, such as Newhouse (1992), believe that new medical technologies, income, and an aging population account for most of the growth in health care spending. Incorporating these factors as much as possible, the equation, predicting the health care labor-share adjusted growth of health care expenditures, HCE, is specified as follows:

$$\left(\frac{L_H}{L_T}\right)_{s,t} \cdot \Delta \log(HCE_{s,t}) = \beta_0 + \beta_1[\Delta \log(W_{s,t}) - \Delta \log(Y_{s,t})]$$

$$+ \alpha \Delta \log(Z_{s,t}) + \eta_s + \tau_t + t \cdot \eta_\varepsilon + \varepsilon_{s,t} \qquad (6.63)$$

with L_H/L_T representing the share of total labor employed in the health care sector. Notice that Equation 6.63 bears a close resemblance to Equation 6.62 except now it formally allows for variation in the various variables across states and over time, hence the subscripts of s and t. Also, a vector Z of continuous variables is included to control for some of the previously mentioned factors affecting health care spending growth along with its vector of slope parameters α, state η_s, and time τ_t; fixed effects are also specified in the estimation equation. The state fixed effects help to control for any unobservable factors affecting the growth of health care spending, such as the proclivity to adopt new medical technologies in each state, and thereby help to reduce the bias normally associated with unobservable heterogeneity. The time-fixed effects capture changes common to all states over time, such as new medical technologies and price inflation. Also, individual state time trends $t \cdot \eta_s$ are specified in the estimation equation, which allows health care expenditures to grow at different rates in different states. The individual state time trends may also control for any other factors that are trending over time within the various states and which may simultaneously affect both the Baumol variable and health care costs. The only other variable not yet identified is the error term $\varepsilon_{s,t}$.

Notice that Equation 6.63 is estimated in first difference form after transforming the variables into logarithms. Hartwig (2011) pointed out that Gerdtham and Jonsson (2000) recommended macroeconomic variables specified as growth rates when conducting regression analysis because of unit roots possibly present in the data. One unit root means that a variable passes through its mean only once. The presence of unit roots in the data can lead to spurious regression results (driven by a high degree of correlation between two variables but lacking any causality) with high adjusted R^2 and t-statistics and little, if any, economic interpretation. First differenced data series typically pass through their mean values much more often than once. Both Hartwig (2008) and Colombier (2010) estimated their equations in first difference form after transforming the variables into logarithms. Specific variables included in vector Z for the empirical test are gross state product (GSP) per capita, the percentage of the population 65 years of age and older, and the poverty, union coverage, and unemployment rates. These specific control variables are included in the empirical model because consistent data are available for them across states and over time. Admittedly, these control variables are ad hoc in nature, but they may control for some potentially important factors. For example, aging, income, and insurance coverage can be mentioned as main drivers of health care spending growth. The poverty (i.e., percentage of the population with incomes below the federal poverty level), union coverage, and unemployment rates serve as proxy variables for the latter variable given that a more direct measure of insurance coverage is unavailable.

Data for nominal health care costs per capita, HCE, in each state over time come from the Bureau of Economic Analysis. These cost figures reflect the amount of GSP generated by the private ambulatory care, hospital, and nursing and residential care industries (i.e., NAICS categories 621–623) in each state over time. These figures are used instead of the spending data from the health care accounts at the Center for Medicare and Medicaid Services (CMS) because corresponding data for employment in the health care sector are also necessary to conduct the empirical test. Corresponding data for health care costs and employment can only be obtained from the BEA for those three health care industries. The GSP generated by these three industries represent about 78% of all personal health care spending in the United States. The difference between the two measures is that the GSP generated from other health care goods and services, such as health insurance, prescription drugs, and medical devices, and the public sector production of medical care are not included in these BEA estimates. Nevertheless, based upon the state-year observations used in this study, the two types of health care costs are highly correlated over time ($r = 0.973$).

The overall nominal wage per worker, W, in Equation 6.63 is measured by dividing total nominal wages and salaries (including sponsored benefits) by the total number of persons employed, including the self-employed, for each state-year observation. Economy-wide output per worker, Y, is calculated by dividing the real GSP by the total number of persons employed, including the self-employed, for each state-year observation. Data for the nominal wage per worker and real GSP in each state are obtained from the U.S. Bureau of Economic Analysis. Figures for the unemployment rates are found from the U.S. Bureau of Labor Statistics. Data for population, the poverty rate, and the percent 65 years of age and older are gathered from the U.S. Bureau of the Census. Finally, the union coverage data are obtained from unionstats.com and are described by Hirsch and Macpherson (2003).

6.5.3 Regression Results

When estimating Equation 6.63, standard errors are made fully robust against arbitrary heteroskedasticity and serial correlation by clustering them at the state level. Equation 6.63 is initially estimated by using the OLS technique. The OLS findings are reported in Table 6.7. As a robustness test, the specification behind the results in column 3 includes individual state time trends whereas those in column 2 do not.

Notice from Table 6.7 that roughly two-thirds of the variation in the growth of health care costs can be explained by the right-hand side variables in the two multiple regression equations. Given the statistically significant coefficient estimates on the growth of GSP per capita, the results suggest that health care is a normal good as many other studies have also confirmed. In addition, according to these OLS results, health care costs grow more rapidly when the elderly take on an increasing share of the population. Lastly, the findings imply that health

Table 6.7 OLS Results

	Estimated Coefficient (Absolute Value of t-Statistic)	
Constant	0.003*** (20.75)	−0.053(1.22)
Baumol variable	0.011*** (4.52)	0.009*** (3.42)
Growth of nominal GSP per capita	0.014*** (6.82)	0.010*** (4.85)
Growth of percent old	0.013*** (2.89)	0.007* (1.35)
Growth of the unemployment rate	0.001*** (2.78)	0.001*** (2.29)
Growth of the union membership rate	−0.00001 (0.03)	−0.00006 (0.18)
Growth of the poverty rate	−0.00009 (0.53)	−0.0001 (0.67)
Time-fixed effects	Yes	Yes
State-fixed effects	Yes	Yes
Individual state time trends	No	Yes
Adjusted R^2	0.647	0.661
D.W. statistic	1.70	1.81
Number of observations	1450	1450

Note: (1) Standard errors clustered at the state level. (2) Panel data ordinary least-squares estimation. Here, * implies statistical significance at the 10% level, ** implies statistical significance at the 5% level, and *** implies statistical significance at the 1% level.

care costs rise more quickly when the unemployment rate grows faster but that the growth of health care costs is unrelated to both the growth of the union and the rate of poverty. The direct relationship between the health care costs and growth rates of unemployment may reflect the increased spending on Medicare that takes place during a recession. Of more importance for the main hypothesis, the estimated coefficients of the Baumol variable are positive and statistically significant in both specifications. In addition, the magnitudes of the two estimated coefficients of the Baumol variable are reasonably close. Individual state time trends often dominate the explanatory power of an estimation equation, so the robustness of the Baumol variable in the presence of individual state time trends is quite remarkable.

Before regression, let us look at the descriptive statistics of the data. Baumol collected data in this case study from 11 different indexes, which are, respectively,

1. Share of labor in the health care sector
2. Growth of nominal health care costs per capita
3. Growth of overall nominal wages per worker
4. Growth of real GSP per worker
5. Baumol variable
6. Growth of nominal GSP per capita
7. Growth of fraction old
8. Growth of unemployment rate
9. Growth of union coverage rate
10. Growth of poverty rate
11. Growth of housing prices in the previous year

At the viewpoint of mean, the maximum value of mean is growth of nominal health care costs per capita. Its value is equal to 0.068. The minimum value of mean is growth of union coverage rate. Its value is equal to −0.024. At the viewpoint of deviation, the maximum value of standard deviation is growth of fraction old. Its value is equal to 0.012. The minimum value of standard deviation is growth of unemployment rate. Its value is equal to −0.024.

Notice that roughly two-thirds of the variation in the growth of health care costs can be explained in the two multiple regression equations. If the average coefficient estimate of 0.010 on the Baumol variable is divided by the average health care labor share of 0.68, the results can be compared to the estimates reported by Hartwig (2008) and Colombier (2010). In any case, much like their results for OECD countries, this OLS analysis implies that the U.S. health care economy is afflicted by Baumol's cost disease. One potential problem with these OLS results, however, is that the estimated coefficients on the Baumol variable may be influenced by endogeneity bias. In particular, a shock leading to a change in the amount of GSP produced in the health care sector may affect the Baumol variable through its impact on wages and productivity in the overall economy. Stated alternatively, instead of contracting the Baumol disease, health care industries may cause the Baumol disease. If so, this reverse causality means that only inferences regarding association, rather than causation, between the Baumol variable and the growth of health care costs can be drawn from the OLS results.

To identify if a causal relationship holds true, the two-stage least-squares (2SLS) technique is used to reestimate Equation 6.63. In the first stage equation that predicts the current-year Baumol variable, the nominal housing price growth rate during the previous year is specified as an instrumental variable. Recall that the growth rate of nominal wages per worker shows up as one of the two components that make up the Baumol variable. The plausibility of specifying the housing-price growth rate as an instrument rests on the notion that faster-growing housing prices lead to a greater cost of living in a state. The greater cost of living, in turn, translates into higher nominal wages because employers are required to offer a compensating wage differential to employees or face the prospect of losing workers to states with slower-growing housing prices. As conditioned on the other covariates, the correlation between the previous-year housing price growth rate and the current-year wage growth rate should offer a suitable identification strategy.

The first stage of the 2SLS procedure tests if the previous-year housing price growth rate does indeed correlate strongly with the current-year Baumol variable. The average housing price data for each state-year observation come from the Lincoln Land Institute. The first stage results are reported in Table 6.8 for specifications with and without individual state time trends. As expected, for both specifications, the estimated coefficients are positive and statistically significant on the past growth rate of housing prices. A Wald test, which restricts the coefficients on housing price growth to be equal to zero, confirms a strong correlation between the previous-year housing price growth rate and the current-year Baumol variable, producing F-statistics of 51.7 and 36.9, respectively. These F-statistics exceed the threshold value of 10 suggested by Stock and Watson (1993) for detecting weak instruments given one endogenous regression.

Table 6.9 reports the second stage results. Qualitatively, the 2SLS results are very similar to the OLS findings. However, quantitatively, the estimated coefficients of the Baumol variable are now larger in magnitude. Another difference is that the coefficients are estimated with less precision. Nevertheless, they remain greater than zero at conventional levels of statistical significance (especially for a one-tailed test which is appropriate). These 2SLS results provide further evidence that Baumol's cost disease inflicts the inpatient and outpatient sectors of the U.S. health care system. Initial specifications also included population growth as another covariate in the estimation equation, but a referee recommended against its inclusion. Population growth controls for a host of reasons why health care cost growth may vary across states including density, agglomeration economies, and population heterogeneity. When included in the estimation equation,

Table 6.8 First-Stage Results (Dependent Variable: Baumol Variable)

	Estimated Coefficient (Absolute Value of t-Statistic)	
Constant	0.056*** (59.0)	−0.135 (0.26)
Growth of housing prices	0.066*** (7.19)	0.054*** (6.08)
Growth of nominal GSP per capita	−0.677*** (56.17)	−0.710*** (59.35)
Growth of percent old	−0.270*** (5.60)	−0.361*** (6.93)
Growth of unemployment rate	−0.046*** (14.63)	−0.045*** (14.72)
Growth of union membership rate	−0.004 (1.17)	−0.004 (1.15)
Growth of the poverty rate	−0.001 (0.64)	−0.001 (0.69)
Time-fixed effects	Yes	Yes
State-fixed effects	Yes	Yes
Individual state time trends	No	Yes
Adjusted R^2	0.792	0.802
D.W. statistic	1.82	1.89
Number of observations	1450	1450

Note: (1) Standard errors clustered at the state level; (2) panel data ordinary least-squares estimation. Here, *** implies statistical significance at the 1% level.

***$p < 0.01$.

the coefficient on population growth turns out to be negative and statistically significant in all four estimated equations. The only difference in the overall results is that the estimated coefficient on the growth of percent old ends up being no different from zero and the estimates improve on the Baumol variable (e.g., for the weaker fourth specification, it increases to 0.050 with a t-statistic of 2.36). Apparently, greater population growth is picking up an increasing percentage of younger people, which swamps the effect of percent old, but also captures some other effects associated with population growth besides a changing age distribution.

Before closing, we should briefly discuss the empirical findings associated with a falsification test involving this same panel data set of U.S. states. The falsification test basically replaces in Equation 6.63 the labor-share adjusted health care cost growth rate with the labor-share adjusted manufacturing cost growth rate. If the previous results are actually capturing some systematic trends in the data over time rather than the hypothesized Baumol cost disease

Table 6.9   Results of 2SLS

	Estimated Coefficient (Absolute Value of t-Statistic)	
Constant	0.002** (2.37)	−0.052 (1.14)
Baumol variable	0.037*** (2.79)	0.028** (1.79)
Growth of nominal GSP per capita	0.031*** (3.45)	0.024** (2.15)
Growth of percent old	0.024*** (3.02)	0.016 (1.63)
Growth of unemployment rate	0.002*** (3.00)	0.002** (2.07)
Growth of union membership rate	0.00008 (0.22)	0.00001 (0.03)
Growth of the poverty rate	−0.00004 (0.23)	−0.00007 (0.43)
Time-fixed effects	Yes	Yes
State-fixed effects	Yes	Yes
Individual state time trends	No	Yes
Adjusted R^2	0.622	0.647
D.W. statistic	1.73	1.83
Number of observations	1450	1450

Note: (1) Standard errors clustered at the state level; (2) panel data two-stage least-squares estimation. Here, ** implies statistical significance at the 5% level, and *** implies statistical significance at the 1% level.

**p < 0.05 and ***p < 0.01.

relationship, the results for the manufacturing industry with respect to the Baumol variable should resemble those associated with the health care industry. That is, the Baumol variable should also directly relate to manufacturing costs. The same four equations are estimated as those shown in Tables 6.7 and 6.9. In only one case, the basic OLS specification without individual state time trends, is the coefficient estimate both positive and statistically significant (estimate of 0.146 with *t*-statistic of 1.66) on the Baumol variable in the manufacturing cost equation. For the other three specifications, however, and in agreement with the underlying theory, the estimated coefficient on the Baumol variable is no different from zero at conventional levels of statistical significance. Taken as a whole, it seems that the proper conclusion to draw from all of these tests is that the findings reflect the presence of Baumol's cost disease rather than some spurious correlation among the data.

6.5.4 Conclusions

In this section, we use a set of panel data of U.S. states and an empirical strategy developed by Hartwig (2008) and Colombier (2010) to test whether or not the health care sector in the United States suffers from Baumol's cost disease. A set of panel data of U.S. states over the time period from 1980 to 2009 is used in the empirical analysis. The empirical test allows for time- and state-fixed effects of individual state time trends, and 2SLS estimation. Overall, a reasonably confident YES is the answer to the question raised by the title in this section.

The major implication is that Baumol's cost disease cannot be ignored as a valid reason for rising health care costs in the United States in addition to other causes such as higher income, an aging population, and new medical technologies. Interestingly, Newhouse (1992) remains doubtful about Baumol's cost disease "because I do not believe we have any good empirical basis for the medical care expenditure into increases in price and increases in quantity." Thankfully, the methodology developed by Hartwig (2008) and Colombier (2010) allows us to sidestep that difficult challenge and test for Baumol's cost disease in an alternative way.

If corresponding data become available for both employment and costs for other health care industries, it may be insightful to examine how different health care industries respond to Baumol's cost disease over time. For instance, costs in the medical device and pharmaceutical retailing industries may not react to Baumol's cost disease to the same degree as the outpatient and inpatient care industries do. Even among various types of inpatient cares, different reactions to the disease may be observed. For example, costs in the nursing home industry may respond differently to Baumol's cost disease than costs in the hospital services industry. At the very least, this study offers a useful template for conducting a study of that kind when those data become available for the United States or other countries.

6.6 Sail against the Current: Held Currencies in Own Hands

6.6.1 Background

The real exchange rate (*RER*) does not play a central role in traditional growth theory. Both the canonical Solow–Swan growth model and endogenous growth models feature closed economies. However, Ricardo's and Lewis's theories of economic growth suggest a more important role for the *RER*. As nations develop, the *modern* manufacturing sector absorbs *surplus labor*, which directly translates into higher national output. The *RER*, which creates an incentive to allocate resources to the modern manufacturing sector, is therefore of first-order importance to economic growth. The question of what is the optimal relative price of traded goods arises. There are two opposing views on the answer to that question. The aim of this section is to shed additional light on this debate.

The *Washington Consensus*, articulated by Williamson (1990), acknowledges a crucial role of the *RER* in the growth process. According to this view, an appropriate RER should be consistent with macroeconomic objectives in the medium run and *sufficiently competitive* such that exports grow at a rate consistent with external balance. However, an overly competitive *RER* is not appropriate because it would fuel inflation and curb resources available for investment. Underlying this view is the notion that there exists an equilibrium real exchange rate (*ERER*) that satisfies external and internal balances (Nurkse, 1947). Seen in this light, any deviation from the *ERER* will hamper economic growth.

The opposing view, with Rodrik (2008) at the forefront, maintains that *RER* overvaluation harms growth and undervaluation promotes it. This stance is in part due to the success story of export-led growth in conjunction with apparently undervalued currencies in East-Asian countries. But there are also other plausible explanations for why real undervaluation is good for growth. In the export-oriented growth literature, it is often argued that the manufacturing sector is special because positive externalities (learning-by-doing effects, technology spillovers) are more pronounced for export-linked activities than other sectors of the economy. Another explanation is that an undervalued *RER* encourages higher savings and investment (Levy-Yeyati and Sturzenegger, 2007). Finally, Rodrik (2008) conjectured that the manufacturing sector in developing countries is disproportionately subject to distortions; hence, it is below its optimal size in equilibrium. Because removing those distortions proves difficult in practice, an undervalued *RER* serves as a *more practical* second-best mechanism to optimally reallocate resources toward the manufacturing sector (Rodrik, 2008).

However, there is little systematic evidence supporting any of these views. The nature and prevalence of those positive externalities associated with exporting remain obscured (Harrison and Rodríguez-Clare, 2009). Rodrik (2008) was unable to empirically verify that the manufacturing sector is disproportionately subjected to distortions in developing countries. In addition, all of these propositions seem to ignore the distortion cost associated with real undervaluation in the form of reduced aggregate demand (Corden, 1981). It is therefore not clear if the gain in exports outweighs the loss in absorption, especially over longer time horizons. Finally, according to Edwards (1989, 1995), *RER* distortions can lead to resource misallocation across sectors as economic agents base their investment decisions on a relative price in disequilibrium. Because the *RER* tends to adjust to equilibrium over time, real undervaluation may induce investments in short-lived projects.

The early empirical literature identifies a negative impact of *RER* overvaluation on growth but does not address *RER* undervaluation (Ghura and Grennes, 1993).

However, recent empirical studies unanimously reject the Washington Consensus view in the sense that they find a positive effect of *RER* undervaluation on economic growth. The most prominent example is Rodrik (2008), whose empirical findings suggest that higher medium-term growth is systematically associated with undervalued exchange rates in developing countries. While Rodrik defines purchasing power parity (PPP) adjusted for the Balassa–Samuelson effect as the *ERER*, there is also a sizable number of empirical studies estimating ERERs consistent with internal and external balances that broadly reach the same conclusion (Berg and Miao, 2010; MacDonald and Vieira, 2010). Since these two concepts vastly differ from one another and are not directly comparable, this section considers both *ERER* definitions but with the prime focus being on *RER* misalignment in the sense of Nurkse (1947).

There are two important sources of inconsistencies driving previous results and the bulk of the literature suffers from at least one of these. Firstly, relying on conventional panel data techniques to estimate ERERs imposes strong homogeneity assumptions on cross-country long-run *RER* behavior. This approach does not conform to the economic theory underlying the *ERER* and therefore generates misleading results. Secondly, the objective to infer the effect on the growth of two variables (real over- and undervaluation) from a single continuous variable (*RER* misalignment) introduces a number of pitfalls, which can lead to growth regression misspecification.

This section explicitly takes into account heterogeneity in long-run *RER* behavior across countries by individually estimating *RER* misalignments for 63 developing countries over the period from 1970 to 2007. It then empirically analyzes how *RER* over- and undervaluation affect economic growth. To this end, the study employs system-generalized method of moments (SGMMs) developed by Blundell and Bond (1998). To ensure robust inference, various measures of *RER* misalignment are used.

The empirical results provide evidence in favor of the Washington Consensus view and reject the notion that *RER* undervaluation is an expedient development policy tool. This means that the optimal growth promoting relative price of traded goods is the value of the ERER. The study in this section also shows that the identified inconsistencies drive previous results, rather than differences in estimation methods or data sets.

6.6.2 Model

Before the relationship between *RER* distortions and economic growth can be analyzed, deviations of the actual *RER* from its equilibrium value need to be estimated. The problem, which any empirical study on this subject faces, is that the *ERER* is not directly observable. The starting point to resolve this issue is to define the *RER* and the *ERER*.

The RER is defined as the domestic relative price of traded to nontraded goods. That is, $RER = E \times PT/PN$, where E denotes the nominal exchange rate (measured as domestic currency per foreign currency). And PT and PN refer to the price of tradables and nontradables, respectively. Note that an increase in RER indicates depreciation. The equilibrium real exchange rate ($ERER$) in the sense of Nurkse (1947) is defined as that value of the RER that results in the simultaneous attainment of both internal and external equilibriums, given sustainable values of relevant variables achieving this objective. Nurkse's definition directly implies that the $ERER$ is determined by a set of macroeconomic fundamentals. Based on Montiel (1999b, 2007), the $ERER$ is a function of the following variables:

$$ERER = ERER(TOT, \phi, \xi, G_N, G_T, I, NFA) \tag{6.64}$$

where
  TOT refers to the terms of trade
  ϕ is a measure of trade policy
  ξ captures productivity differentials (Balassa–Samuelson effect)
  G_N and G_T are the government consumption on nontradables and tradables
  I refers to investment
  NFA refers to the net foreign asset position

Importantly, theoretical priors point to an ambiguous effect of some fundamentals on the $ERER$, as shown by the signs of the partial derivatives below. In case a country faces a binding credit constraint, the trade surplus will depend on exogenous foreign aid flows (Baffes et al., 1999). Therefore, Equation 6.64 takes a modified form:

$$ERER = ERER(TOT, \phi, \xi, G_N, G_T, I, TS) \tag{6.65}$$

where the net foreign asset position has been replaced with the trade surplus TS. These two specifications differ fundamentally with regard to the underlying assumption of how the stock of net international indebtedness feeds back on net capital inflows and the $ERER$. The former conditions the $ERER$ on given (sustainable) values of the stock of net international indebtedness, which also affects the nonexogenous component of net capital inflows (Montiel, 1999a). The latter on the other hand specifies the $ERER$ as a function of exogenous (sustainable) net capital inflows only, with no feedback from the accumulated stock of net foreign assets. Therefore, the concept of external balance is a *stock-flow* approach in Equations 6.64 and 6.65 (Faruqee, 1995; Elbadawi, 2012) and a *flow* approach in Equations 6.64 and 6.65 (Montiel, 1999a). As for the question, which of the two specifications is relevant for a given country, the answer will depend on the country's economic structure.

Intuitively, the flow approach is suitable in foreign aid–receiving low-income countries, whereas the stock-flow approach fits better for middle-income countries. However, rather than imposing possibly restrictive assumptions in what follows, the approach will be to let the data *choose* the appropriate specification. The approach adopted to empirically estimate ERERs in this section is the single-equation approach developed by Baffes et al. (1999). It comprises three steps:

1. Estimate the long-run equilibrium relationship between the *ERER* and its fundamentals.
2. Derive sustainable values of those fundamentals that explain long-run *RER* behavior.
3. Calculate the *ERER* and the degree of misalignment.

According to this approach, the empirical equivalent of Equations 6.64 and 6.65 under the assumption of linearity in the theorized long-run relationship takes the following form: uncovering β involves estimating some form of the empirical model in Equation 6.66 except that *ERER* and *FS* have to be replaced with their observable counterparts, the actual *RER* and actual values of the fundamentals, respectively:

$$\ln ERER_{it} = \beta_i^T FS_{it} + \varepsilon_{it} \qquad (6.66)$$

where
   subscript *i* refers to the country in question
   FS is the vector of the set of fundamentals at sustainable values
   the vector β contains the long-run parameters to be estimated
   the error term ε_{it} is assumed to be stationary with zero mean

Once β and *FS* are derived, the degree of misalignment (mis_{it}) can be calculated with the following formula:

$$mis_{it} = \frac{ERER_{it} - RER_{it}}{RER_{it}} \qquad (6.67)$$

where positive (negative) values of mis_{it} indicate overvaluation (undervaluation).

The approach in this section is to estimate ERERs for each country individually, and panel data imposing the homogeneity assumption will also be employed for comparison purposes.

The first step is to determine the order of integration of the variables using Augmented Dickey–Fuller (ADF) as well as Phillips–Perron (PP) tests. In case the unit root null is rejected, the *RER* follows a stationary process, providing evidence for relative PPP to hold. The *ERER* can then be set at the sample mean.

The second step is to estimate the long-run parameters (β_i). Estimating the full joint distribution of the *RER* and its fundamentals would be desirable, but small samples with at most 38 observations and 6 possible fundamentals render system estimation practically infeasible (Baffes et al., 1999). This motivates a single-equation setting. In principle, the static OLS estimator (SOLS) can be used to estimate β in Equation 6.67. While SOLS is superconsistent (the rate of convergence is proportional to the sample size) and there is no asymptotic bias from simultaneous equations or measurement error (Phillips and Durlauf, 1986), SOLS performs poorly for small samples (Banerjee et al., 1993) and the nonstandard distribution of the t-statistics prevents valid inference. Stock and Watson (1993) propose a dynamic OLS estimator that can handle these issues.

The augmented version of Equation 6.67 with m_1 leading and m_2 lagging takes the following form:

$$\ln RER_{it} = \beta'_{i1} F_{it} + \beta'_{i2} Z_{it} + \sum_{s=-m_2}^{m_1} \gamma'_{is} \Delta F_{it+s} + v_{it} \qquad (6.68)$$

where the vectors F and Z contain the fundamentals that are $I(1)$ and $I(0)$, respectively (Hansen, 1992).

To test for cointegration, it needs to employ the ADF-cointegration test and the Johansen method (Johansen, 1992). Since shocks, such as trade or capital account liberalization, may shift the equilibrium relationship between the *ERER* and its fundamentals, it also needs to test for parameter's stability by relying on Andrews' parameter stability tests.

Now, let us consider how to examine the impact of *RER* over- and undervaluation on economic growth. To control for cyclical variations, it is necessary to split the sample period 1970–2007 into nonoverlapping 5-year periods.

The empirical growth equation is derived from the Solow–Swan growth model:

$$g_{it} = y_{it} - y_{i,t-1} = \alpha + \beta y_{i,t-1} + \gamma' x_{it} + \psi m_{it} + \mu_i + \lambda_t + \varepsilon_{it} \qquad (6.69)$$

In this equation, g_{it} reflects the real GDP per capita growth rate for the 5-year period and $y_{i,t-1}$ the logarithm of real GDP per capita at the beginning of the period. The vector x_{it} contains the logarithm of the investment ratio and other growth determinants, each averaged over the 5-year period. The term mis_{it} represents the variable used to investigate the relationship between *RER* under- and overvaluation on growth. The last three components on the right-hand side represent unobserved country-specific effects (μ_i), time-specific effects (λ_t), and the idiosyncratic error term (ε_{it}), respectively.

However, averaging mis_{it} with the intent to obtain the 5 year average degree of misalignment generates a misleading time series. To deal with this problem, we split mis_{it} into two variables: one taking negative values when the *RER* is undervalued, zero otherwise, and another taking positive values when the *RER* is overvalued, and zero otherwise. Then, it needs to average these two variables over the 5-year periods. A negative signed coefficient on both variables supports the hypothesis that *RER* undervaluation (*RER* overvaluation) fosters (harms) economic growth.

So far, the previous discussion has examined the growth effect of deviations of *RER* from the level consistent with internal and external balances. However, the results are not comparable with those studies of Rodrik (2008) and others that have measured *RER* misalignment following the (Balassa–Samuelson adjusted) PPP approach (Balassa, 1964). It could well be that *RER* undervaluation from the point of view of internal and external equilibriums retards growth, whereas undervaluation relative to PPP is growth promoting. Rather than further discussing the relative merits of the two approaches to measuring *RER* misalignment, in this subsection, we probe whether or not the results of Rodrik (2008) stand up to scrutiny.

Rodrik used the exchange rate (*XRAT*) and PPP conversion factors to compute the *RER* as $\ln RER_{it} = \ln(XRAT/PPP)_{it}$, with t indexing 5-year averages over the period 1950–2004. To adjust for the Balassa–Samuelson effect, Rodrik (2008) regressed the *RER* on GDP per capita (*RGDPCH*):

$$\ln RER_{it} = \alpha + \beta \ln RGDPCH_{it} + f_t + u_{it} \tag{6.70}$$

where
f_t is a time fixed effect
u_{it} the error term

The latter is also the estimate of adjusted PPP misalignment (*MISPPP*). Note that this specification implicitly assesses that the magnitude of the impact of productivity gains on *RER* (operating via wages) is uniform across countries. Thus, it ignores differences in labor market distortions or *surplus labor* situations across countries.

To examine the sensitivity of the results of Rodrik (2008) to this restrictive assumption, we may follow his approach to estimate *MISPPP* and reestimate Equation 6.71 by adding country slope dummies to obtain *MISPPP* measures that account for parameter heterogeneity. The future task is to compare the growth effects of the two alternative misalignment indicators by estimating the baseline growth specification of Rodrik (2008):

$$g_{it} = \alpha + \beta \ln RGDPCH_{i,t-1} + \delta MISPPP_{it} + f_t + f_i + u_{it} \tag{6.71}$$

where
g_{it} is the annual growth rate
f_i country-specific effects

6.6.3 Data

The absence of readily available indices for the actual *RER* and some of the fundamentals imposes a considerable obstacle to the empirical estimation of *ERER*s. The fundamentals for which reliable time series are available are investment, the terms of trade, the net foreign asset position, and the trade balance. Proxies have to be used for the actual *RER* and the other fundamentals.

The *RER* is an incentive measure for both producing and consuming tradable and nontradable goods. Constructing corresponding indexes is infeasible due to conceptual problems and data constraints in low income countries (Hinkle and Nsengiyumva, 1999). Therefore, the *RER* is measured as the ratio of trade-weighted foreign consumer price indexes (CPI) converted at the official exchange rate relative to the domestic CPI.

There is also no direct measure for trade policy. The bulk of the empirical literature proxies this variable through the ratio of total exports plus imports to GDP under the assumption that countries with more liberal trade regimes have higher trade volumes. Three proxies are considered: the ratio of total exports plus imports to GDP at current (OPEN) and constant (OPEN1) prices as well as the ratio of current imports relative to current GDP (OPEN2). To capture the Balassa–Samuelson-effect, the variable PROD is constructed, which is equal to the ratio of the home country's GDP per capita to the OECD average GDP per capita. This proxy directly incorporates Balassa's assumption (Balassa, 1964) that productivity gains are associated with higher growth rates.

Finally, there are no data on the composition of government consumption. Data are, however, available on total government consumption as a share of GDP (GEXP). Empirically, government consumption tends to fall disproportionately on nontraded goods (Edwards, 1989). Therefore, GEXP serves as a proxy for GN. However, this need not be true for all countries (Elbadawi and Soto, 1997). Thus, to avoid imposing more restrictions than necessary, the approach is to let the data decide whether GEXP proxies either GN or GT.

Under the criterion of at least 20 consecutive yearly observations within the time span from 1970 to 2007, the final sample consists of 63 developing countries, excluding outliers. While the model is initially estimated by OLS, this procedure does not address the issue of endogeneity. Nickell (1981) showed that the standard fixed-effects estimator of a dynamic panel data model is inconsistent when T (the number of time series observations) is fixed even as $N \to \infty$ (the number of cross-sectional units). In addition, growth is likely to affect the independent variables, which gives rise to inconsistent estimates.

6.6.4 Conclusions

The purpose of this section is to contribute to the debate on the impact of RER misalignment on economic growth with particular emphasis on the inference of some recent studies that real undervaluation promotes growth. While the traditional position on this issue (the *Washington Consensus*) advocates for the *RER* being close to its equilibrium level, the recent theoretical and empirical literature emphasizes the economic benefits of real undervaluation. This study estimated *RER* misalignments for 63 developing countries and analyzed the impact of *RER* over- and undervaluation on economic growth. In accordance with the Washington Consensus, the results suggest that any deviation of the *RER* from the level that is consistent with external and internal equilibriums lowers economic growth. Previous results seem to be driven by two inconsistencies:

First, the strong homogeneity assumption on long-run *RER* behavior across countries produces misleading results and is inconsistent with economic theory.

Second, the objective to infer the effect on growth of two variables (real over- and undervaluation) from a single continuous variable (*RER* misalignment) can lead to model misspecification.

The case study in this section also revisited the claim that PPP undervaluation promotes growth in developing countries. Again, when the two problems are taken into account, the results suggest that deviations from adjusted PPP do not impact on the growth performance of developing countries.

Thus, despite recent criticisms, the Washington Consensus still has valuable policy guidelines to offer. Developing countries should aim to keep the *RER* close to its equilibrium level in the sense of Nurkse (1947), which reinforces another policy guideline of the Washington Consensus: sound macroeconomic policies. In addition, countries with fixed but adjustable exchange rate regimes should closely review their current pegs if there are movements in anchor currencies. However, while it is true that the fastest-growing countries tend to have avoided excessive *RER* distortions in either direction, it is not a sufficient condition for growth to take off. For example, the *RER* rarely diverged from equilibrium in Central African Republic, but the country did not experience fast growth. Therefore, "the real exchange rate is best thought of as a facilitating condition," as concluded in Eichengreen (2008).

6.7 Nowhere to Hide: Financial Contagion Effects

6.7.1 Background

Since the 1990s, with the expansion of international hot money, the frequent outbreaks of financial crises, greatly enhanced crises transmissions, and profound impacts on the order of the global economic development have caught great attention of most governments and scholars. Specially, there have appeared some new features in the global financial crisis that was initially triggered by the U.S. subprime mortgage crisis. For example, a nonsystematic risk that triggers a regional financial crisis in one country or area has evolved into a risk of the global financial system along the paths of *globalization, integration,* and/or *liberalization.* So the breadth and depth of the financial crisis' impact has been much more than those of the previous financial crises. Although as of this writing the global financial contagion has been controlled, the subsequent crises have added the uncertainty of world economic environment, such as Dubai debt crisis, the Greek economic crisis, the rise of asset bubbles, and the recent European debt crisis. That has been why it is more difficult than before for a national economy to recover and to maintain financially sustainable and healthy development. To this end in this section, we will focus on testing the financial crisis contagion effects of the current global financial crisis, and we will especially compare the difference among multinational markets with different degrees of capital account liberalization under the brewing, outbreak, and restore stage of a crisis.

There is great amount of work published addressing this recent financial crisis and its contagion effect. But the relevant empirical results have been different under varied definitions of financial crisis contagion. Even under a similar framework of concepts, the results are still not entirely consistent. For example, by defining financial crisis contagion as significantly enhancing correlation between financial markets after the crisis, some studies showed that there existed contagions in cross-border capital markets during several major financial crises (Rosenberg, 1982; Hodrick and Prescott, 1997). However, by strictly defining the contagion and interdependence and considering the conditional variance, Forbes and Rigobon (2002) did not find any evidence of correlation between various markets during and after several major financial crises. Embrechts et al. (1999) pointed out that the correlation coefficient only described a linear quantitative relationship between two financial markets. So it is not suitable for testing nonlinear changes in the financial markets. And Boyer et al. (1999) found that the obtained results would be biased if the conditional heteroskedasticity were not considered.

But after comparing the similarities and differences of empirical research methods to a variety of financial contagions, Dunger et al. (2004) argued that due to the fact that some models included all of the information while others only part of the information, the difference of these slightly different concepts and relevant tests was smaller, and even does not exist under certain conditions. It seems reasonable to expect that the effective financial contagion model should not depend on which conceptual framework is employed. Instead it should only relate to whether linear information or nonlinear information is used. To this end, one can select one or some of the recent research methods, such as the minimum spanning tree method, Copula function, binary probit model, GARCH model, symbolic time series analysis, dynamic factor model, and differential dynamics model (Roudet et al., 2007; Mongardini and Rayner, 2009; Roodman, 2009) to deal with the nonlinear characteristics of financial contagion.

In addition, in order to test the dynamic effects and impact intensity of financial contagion, some scholars used the vector autoregression (VAR) model to conduct their empirical research. However, there might be some strong correlations within the new VAR information (innovations) that make the economic meaning of the impulse response function unclear. To resolve this problem, the structured vector autoregression (SVAR) model is often used to obtain structural shocks of the impulse response function. Therefore, we choose SVAR model to test the effect of financial crisis contagion. The rising taxes and the introduction of rigorous quantitative restrictions may be used to explain the major change that occurred in 2009. In this change, the activity of the sector's entrepreneur unions must have played an important part, carrying out an active plan of action, spreading information, and posting the problem in the media. These actions worked to face the *rational ignorance* problem in the political market assuming that the absence of voluntary cooperation is lower in the context of repeated games. The high costs—in terms of lost earnings—of the measures against the sector encouraged voters to have a rational behavior by assuming the cost of the plans of action and of obtaining information.

6.7.2 SVAR Modeling

By taking into account both the effects and the lagged effects of multiple time series, SVAR model may be imposed with short-term or long-term constraints as needed by economic theory to identify the impact of different sequences. The global financial crisis is started from the U.S. subprime mortgage crisis, and financial connection is the initial and main channels of transmission. Thus, we can study financial contagion by observing the changes in asset prices, such as stock price index, stock returns, and exchange rates.

Let $y_t = (y_{1t}, y_{2t}, \ldots, y_{kt})$, ($t = 1, 2, \ldots, T$), where y_{kt} is a time series vector of asset price (or return on assets) of k countries or financial markets. The general form of SVAR model can be constructed as follows:

$$B_0 y_t = B_1 y_{t-1} + B_2 y_{t-2} + \cdots + B_p y_{t-p} + u_t \tag{6.72}$$

or

$$B(L)y_t = (B_0 - B_1 L - B_2 L^2 - \cdots - B_p L^p)y_t = u_t \tag{6.73}$$

where

B_0 is the coefficient matrix of y_t, a nonsingular matrix with diagonal elements equal to 1, which reflects the effect of the same time period

B_i is the coefficient matrix of y_{t-i}, which reflects the lagged effects, $i = 1, 2, \ldots, p$

$B(L)$ is a P-order lag operator

u_t is a vector of structural shocks with mean equal to zero and no serial correlation such that its covariance is diagonal matrix, that is, $E(u_t u_t') = \Lambda$

Then the reduced-form VAR model of SVAR model is as follows:

$$y_t = A_1 y_{t-1} + A_2 y_{t-2} + \cdots + A_p y_{t-p} + \varepsilon_t \tag{6.74}$$

or

$$A(L)y_t = (I - A_1 L - A_2 L^2 - \cdots - A_p L^p)y_t = \varepsilon_t \tag{6.75}$$

where
 A_i is the coefficient matrix of y_{t-i}, reflecting the lagged effects of the sequences, for $i = 1, 2, \ldots, p$
 ε_t is a vector of shocks in the reduced-form VAR model

From Equations 6.72 through 6.75, we can obtain the linear relation of the simple shocks ε_t and structural shock u_t as follows:

$$A(L)y_t = B_0^{-1} * B(L)y_t \tag{6.76}$$

$$\varepsilon_t = B_0^{-1} u_t \tag{6.77}$$

where $E(\varepsilon_t \varepsilon_t') = B_0^{-1} E(u_t u_t')(B_0^{-1})' = B_0^{-1} \Lambda (B_0^{-1})'$.

By left multiplying $A(L)^{-1}$ to (6.74) and (6.75), Equations 6.76 and 6.77 imply that the structural vector moving average (SVMA) model can be deduced as follows:

$$y_t = \sum_{i=1}^{\infty} \psi_i B_0^{-1} L^i u_t \tag{6.78}$$

Then, based on the estimation of VAR (P) model, SVAR (P) model can be estimated with short-term or long-term constraints of B_0. It should be noted that SVAR model with k variables needs $k(k-1)/2$ short-term constraints to determine the structural impact. Theoretical and empirical analyses show that financial contagion often spread in one way, or two ways, or both forms. And its transmission direction can be determined by using the degree of the relevant economic and financial links, geographical relationships, trade relations, political relationships, degrees of the capital account openness, financial integration or globalization, and so on (Li, 2004). However, with economic and financial integration and improvement in the technology of electronic dissemination of asset transactions, financial market volatility and contagion become stronger and faster. So, we pay additional attention on the effect of asset price or return series in the same time period, and only discuss the short-term constraints of B_0.

6.7.3 Regression

Firstly, let us look at the sample selection. To compare the contagion effect of the U.S. subprime mortgage crisis on the markets with different degrees of capital account openness, we select two groups: the developed Eurasian countries that are used as states with high degrees of capital account openness, and the BRIC countries that are used as states with moderate degrees of capital account openness. The first sample group includes the United States, Germany, France, Britain, and Japan. And the second one contains Russia, Brazil, India, and China. The international index data of Morgan Stanley Capital of the time period from January 4, 2006, to January 4, 2010, are used in our empirical analysis by fitting SVAR model. The data can be downloaded from www.msci.com.

Before the outbreak of the said crisis, the U.S. subprime mortgage risk had been gradually emerging. For example, in February 13, 2007, New Century Financial Corporation issued a profit warning of the fourth quarter of 2006. And Bear Stearns Companies announced that its hedge funds would be in bankruptcy procedure on July 19. Then the U.S. MSCI index started to decline from 1467 on that day. And after reaching a new high point of 1478 on October 12, it entirely entered a 5-month decline channel. That marked the outbreak of the subprime mortgage crisis. During the crisis, BNP Paribas, the biggest bank in France, announced its involvement in the U.S. subprime on August 9, 2007. Then most of the global stock indexes started to fall, and metal and crude oil futures and the spot price of gold fell sharply, showing a strong transmission of the U.S. subprime mortgage crisis throughout the world. After the European Central Bank took the *bailout* measures on August 10, 2007, many countries also introduced policies to stabilize their financial environment and stimulate economic development. Then on March 9, 2009, the U.S. MSCI index began to rise from 645, signaling a weakening impact of the U.S. subprime mortgage crisis. That was followed by a period of recovery. Therefore, correspondingly, we test respectively the transmission effects of the U.S. subprime mortgage crisis during the gestation period from January 4, 2006 to July 19, 2007, the deepening period from July 19, 2007 to March 9, 2009, and the recovery period from March 9, 2009 to January 4, 2010, see Figure 6.4.

Secondly, let us discuss the short-term constraints. Sims (1980) imposed short-term constraint by using the Cholesky factorization of residual covariance matrix to orthogonalize the innovations of the VAR model. When the residual covariance matrix is diagonal, the variable order does not affect the results of structural impact. In this case, we can use Wold causal chain, a form of recursive system, to define directly the upper triangular elements as being 0. There are 10 short-term constraints. However, these constraints do not agree with the actual performance of market volatility and contagion. Therefore, we sort the five sequences according to their related degrees in economy, finance, trade, location, and political relationship from the strongest to the weakest, and determine the short-term constraints by jointly using theory and experience.

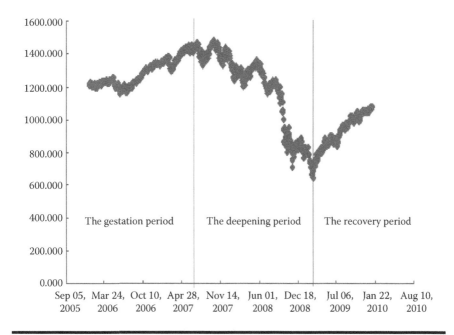

Figure 6.4 The U.S. MSCI index trends.

In order to investigate the transmission dynamics of the U.S. subprime mortgage crisis, the U.S. MSCI index is listed in the first column of B_0, recorded as R0. The order of other four sequences in the first sample is Britain, France, Germany, and Japan, recorded, respectively, as R1, R2, R3, and R4 (see Figure 6.5). The order in the second sample is India, Russia, China and Brazil, recorded as R5, R6, R7, and R8 (see Figure 6.6).

According to experience, B_0 is identified as a 5×5 structure factor matrix with diagonal elements equal to 1. First, to examine the dynamics of the crisis contagion with the United States as the center, we assume that the subprime crisis was originated in the United States and had only one-way impact on the other four countries in the same period. Then the latter four elements in the first row of B_0^1 are 0. Secondly, we assume that Britain, France, and Germany did not have any impact on Japan in the same period due to their relationship of location, policy, trade, and economic and financial integration. So the third and fourth elements in the fifth row of B_0^1 are 0. Thirdly, we assume that Japan did not have any impact on France and Germany in the same period due to their political relationships. So, the third and fourth elements in the fifth column of B_0^1 are 0. Fourthly, by taking into account the location and economic development, we assume that Germany did not have any impact on United Kingdom in the same period. So, the second element in the fourth column is 0. Based on all these assumptions, the short-term constraint matrix B_0^1 of the first sample is

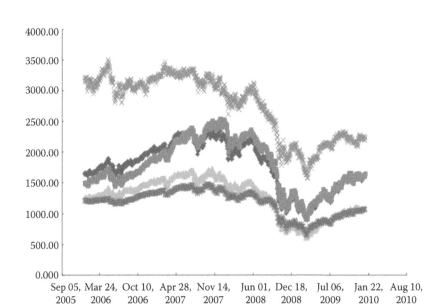

Figure 6.5 The MSCI index trends of the first sample.

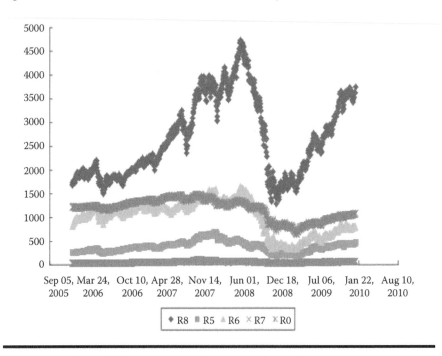

Figure 6.6 The MSCI index trends of the second sample.

$$\begin{pmatrix} 1 & 0 & 0 & 0 & 0 \\ b_{21}^1 & 1 & b_{23}^1 & 0 & b_{25}^1 \\ b_{31}^1 & b_{32}^1 & 1 & b_{34}^1 & 0 \\ b_{41}^1 & b_{42}^1 & b_{43}^1 & 1 & 0 \\ b_{51}^1 & b_{52}^1 & 0 & 0 & 1 \end{pmatrix} \tag{6.79}$$

By doing a similar analysis, we get the short-term constraint matrix B_0^2 of the second sample as follows:

$$\begin{pmatrix} 1 & 0 & 0 & 0 & 0 \\ b_{21}^2 & 1 & b_{23}^2 & b_{24}^2 & 0 \\ b_{31}^2 & b_{32}^2 & 1 & b_{34}^2 & 0 \\ b_{41}^2 & b_{42}^2 & b_{43}^2 & 1 & b_{45}^2 \\ b_{51}^2 & 0 & 0 & b_{54}^2 & 1 \end{pmatrix} \tag{6.80}$$

These short-term constraints are also used to test the effects in the gestation and recovery periods of the crisis. Firstly, the level of economic development in the United States has been No. 1 in the world, while maintaining its economic influence on other countries for a long time. Secondly, although there have been a long-term appreciation of the yen, the sudden emergence of the euro, and the impact of the most recent global financial crisis, the U.S. dollar, as the *world currency*, can still not be replaced as the world currency. In addition, for the purpose of comparing the impacts of the outbreak of the financial crisis in the United States on the other countries during these three time periods, it is necessary for us to assume the same short-term constraints as what are given earlier.

6.7.4 Financial Contagion Effect between Markets with High Capital Account Openness

First, let us look at the data stationarity test.

The results of ADF stationarity test in EViews6.0 show that the five logarithmic series of MSCI index in the first sample are unstable. But their difference series are stable. That means that the asset return series are stable.

Secondly, let us estimate the structural factor matrix B_0^1.

Asset return series of the five countries are used to build the reduced-form VAR models, respectively, for the gestation, deepening, and recovery period of the U.S. subprime mortgage crisis. The optimal lag orders of the three models are 2, 6, and 2, respectively, in accordance with the AIC principle. The lag structure test of the three models shows that the reciprocal of the root of the AR character-istic polynomial is in the unit circle. So the three models are stable. Hence, the coefficient matrix B_0^1 can be estimated from three models. Table 6.10 shows the results.

From Table 6.10, we can get the following results:

1. There are positive impacts in the same period during the three stages, such as the United States on Britain and France, Britain on France, France on Germany, and Japan on Britain. There are also negative effects between the other countries in the same period. It implies that there may be a lag infection of the U.S. subprime mortgage crisis in some countries.
2. The impact strength in the same period of the United States on France and Japan is increasing from the gestation, deepening to recovery period. But the impacts of others are smallest in the deepening period. This means that the recovery period from the crisis will be relatively long. And any change from the United States or a primarily infected country may be transmitted more greatly and fast to other countries and then become a main obstacle of the national economic recovery in the postfinancial crisis era. So governments must pay enough attention to this expectation.

Thirdly, let us analyze the impulse response.

The impulse response analysis is employed to reflect the changes in the current and future values of all endogenous variables as caused by one-time impact of an endogenous variable. By using structural factorization impulse response analysis in the three stages, we find that the results are not obvious. So the generalized impulse analysis need to be used in future research.

6.7.4.1 Response of Other Countries to the U.S. Shocks

Figure 6.7 shows the following results:

1. The impulse responses of other countries to the shock of the United States are weaker during the crisis gestation period. The maximum of all fluctuations is less than 0.006, and the duration is up to 9 days. But their responses during the crisis deepening and recovery period are stronger. The largest response strengths average over 0.01, and the duration is at least 9 days. The average duration in the deepening period is more than 20 days. This indicates that the U.S. subprime mortgage crisis was strongly infectious in the first sample. Even in the recovery period, the impact of the United States on the other countries in the sample also remains strong.

Table 6.10 Results of B_0^1

Impacts in the Same Period	Crisis Gestation Period	Crisis Deepening Period	Crisis Recovery Period
United States → Britain	6.877***	4.517***	8.732***
	(20.713)	(6.147)	(4.848)
United States → France	2.658***	4.701***	13.962***
	(8.694)	(5.923)	(8.463)
United States → Germany	−1.703***	−1.355*	−13.117***
	(−5.733)	(−1.74)	(−8.005)
United States → Japan	−5.789***	15.986***	22.064***
	(−27.731)	(29.025)	(20.657)
Britain → France	219.058***	79.682***	107.576***
	(22.287)	(20.556)	(15.475)
Britain → Germany	−237.624***	−109.923***	−120.765***
	(−22.844)	(−23.402)	(−16.113)
Britain → Japan	−121.184***	−57.512***	−68.772***
	(−28.161)	(−29.163)	(−20.748)
France → Britain	−37.451***	−12.515***	−10.412**
	(−6.841)	(−4.563)	(−2.245)
France → Germany	234.04***	121.397***	137.6***
	(28.248)	(29.256)	(20.784)
Germany → France	−204.219***	−94.27***	−117.7304***
	(−28.248)	(−29.255)	(−20.784)
Japan → Britain	103.371***	66.448***	84.615***
	(28.246)	(29.251)	(20.782)

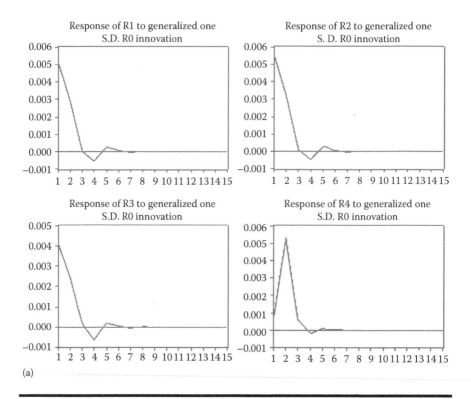

Figure 6.7 Response of developed countries to the United States. (a) The crisis gestation period *(Continued)*

2. In the three periods, the responses of the United States to the shocks of Britain, France, and Germany are similar, and their transmission strength and duration are also similar. But Japan's response is relatively unique. This implies that the infectious characteristics of the U.S. subprime mortgage crisis are felt differently by different countries. It can be seen as an indirect evidence that geopolitical relation is one of the paths of transmission.

6.7.4.2 Response of the United States to the Shocks of Other Countries

By comparing the response of the United States to shocks of other countries in the three periods, we found from Figure 6.8 similar situations as those existing in Figure 6.7. That is, the response of the United States is stronger during the crisis gestation and recovery period, and their durations are also longer. Especially, the responses of United States to England, France, and Germany are similar in strength, duration,

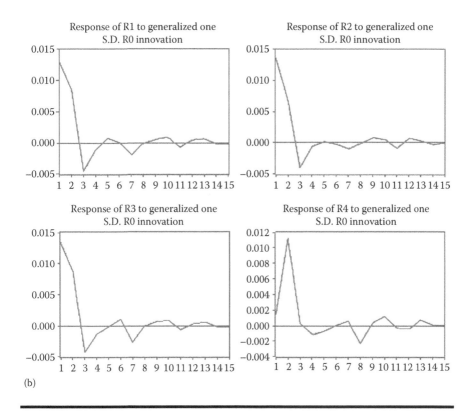

(b)

Figure 6.7 (Continued)   Response of developed countries to the United States. (b) the crisis deepening period *(Continued)*

and performance. And its response to Japan is much smaller, not exceeding ±0.02. This is totally different from the situation with the other three countries. But all the durations are similar. This fact indicates that European countries that were infected by the U.S. subprime mortgage crisis also had strong impacts on the United States. So, that constitutes a piece of indirect evidence showing the fact that geopolitical relationship is one of the contagion paths of crises.

6.7.4.3 Response between Other Countries

Comparing the impulse responses between the other countries in Figures 6.9 through 6.11, we find the following:

1. In the three specified periods, the impulse responses between Britain, France, and Germany are stronger than those between Japan and these three countries. But all their durations are similar.

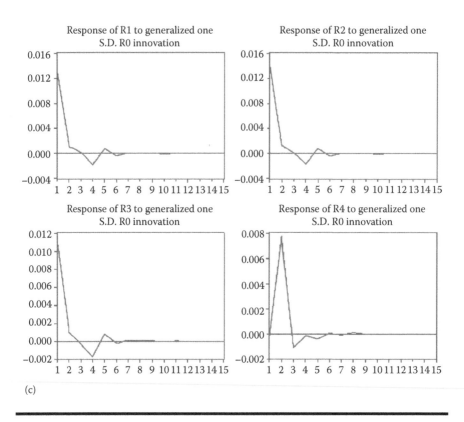

(c)

Figure 6.7 (Continued)   Response of developed countries to the United States. (c) the crisis recovery period.

2. From the curve of impulse response, it follows that Britain, France, and Germany have similar performance. But Japan is unique. This indicates that there are two-way infections between the other four sample countries during the crisis gestation and recovery period. And there are three types of infection status. The first one is in Britain, France, and Germany. The second is the response of Japan to these three countries. The third is the response of other three countries to Japan.

6.7.5 Financial Contagion Effect between Markets with At Least Moderate Capital Account Liberation

First, let us consider data stationarity test and estimation of the structural factor matrix B_0^2.

Similar to what has been discussed earlier, the results of ADF stationarity test in EViews6.0 show that the five difference series of logarithmic MSCI indexes in the

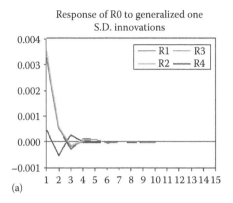

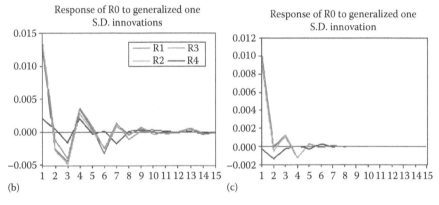

Figure 6.8 **The response of the United States to shocks of developed countries. (a) The crisis gestation period, (b) the crisis deepening period, and (c) the crisis recovery period.**

second sample are stable. And the optimal lag orders of the three models developed for the gestation, deepening, and recovery period are 2, 6, and 2, respectively, in accordance with the AIC principle. And these three models are stable. The results of the coefficient matrix B_0^2 are shown in Table 6.11.

Table 6.11 indicates that there are seven same-period positive impacts during the crisis deepening and recovery period, including the United States to India and China, Russia to China, India to Russia and China, and China to India and Brazil. And all other impacts are negative. It also shows that there may be lagging transmissions between the United States and the BRIC during the subprime mortgage crisis. And what is employed is the analysis of structural impulse responses, which are very different from those between the United States and developed countries with high degree of capital account liberalization. For example, the United States does not impact on Russia and Brazil in the same periods. It seems that the geopolitical relation is not one of the main transmission channels.

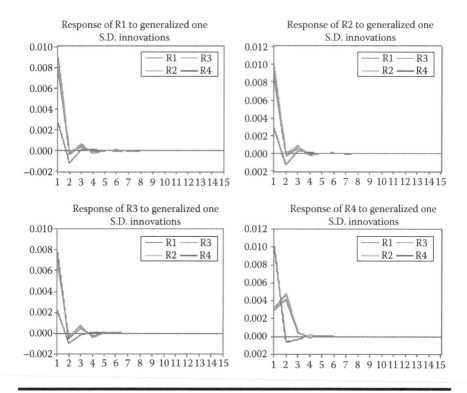

Figure 6.9   Response between developed countries (except United States) during the crisis deepening period.

In fact, a new transmission characteristic of the U.S. subprime mortgage crisis in 2008 is to make the nonsystematic risk of the financial crisis induced by a single country or local area into a global financial system risk along the paths of *globalization, integration,* and *liberalization.* It impacts at first on the developed countries with higher degree of capital account openness and spreads quickly among such countries. Then it spills over to emerging market countries or developing countries with moderate degree of capital openness. Thus, the breadth and depth of the transmission during the U.S. subprime mortgage crisis depended first on the degree of capital account liberalization rather than geographical distances to the United States. Clearly, the results of Tables 6.10 and 6.11 support this analysis.

Second, let us discuss the impulse response.

6.7.5.1 Response of the BRIC to the U.S. Shocks

Overall, the strength of response of the BRIC to the United States in the crisis gestation period is less than in the crisis deepening and recovery period.

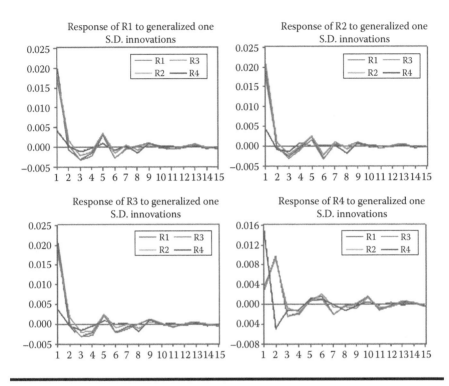

Figure 6.10 Response between developed countries (except United States) during the crisis deepening period.

Their response curves are very different during the three periods. But the difference of their durations is smaller. The duration is 6 or 7 days in the crisis gestation and recovery period, but more than 20 days on average in deepening period (see Figure 6.12). This fact indicates that the United States also has a strong impact on the BRIC countries during the subprime mortgage crisis. But the transmission status between these countries is different. First, the Russian and Chinese response to the U.S. shocks is similar in response curve during the gestation period, but Russia's largest positive response strength is less than China's during the crisis deepening period. Secondly, the response strength of India to the United States is the smallest of the BRICs in the crisis gestation period. But its maximum positive response strength is close to 0.02 in the crisis deepening period, ranking No. 2, and its response curve is similar to that of Russia. In addition, its response strength is the largest in the crisis recovery period, and its curve is similar to that of Brazil. Thirdly, in the three periods, the response curve of Brazil to the United States is different from those of other three countries. Though there are a lot of factors affecting the crisis contagion status and strength and no clear clues to be

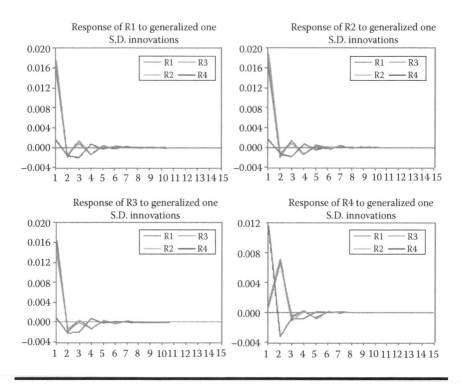

Figure 6.11 Response between developed countries (except United States) during the crisis recovery period.

investigated in the impulse response results, by combining the analyses of these two samples, it can be inferred that the capital account openness is superior to geopolitical factor in the impacts on contagion.

6.7.5.2 Response of the United States to the BRIC Countries' Shocks

Figure 6.13 shows that in the three periods, the response strength of Brazil to the United States is the largest, and its largest positive response strengths are, respectively, 0.0045, 0.016, and 0.012, being at least two times greater than the others. And its response curve is also different from the others. However, the responses of the United States to the other three shocks are similar in terms of status, strength, and duration during the three periods. Thus, there are two-way contagions between the BRIC countries and the United States. And there is the similar indirect evidence that geopolitical relation is one of the financial contagion paths.

Table 6.11   Results of B_0^2

Impacts in the Same Period	Crisis Gestation Period	Crisis Deepening Period	Crisis Recovery Period
United States → Russia	2.953*	−4.834*	−1.354*
	(11.860)	(−11.440)	(−11.120)
United States → India	−0.616*	8.783*	5.6*
	(−2.715)	(18.209)	(19.661)
United States → China	2.477*	18.295*	3.34*
	(9.655)	(21.057)	(8.034)
United States → Brazil	4.43*	−7.324*	−0.906*
	(27.055)	(−26.661)	(−10.749)
Russia → India	−56.064*	−31.812*	−35.451*
	(−28.242)	(−29.202)	(−20.724)
Russia → China	30.668*	16.786*	30.363*
	(10.792)	(10.73)	(11.232)
India → Russia	75.861*	47.612*	50.333*
	(28.244)	(29.212)	(20.729)
India → China	11.690*	5.579*	10.189*
	(3.274)	(2.821)	(3.349)
China → Russia	−33.842*	−23.674*	−34.991*
	(−7.848)	(−12.192)	(−8.015)
China → India	34.229*	15.69*	25.714*
	(7.931)	(8.534)	(6.118)
China → Brazil	82.892*	36.485*	59.121*
	(28.368)	(29.422)	(20.877)
Brazil → China	−66.174*	−40.529*	−69.904*
	(−28.244)	(−29.208)	(−20.733)

Note: *Significant at 1% level. Z statistics in brackets.

6.7.5.3 Response between the BRIC Countries

Comparing the impulse responses between BRIC countries in Figures 6.14 through 6.16, we found the following:

1. In the three periods, the largest positive response strength of Russia to the other three countries' shocks is the smallest. In the crisis deepening period, Brazil is the largest. In the crisis recovery period, China is the smallest.
2. From the curves, the responses of Russia, India, and Brazil to China are more similar in the crisis gestation period than in other periods, and there are a few differences in the crisis deepening and recovery period. Although it is difficult to classify the other response curves in the three periods, there is evidence of two-way transmission between the BRIC countries.

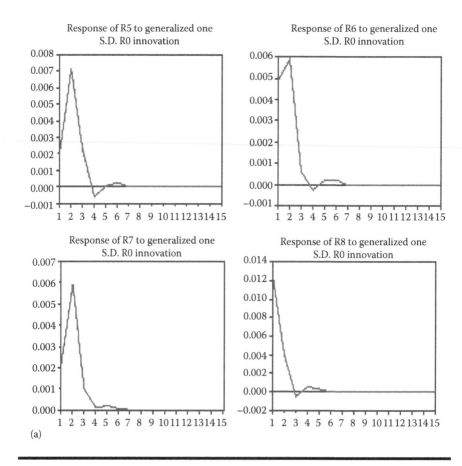

(a)

Figure 6.12 The response of the BRIC countries to the United States. (a) The crisis gestation period *(Continued)*

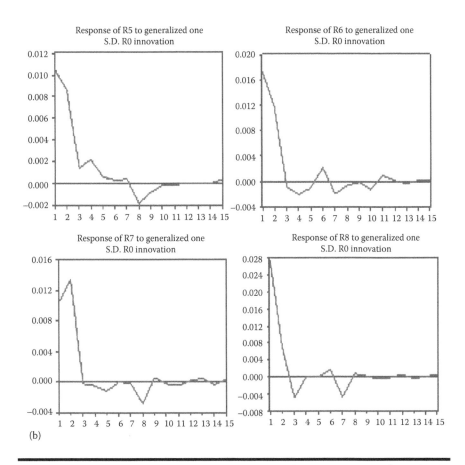

(b)

Figure 6.12 (Continued)   The response of the BRIC countries to the United States. (b) the crisis deepening period *(Continued)*

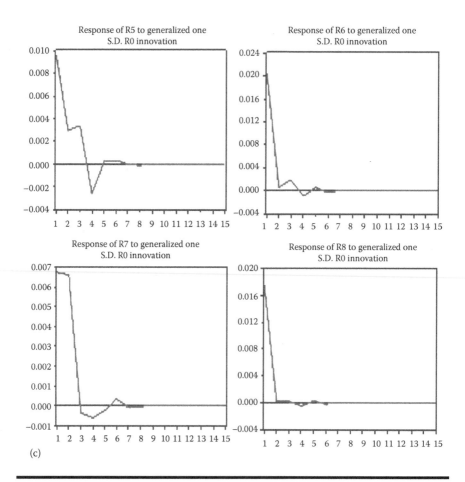

(c)

Figure 6.12 (Continued)   The response of the BRIC countries to the United States. (c) the crisis recovery period.

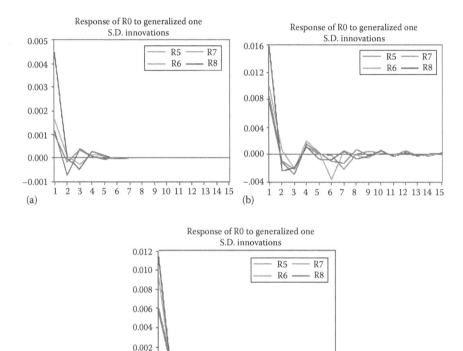

Figure 6.13   The response of the United States to the BRIC countries' shock. (a) The crisis gestation period, (b) the crisis deepening period, and (c) the crisis recovery period.

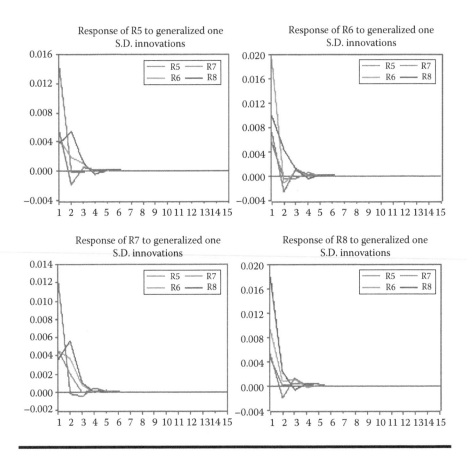

Figure 6.14   The response between the BRIC countries (except the United States) during the crisis gestation period.

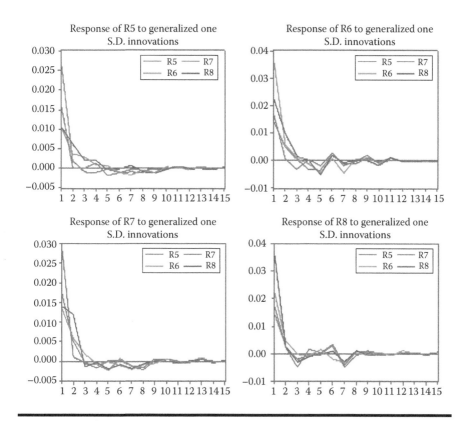

Figure 6.15   The response between the BRIC countries (except the United States) during the crisis deepening period.

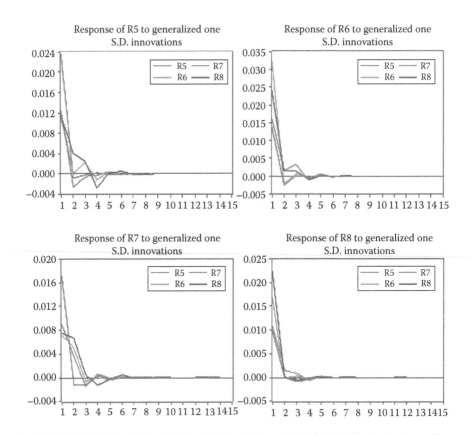

Figure 6.16   The response between the BRIC countries (except the United States) during the crisis recovery period.

References

Aït-Sahalia, Y. and Lo, A. W. (2000). Nonparametric risk management and implied risk aversion. *Journal of Econometrics*, 94, 9–51.

Alfarano, S., Lux, T., and Wagner, F. (2008). Time variation of higher moments in a financial market with heterogeneous agents: An analytical approach. *Journal of Economic Dynamics and Control*, 32(1), 101–136.

Alfaro, L. and Hammel, E. (2007). Capital flows and capital goods. *Journal of International Economics*, 72(1), 128–150.

Alfaro, L., Kalemli-Ozcan, S., and Volosovych, V. (2003). Why doesn't capital flow from rich to poor countries? An empirical investigation. *Mimeo*, 32, 363–394.

Amin, K. I. and Jarrow, R. A.(1991). Pricing foreign currency options under stochastic interest rates. *Journal of International Money and Finance*, 10, 310–329.

Anderson, K., McRae, C., and Wilson, D. (eds.) (2001). *The Economics of Quarantine and the SPS Agreement*. Centre for International Economic Studies, AFFA Biosecurity Australia, Adelaide/Canberra, Australia.

Andolfatto, D. (1994). Business cycles and labor market search. Manuscript, University of Waterloo, Waterloo, Ontario, Canada.

Andrews, D. W. K. (1991). Heteroskedasticity and autocorrelation consistent covariance matrix estimation. *Econometrica*, 59, 817–858.

Arthur, W. B. (1989). Competing technologies, increasing returns, and lock-in by historical events. *The Economic Journal*, 99, 116–131.

Aseda, T., Semmler, W., and Novak, A. (1995). Endogenous growth and balanced growth equilibrium. Technical Report, pp. 95–102.

Ayadi, M. A. and Kryzanowski, L. (2008). Portfolio performance sensitivity for various asset-pricing kernels. *Computers & Operations Research*, 35(1), 171–185.

Baffes, J., Elbadawi, I. A., and O'Connell, S. A. (1999). Single-equation estimation of the equilibrium exchange rate. In: Hinkle, L. E. and Montiel, P. J. (eds.), *Exchange Rate Misalignment: Concepts and Measurement for Developing Countries*. A World Bank Research Publication, Oxford University Press, Oxford, U.K., pp. 405–464.

Baig, T. and Goldfajn, I. (1998). Financial market contagion in the Asian crisis. IMF Working paper no. 98/155.

Balassa, B. (1964). The purchasing-power parity doctrine: A reappraisal. *Journal of Political Economy*, 72, 584–596.

Banerjee, A., Dolado, J. J., Galbraith, J. W., and Hendry, D. F. (1993). *Co-Integration, Error Correction, and the Econometric Analysis of Non-Stationary Data*. Oxford University Press, Oxford, U.K.

Barbier, E. B. et al. (2008). Coastal ecosystem-based management with nonlinear ecological functions and values. *Science*, 319, 321–323.

Barbiero, A. (2014). An alternative discrete skew Laplace distribution. *Statistical Methodology*, 16, 47–67.

Batra, G., Kaufmann, D., and Stone, A. H. (2003). The firms speak: What the world business environment survey tells us about constraints on private sector development. World Bank working paper.

Baumol, W. J. (1967). Macroeconomics of unbalanced growth: The anatomy of urban crisis. *American Economic Review*, 57(3), 415–426.

Baye, M. R., Morgan, J., and Scholten, P. (2006). Information, search, and price dispersion. In: Hendershott, T. (ed.), *Handbook on Economics and Information Systems*. Elsevier, Amsterdam, the Netherlands.

Bayraktar, E. and Xing, H. (2011). Pricing Asian options for jump diffusion. *Mathematical Finance*, 21(1), 117–143.

Bellman, R. and Cooke, K. (1963). *Differential Difference Equations*. Academic Press, New York.

Benhabib, J., Schmitt-Grohé, S., and Uribe, M. (2001). Monetary policy and multiple equilibria. *American Economic Review*, 91, 167–186.

Bergman, U. M. and Jellings, M. (2010). Monetary policy during speculative attacks: Are there diverse medium term effects? *The North American Journal of Economics and Finance*, 21(1), 5–18.

Beretta, A., Roman, E., and Raicich, F. (2005). Long-time correlation of sea-level and local atmospheric pressure fluctuations at Trieste. *Physica A: Statistical Mechanics and Its Applications*, 347(1), 695–703.

Berg, A. and Miao, Y. (2010). The real exchange rate and growth revisited: The Washington consensus strikes back? IMF working paper no. WP/10/58.

Berger, A. and De Young, R. (1997). Problem loans and cost efficiency in commercial banks. *Journal of Banking and Finance*, 21, 849–870.

Bernard, D., Peter, J. L., and Bertill, N. (1995). Siegel's paradox and the pricing of currency options. *Journal of International Money and Finance*, 2, 213–223.

Bershadskii, A. (1999). Multifractal critical phenomena in traffic and economic processes. *European Physical Journal B*, 11, 361–404.

Black, F. and Cox, J. C. (1976). Valuing corporate securities: Some effects of bond indenture provisions. *Journal of Finance*, 31(2), 351–367.

Black, F. and Litterman, R. (1992). Global portfolio optimization. *Financial Analysts Journal*, 48(5), 28–43.

Blackburn, K., Bose, N., and Capasso, S. (2012). Tax evasion, the underground economy and financial development. *Journal of Economic Behavior and Organization*, 83, 243–253.

Blair, B. J., Poon, S. H., and Taylor, S. J. (2001). Forecasting S&P 100 volatility: The incremental information content of implied volatilities and high frequency returns. *Journal of Econometrics*, 105, 5–26.

Blanchard, O. J. and Diamond, P. A. (1989). The Beveridge curve. *Brookings Papers on Economic Activity*, 1, 1–60.

Blundell, R. and Bond, S. (1998). Initial conditions and moment restrictions in dynamic panel data models. *Journal of Econometrics*, 87, 115–143.

Boldrin, M. and Rustichini, A. (1994). Growth and indeterminacy in dynamic model with externalities. *Econometrica*, 62(2), 323–342.

Bordo, M. D. and Choudhri, E. U. (1982). Currency substitution and the demand for money: Some evidence for Canada. *Journal of Money, Credit and Banking*, 14(2), 48–57.

Bose, N., Capasso, S., and Wurm, M. (2012). The impact of banking development on the size of the shadow economy. *Journal of Economic Studies*, 39(6), 620–628.

Boyer, B. H., Gibson, M. S., and Loretan, M. (1999). Pitfalls in tests for changes in correlations. Provided by Board of Governors of the Federal Reserve System (U.S.) in its series International Finance Discussion Papers, No. 597.

Brennan, M. J. and Subrahmanyam, A. (1996). Market microstructure and asset pricing: On the compensation for illiquidity in stock returns. *Journal of Financial Economics*, 41, 441–464.

Brida, J. G., Gómez, D. M., and Risso, W. A. (2009). Symbolic hierarchical analysis in currency markets: An application to contagion in currency crises. *Expert Systems with Applications*, 36(4), 7721–7728.

Calvo, S. and Reinhart, C. (1996). Capital flows to Latin America: Is there evidence of contagion effects? WPS1619, The World Bank.

Cecen, A. and Erkal, C. (1996). Distinguishing between stochastic and deterministic behavior in high frequency foreign exchange rate returns: Can nonlinear dynamics help forecasting. *International Journal of Forecasting*, 12(4), 465–473.

Cetin, U., Jarrow, R. A., and Protter, P. (2002). Option pricing with liquidity risk. Working paper. Cornell University, Ithaca, NY.

Cetin, U., Jarrow, R., A., and Protter, P. (2003). Liquidity risk and arbitrage pricing theory. Working paper. Cornell University, Ithaca, NY.

Chaloupka, F. (2011). Lessons for obesity policy from the tobacco wars. In: Cawley, J. (ed.), *The Oxford Handbook of the Social Science of Obesity*. Oxford University Press, Oxford, U.K.

Chavarriga, J. and Garcia, I. A. (2004). The Poincaré problem in the non-resonant case. Preprint, Universitat de Lieida, Lieida, Spain.

Chavarriga, J., Garcia, I. A., and Sorolla, J. (2005). Resolution of the Poincare problem and nonexistence of algebraic limit cycles in family (I) of Chinese classification. *Chaos Solitons and Fractals*, 24(2), 491–499.

Chavarriga, J., Giacomini, H., and Grau, M. (2003). Necessary conditions for the existence of invariant algebraic curves for planar polynomial systems. Preprint, University of Lieida, Lieida, Spain.

Chavarriga, J., Llibre, J., and Sorolla, J. (2004). Algebraic limit cycles of degree 4 for quadratic systems. *Journal of Differential Equations*, 200, 206–244.

Chen, C., Li, G., Zhou, J., and Wang, W. (2008). *Mathematical Physics Equations*. Science Press, Beijing, China.

Chen, K. and Ying, Y.-R. (1910). Thomas constraint in currency substitution. *Proceedings of SSMSSD 2010*, Shanghai, China.

Chen, P. (1988). Empirical and theoretical economic chaos. *System Dynamic Review*, 4, 81–108.

Chen, X.-H. and He, P. (2006). Listing corporation size and growth rates of probability distribution of the contrast experimental study. *System Engineering*, 24(2), 1–10.

Chiou, L. and Muehlegger, E. (2008). Crossing the line: Direct estimation of cross-border cigarette sales and the effect on tax revenue. *The B.E. Journal of Economic Analysis and Policy*, 8(1), 1–41.

Choi, H. S. (2011). Monetary policy and endowment risk in a limited participation model. *Economic Inquiry*, 49(1), 89–93.

Chunbo, M. and David, I. (2008). China's changing energy intensity trend: A decomposition analysis. *Energy Economics*, 30(3), 1037–1053.

Cipollini, A. and Kapetanios, G. (2009). Forecasting financial crises and contagion in Asia using dynamic factor analysis. *Journal of Empirical Finance*, 16(2), 188–200.

Coeurdacier, N. and Martin, P. (2009). The geography of asset trade and the euro: Insiders and outsiders. *Journal of the Japanese and International Economies*, 23(2), 90–113.

Cogley, T. and Nason, J. M. (1995). Output dynamics in real-business-cycles models. *American Economic Review*, 85, 492–511.

Colombier, C. (2010). Drivers of health care expenditures: Does Baumol's cost disease loom large? Papers Presented at *the 66th Congress of the International Institute of Public Finance*, Uppsala, Sweden, August 2010.

Concha, A., Galindo, A. J., and Vasquez, D. (2011). An assessment of another decade of capital controls in Colombia: 1998–2008. *The Quarterly Review of Economics and Finance*, 51(4), 319–338.

Corden, M. W. (1981). Exchange rate protection. In: Cooper, R. N. and Triffin, R. (eds.), *The International Monetary System under Flexible Exchange Rates: Global, Regional and National. Essays in Honor of Robert Triffin*. Ballinger, Cambridge, MA, pp. 17–34.

Cruz-Baez, D. I. and Gonzalez-Rodrıguez. J. M. (2008). A different approach for pricing Asian options. *Applied Mathematics Letters*, 21, 303–306.

Daheng, P. (2007). Pricing of perpetual American put with fractionl O-U process. *Acta Mathematica Scientia*, A(6), 1141–1147.

Dai, M., Li, P., and Zhang, J. (2010). A lattice algorithm for pricing moving average barrier options. *Journal of Economic Dynamics and Control*, 34, 542–554.

Darrell, D., Jarrow, R., Purnanandam, A., and Yang, W. (2003). Market pricing of deposit insurance. *Journal of Financial Services Research*, 24(2–3), 93.

de Jong, E., Verschoor, W. F. C., and Zwinkels, R. C. J. (2009a). A heterogeneous route to the European monetary system crisis. *Applied Economics Letters*, 16, 929–932.

de Jong, E., Verschoor, W. F. C., and Zwinkels, R. C. J. (2009b). Behavioural heterogeneity and shift-contagion: Evidence from the Asian crisis. *Journal of Economic Dynamics and Control*, 33(11), 1929–1944.

DeCicca, P., Kenkel, D. S., and Liu, F. (2013a). Who pays cigarette taxes? The impact of consumer price search. *Review of Economics and Statistics*, 95(2), 516–529.

DeCicca, P., Kenkel, D. S., and Liu, F. (2013b). Excise tax avoidance: The case of state cigarette taxes. *Journal of Health Economics*, 32(6), 1130–1141.

Deelstra, G., Petkovic, A., and Vanmaele, M. (2010). Pricing and hedging Asian basket spread options. *Journal of Computational and Applied Mathematics*, 233, 2814–2830.

Deng, S. (2009). The formation cause and treatment of non-performing loans to commercial bank. *Times Finance*, 2, 23–24.

Dewynne, J. N. and Shaw, W. T. (2008). Differential equations and asymptotic solutions for arithmetic Asian options: 'Black-Scholes formulae' for Asian rate calls. *European Journal of Applied Mathematics*, 19, 353–391.

Dewynne, S. (2008). Differential equations and asymptotic solutions for arithmetic Asian options: 'Black-Scholes formulae' for Asian rate calls. *European Journal of Applied Mathematics*, 19, 353–391.

Ding, J., Lu, Y.-Z., and Chu, J. (2013). Studies on controllability of directed networks with extremal optimization. *Physica A: Statistical Mechanics and Its Applications*, 392(24), 6603–6615.

Doraisami, A. (2004). From crisis to recovery: The motivations for and effects of Malaysian capital controls. *Journal of International Development*, 16, 241–254.

Drasson, A. and Masson, P. R. (1994). Credibility of policies versus credibility of policy makers. *Quarterly Journal of Economics*, 109(3), 735–754.

Dreher, A. and Schneider, F. (2010). Corruption and the shadow economy: An empirical analysis. *Public Choice*, 144, 215–238.

Duke, N. C. et al. (2007). A world with mangroves? *Science* (in Letters), 317, 41–42.

Dunger, M., Fry, R., Gonzalez-Hermosillo, B., and Martin, V. (2004). Empirical modeling of contagion: A review of methodologies. IFM working paper no. 04/78.

Edison, H. J., Levine, R., Ricci, L., and Slok, T. (2002). International financial integration and economic growth. *Journal of International Money and Finance*, 21, 749–776.

Edison, H. J. and Warnock, F. E. (2008). Cross-border listings, capital controls, and equity flows to emerging markets. *Journal of International Money and Finance*, 27, 1013–1027.

Edwards, S. (1989). *Real Exchange Rates, Devaluation, and Adjustment*. The MIT Press, Cambridge, MA.

Edwards, S. (1995). Trade policy, exchange rates, and growth. *Reform, Recovery, and Growth: Latin America and the Middle East*. NBER Chapters. National Bureau of Economic Research, Inc., pp. 13–52.

Edwards, S. (2007). *Capital Controls, Sudden Stops, and Current Account Reversals*. University of Chicago Press, Chicago, IL.

Edwards, S. (2010). The international transmission of interest rate shocks: The Federal Reserve and emerging markets in Latin America and Asia. *Journal of International Money and Finance*, 29(4), 685–703.

Efrati, A. (2007). Cigarette-tax disparities are a boon for border towns. *Wall Street Journal*, March 2.

Ehrlich, P. R. and Holdens, J. P. (1971). The impact of population growth. *Science*, 171, 1212–1217.

Eichengreen, B. (2008). The real exchange rate and economic growth. Working paper no. 4. Commission on Growth and Development.

Eichengreen, B. and Leblang, D. (2002). Capital account liberalization and growth: Was Mr. Mahathir right? National Bureau of Economic Research. Working paper.

Eicker, F. (1963). Asymptotic normality and consistency of the least squares estimators for families of linear regressions. *Annals of Mathematical Statistics*, 34, 447–456.

Elbadawi, I. A. (2012). Aid, real exchange rate misalignment, and economic growth in Sub-Saharan Africa. *World Development*, 40(4), 681–700.

Elbadawi, I. A. and Soto, R. (1997). Real exchange rates and macroeconomic adjustment in sub-Sahara Africa and other developing countries. *Journal of African Economies*, 6, 74–120 (Supplement).

Ellis, M. and Auernheimer, L. (1996). Stabilization under capital controls. *Journal of International Money and Finance*, 15(4), 523–533.

Ellul, A., Jappelli, T., Pagano, M., and Panunzi, F. (2012). Transparency, tax pressure and access to finance. CEPR Discussion Papers 8939.

Embrechts, P., McNeil, A. J., and Straumann, D. (1999). Correlation and dependence in risk management: Properties and pitfalls. In: Dempster, M. (ed.), *Risk Management: Value at Risk and Beyond*. Cambridge University Press, Cambridge, U.K., pp. 176–223.

Evdokimenko, M. D. (2011). Combustibility factors of Baikalian forests. *Geography and Natural Resources*, 32(3), 242.

Farrel, D. (2004). The hidden dangers of the informal economy. *The McKinsey Quarterly*, Vol. 3.

Faruqee, H. (1995). Long-run determinants of the real exchange rate: A stock-flow perspective. *Staff Papers—International Monetary Fund*, 42, 80–107.

Fazio, G. (2007). Extreme interdependence and extreme contagion between emerging markets. *Journal of International Money and Finance*, 26(8), 1261–1291.

Fleenor, P. (2008). Cigarette taxes are fueling organized crime. *The Wall Street Journal*, A17 (May 7, 2008).

Flood, R. and Garber, P. (1984). Collapsing exchange rate regimes: Some linear examples. *Journal of International Economics*, 17(1–2), 1–13.

Food and Drug Administration (2013). Menthol in cigarettes, tobacco products: Request for comments. A proposed rule by the food and drug administration on 07/24/2013. Federal Register. https://www.federalregister.gov/articles/2013/07/24/2013-17805/menthol-in-cigarettes-tobacco-products-request-for-comments.

Forbes, K. J. and Rigobon, R. (2002). No contagion, only interdependence: Measuring stock markets co-movements. *Journal of Finance*, 57(5), 2223–2261.

Forte, J. C., Letrémy, P., and Cottrell, M. (2002). Advantages and drawbacks of the Batch Kohonen algorithm. In: Verleysen, M. (ed.), *Proceedings of the 10th European Symposium on Neural Networks*. Springer-Verlag, Berlin, Germany.

Frankel, J. and Froot, K. (1986). Understanding the U. S. dollar in the eighties: The expectation of chartists and fundamentals. *Economic Record Supplement*, 10, 24–38.

Freitas, M. L. (2004). The dynamics of inflation and currency substitution in a small open economy. *Journal of International Money and Finance*, 23(1), 133–142.

Freitas, M. L. (2006). Currency substitution, portfolio diversification and money demand. *Canadian Journal of Economics*, 39(3), 719–743.

Frentz, M. and Nyström, K. (2010). Adaptive stochastic weak approximation of degenerate parabolic equations of Kolmogorov type. *Journal of Computational and Applied Mathematics*, 234, 146–164.

Frentz, M., Nyström, K., Pascucci, A., and Polidoro, S. (2010). Optimal regularity in the obstacle problem for Kolmogorov operators related to American Asian options. *Mathematische Annalen*, 347, 805–838.

Galiani, S., Heymann, D., and Magud, N. (2010). On the distributive effects of terms of trade shocks: The role of non-tradable goods. IMF working paper, pp. 10–241.

Gallagher, K. (2011). Regaining control? Capital controls and the global financial crisis. University of Massachusetts Amherst, Amherst, MA, working paper series, Number 250.

Garman, M. B. and Kohlhagen, S. W. (1983). Foreign currency option values. *Journal of International Money and Finance*, 2(12), 231–237.

Gerdtham, U. G. and Jonsson, B. (2000). International comparisons of health expenditure: Theory, data and econometric analysis. In: Culyer, A. J. and Newhouse, J. P. (eds.), *Handbook of Health Economics*, Vol. 1, Part A. Elsevier, Amsterdam, Netherlands, pp. 11–53.

Ghashghaie, S., Breymann, W., Peinke J., Talkner, T., and Dodge, Y. (1996). Turbulent cascade in foreign exchange markets. *Nature*, 381, 767–770.

Ghura, D. and Grennes, T. J. (1993). The real exchange rate and macroeconomic performance in Sub-Saharan Africa. *Journal of Development Economics*, 42, 155–174.

Gollin, D. (2002). Getting income shares right. *Journal of Political Economy*, 110, 458–474.

Greiner, A. and Semmler, W. (1996). Multiple steady states, indeterminacy, and cycles in a basic model of endogenous growth. *Journal of Economics*, 63, 79–99.

Groessl, I. and Fritsche, U. (2007). The store-of-value-function of money as a component of household risk management. Discussion Papers, February 7, http://www.gsoep.de/documents/publikationen /73/55726/dp660.pdf.

Guiso, L., Sapienza, P., and Zingales, L. (2004). Does local financial development matter? *Quarterly Journal of Economics*, 119, 929–969.

Guitian, M. (1998). Capital account convertibility and the financial sector. *Journal of Applied Economics*, 1, 209–229.

Gulich, D. and Zunino, L. (2014). A criterion for the determination of optimal scaling ranges in DFA and MF-DFA. *Physica A: Statistical Mechanics and Its Applications*, 397(1), 17–30.

Hall, B. J. and Murphy, K. J. (2000). Optimal exercise prices for executive stock options. *American Economic Review*, 90, 209–214.

Hall, B. J. and Murphy, K. J. (2002). Stock options for undiversified executives. *Journal of Accounting and Economics*, 33, 3–42.

Hansen, B. E. (1992). Tests for parameter instability in regressions with I(1) processes. *Journal of Business and Economic Statistics*, 10, 321–335.

Harandi, S. and Alamatsaz, M. H. (2013). Alpha–Skew–Laplace distribution. *Statistics and Probability Letters*, 83(3), 774–782.

Harrison, A. E. and Rodríguez-Clare, A. (2009). Trade, foreign investment, and industrial policy. MPRA Paper 15561. University Library of Munich, Munich, Germany

Harte, J. and Newman, E. A. (2014). Maximum information entropy: A foundation for ecological theory. *Trends in Ecology and Evolution*, 29(7), 384–389.

Hartwig, J. (2008). What drives health care expenditure? Baumol's model of unbalanced growth revisited. *Journal of Health Economics*, 27, 603–623.

Hartwig, J. (2011). Can Baumol's model of unbalanced growth contribute to explaining the secular rise in health care expenditure? An alternative test. *Applied Economics*, 43(2), 173–184.

Henry, P. B. (2007). Capital account liberalization: Theory, evidence, and speculation. *Journal of Economic Literature*, XLV, 887–935.

Hines, J. R. (2007). Taxing consumption and other sins. *Journal of Economic Perspectives*, 21(1), 49–68.

Hinkle, L. E. and Nsengiyumva, F. (1999). The two-good internal RER for tradables and non-tradables. In: Hinkle, L. E. and Montiel, P. J. (eds.), *Exchange Rate Misalignment: Concepts and Measurement for Developing Countries*. A World Bank Research Publication, Oxford University Press, Oxford, U.K., pp. 113–173.

Hirsch, A. (1989). The origins and implications of South Africa's continuing financial crisis. *Transformation*, 9, 31–50.

Hirsch, B. T. and Macpherson, D. A. (2003). Union membership and coverage database from the current population survey: Note. *Industrial and Labor Relations Review*, 56(2), 349–354.

Hodrick, R. J. and Prescott, E. C. (1997). Postwar U.S. business cycles: An empirical investigation. *Journal of Money, Credit and Banking*, 29, 1–16.

Holmstrom, B. and Tirole, J. (1996). Modeling aggregate liquidity. *American Economic Review*, 86, 187–191.

Horn, S. D., Horn, R. A., and Duncan, D. B. (1975). Estimating heteroscedastic variances in linear models. *Journal of the American Statistical Association* 70, 380–385.

Hou, Y.-J. (2002). The efficiency of the foreign exchange market and fractal market analysis. *The Theory and Practice of Finance and Economics*, 23(120), 75–79.

Hu, Y. and Shimomura, K. (2007). Status-seeking, catching-up, and comparative, statics in a dynamic Heckscher–Ohlin mode. *Review of Development Economics*, 11(2), 258–274.

Hughes, J., Lang, L., Mester, L., and Moon, C. (1995). Recovering technologies that account for generalized managerial preferences: An application to non-risk neutral banks. Working paper. Center for Financial Institutions, The Wharton School, University of Pennsylvania, Philadelphia, PA, pp. 95–16.

Hull, J. C. (1999). *Options, Future and Other Derivatives*. Prentice Hall, Upper Saddle River, NJ, pp. 189–204.

International Monetary Fund (1998). *World Economic Outlook*. International Monetary Fund, Washington, DC.

Janus, T. and Riera.-Crichton, D. (2013). International gross capital flows: New uses of balance of payments data and application to financial crises. *Journal of Policy Modeling*, 35, 16–28.

Jappelli, T., Pagano, M., and Bianco, M. (2005). Courts and banks: Effects of judicial enforcement on credit markets. *Journal of Money, Credit, and Banking*, 37, 223–244.

Jiang, B. K. and Yang, H. (1999). *Currency Substitution Research*. Fudan University Press, Shanghai, China [in Chinese].

Johansen, S. (1992). Co-integration in partial systems and the efficiency of single-equation analysis. *Journal of Econometrics*, 52, 389–402.

Johnson, S. and Mitton, T. (2003). Cronyism and capital controls: Evidence from Malaysia. *Journal of Financial Economics*, 67, 351–382.

Jorge, S. et al. (2008). The political economy of populism and protection in Argentina: A case study for the agricultural sector. *II Congreso Regional de Economía Agraria*, Montevideo, Uruguay.

Jose, K. K., Tomy, L., and Sreekumar, J. (2008). Autoregressive processes with normal-Laplace marginals. *Statistics and Probability Letters*, 78(15), 2456–2462.

Kaminsky, G. L. and Reinhart, C. M. (1999). The twin crises: The causes of banking and balance-of-payments problem. *American Economic Review*, 89, 473–500.

Khalifa, S. and Hurcan, K. I. (2011). Undiscounted optimal growth with consumable capital: Application to water resources. *Journal of Applied Economics*, 14(1), 145–166.

Khan, M. A. and Mitra, T. (2007). Optimal growth in a two sector model without discounting: A geometric investigation. *Japanese Economic Review*, 58, 191–225.

Khanser, M. A. (1999). *Dance of Chaos: The Application of Chaos Theory in the Philippine Foreign Exchange Market*. Khaner Publishing House, Cagayan de Oro, Philippines.

Khwaja, A., Silverman, D., and, Sloan, F. (2007). Time preference, time discounting, and smoking decisions. *Journal of Health Economics*, 26, 927–949.

Kim, S. (2009). On a degenerate parabolic equation arising in pricing of Asian options. *Journal of Mathematical Analysis and Applications*, 351, 326–333.

Kindleberger, C. (1996). *Maniacs, Panics, and Crashes*. Cambridge University Press, Cambridge, U.K.

King, D. T. (1978). Currency portfolio approach to exchange determination: Exchange rate stability and the independence of money policy. *The Monetary Approach to International Adjustment*. Praeger, New York.

King, M. A. and Wadhwani, S. (1990). Transmission of volatility between stock markets. *The Review of Financial Studies*, 3(1), 5–33.

Kitano, S. (2011). Capital controls and welfare. *Journal of Macroeconomics*, 33, 700–710.

Klein, M. W. and Olivei, G. P. (2008). Capital account liberalization, financial depth, and economic growth. *Journal of International Money and Finance*, 27(6), 861–875.

Kohonen, T. (1982). Self-organized formation of topologically correct feature maps. *Biological Cybernetics*, 66, 59–69.

Kohonen, T. (2001). *Self-Organizing Maps*, 3rd edn. Springer-Verlag, Berlin, Germany.

Korinek, A. (2011). Hot money and serial financial crises. *IMF Economic Review*, 59, 306–339.

Krugman, P. (1979). A model of balance-of-payments crises. *Journal of Money, Credit and Banking*, 11(3), 311–312.

Krugman, P. (1998). Fire sale FDI. Working paper, Massachusetts Institute of Technology.

Kukla, Z. E. and Płatkowski, T. (2013). Onset of limit cycles in population games with attractiveness driven strategy choice. *Chaos, Solitons and Fractals*, 56, 77–82.

Lane, P. R. (2001). International trade and economic convergence: The credit channel. *Oxford Economic Papers*, 53, 221–240.

Layard, R., Nickel, S., and Jackman, R. (1991). *Unemployment: Macroeconomic Performance and the Labor Market*. Oxford University Press, Oxford, U.K.

Leroux, A. and MacLaren, D. (2011). The optimal time to remove quarantine bans under uncertainty: The case of Australian bananas. *The Economic Record*, 87(276), 140–152.

Levy-Yeyati, E. L. and Sturzenegger, F. (2007). Fear of floating in reverse exchange rate policy in the 2000s. Mimeo.

Lewellen, J. and Nagel, S. (2006). The conditional CAPM does not explain asset-pricing anomalies. *Journal of Financial Economics*, 82, 289–314.

Lewellen, K. (2006). Financing decisions when managers are risk averse. *Journal of Financial Economics*, 82, 551–589.

Li, C. and Qu, B. (2007). The economic impact of capital account opening and related countermeasures. *China Credit Card*, 12, 50–52.

Li, G. and Li, Z. (2010). Regional difference and influence factors of China's carbon dioxide emissions. *China's Population Resources and Environment*, 20(5), 22–27.

Li, L., Ying, Y.-R., and Dang, X.-Y. (1996). *Significance, Methodology and Application about Infinite*. Press of Northwestern University, Xi'an, China. [in Chinese]

Li, X. (2004). Trade liberalization and real exchange rate movement. *IMF Staff Papers*, 51, 553–584.

Lin, B. and Jiang, Z. (2009). A forecast for China's environmental Kuznets curve for CO_2 emission, and an analysis of the factors affecting China's CO_2 emission. *Management World*, 4, 27–36.

Lin, Y. and Xu, C. (2006). Analysis of a cumulative type financial product pricing. *Modern Management Science*, 1, 102–104.

Ling, J., Li, Y., and Wang, C. (2009). Welfare effects and policy choice of capital account liberalization. *Future and Development*, 6, 17.

Liu, N. (2009). The influence of capital account openness on inflation—The empirical analysis on base of China's data during1990–2008. *Shanghai Finance*, 6, 18–22.

Liu, Q. (2012). An optimal control approach to probabilistic Boolean networks. *Physica A: Statistical Mechanics and its Applications*, 391(24), 6682–6689.

Love, N. (2013). NH convenience stores oppose cigarette tax hike. *Concord Monitor*, Associated Press, April.

Lucio, R. (2008). El conflicto por el régimen de las retenciones y el futuro de la agricultura argentina. *Anales de la XLIII Reunion Anual de la Asociacion Argentina de Economia Politica*, Córdoba, Argentina.

Luo, J. (2008). Capital account liberalization will trigger the financial crisis. *New Finance*, 228, 53–54.

MacDonald, R. and Vieira, F. (2010). A panel data investigation of real exchange rate misalignment and growth. CESifo working paper no. 3061.

Magud, N. E., Reinhart, C. M., and Rogoff, K. S. (2011). Capital controls: Myth and reality—A portfolio balance approach. National Bureau of Economic Research Working Paper Series. National Bureau of Economic Research.

Manlagnit, M. C. V. (2010). Cost efficiency, determinants, and risk preferences in banking: A case of stochastic frontier analysis in the Philippines. *Journal of Asian Economics*, 10(1).

Mansfield, E. (1963). Intra-firm rates of diffusion of an innovation. *The Review of Economics and Statistics*, 45, 348–359.

Markowitz, H. (1952). Portfolio selection. *Journal of Finance*, 7, 77–91.

Meese, R. A. and Rogoff, K. (1983). Empirical exchange rate models of the seventies: Do they fit out of sample. *Journal of International Economics*, 14, 3–24.

Meltzer, A. H. (1994). Heterodox policy and economic stabilization. *Journal of Monetary Economics*, 34(3), 581–600.

Merton, R. C. (1969). Lifetime portfolio selection under uncertainty: The continuous time case. *Review of Economics and Statistics*, 51, 247–257.

Merton, R. C. (1971). Optimum consumption and portfolio rules in a continuous time model. *Journal of Economic Theory*, 3, 373–413; Erratum 6, 1973, 213–214.

Merton, R. C. (1990). *Continuous-Time Finance*. Wiley-Blackwell, Hoboken, New Jersey.

Merz, M. (1995). Search in the labor market and the real business cycle. *Journal of Monetary Economics*, 36, 269–300.

Michael, D. B., Meissner, C. M., and Stuckler, D. (2010). Foreign currency debt, financial crises and economic growth: A long-run view. *Journal of International Money and Finance*, 29(4), 642–665.

Miles, M. A. (1978). Currency substitution, flexible exchange rates, and monetary independence. *American Economic Review*, 68(3), 428–436.

Miller, M. (1998). Asian financial crisis. *Japan and the World Economy*, 10(3), 355.

Minsky, H. (1982). *Can "It" Happen Again?: Essays on Instability and Finance*. M.E. Sharpe, Armonk, NY.

Mongardini, J. and Rayner, B. (2009). Grants, remittances, and the equilibrium real exchange rate in Sub-Saharan African countries. IMF working paper WP/09/75.

Montiel, P. J. (1999a). Determinants of the long-run equilibrium real exchange rate. In: Hinkle, L. E. and Montiel, P. J. (eds.), *Exchange Rate Misalignment: Concepts and Measurement for Developing Countries*. A World Bank Research Publication, Oxford University Press, Oxford, U.K., pp. 264–290.

Montiel, P. J. (1999b). The long-run equilibrium real exchange rate: Conceptual issues and empirical research. In: Hinkle, L. E. and Montiel, P. J. (eds.), *Exchange Rate Misalignment: Concept and Measurement in Developing Countries*. A World Bank Research Publication, Oxford University Press, Oxford, U.K., pp. 219–263.

Montiel, P. J. (2007). Equilibrium real exchange rates, misalignment and competitiveness in the southern cone. CEPAL, Economic Development Division/United Nations, Santiago de Chile.

Morris, K. and McNicholas, P. D. (2013). Dimension reduction for model-based clustering via mixtures of shifted asymmetric Laplace distributions. *Statistics and Probability Letters*, 83(9), 2088–2093.

Morris, S. and Shin, H. S. (1998). Unique equilibrium in a model of self-fulfill in currency attacks. *American Economic Review*, 88(3), 587–589.

Musiela, M. and Rutkowsk, M. (1997). *Martingale Methods in Financial Modeling: Theory and Application*. Springer-Verlag, Berlin, Germany, pp. 88–97.

Narayan, C. P. (2011). Nexus between capital flows and economic growth: The Indian context. *Journal of International Economics*, 2(1), 18–37.

Newhouse, J. P. (1992). Medical care costs: How much welfare loss? *Journal of Economic Perspectives*, 6(3), 3–21.

Nickell, S. (1981). Biases in dynamic models with fixed effects. *Econometrica*, 49, 1417–1426.

Nurkse, R. (1947). International monetary policy and the search for economic stability. *American Economic Review*, 7(2), 569–580.

Obstfeld, M. (1994). The Logic of currency crises. *Cahierséconomiques et monétairs*, 43, 189–213.

Obstfeld, M. (1996). Models of currency crises with self-fulfilling feature. *European Economic Review*, 40, 1037–1047.

Odani, K. (1995). The limit cycle of the van der Pol equation is not algebraic. *Journal of Different Equations*, 115, 146–152.

Onofri, A. (2009). *Impactos económicos y sociales de la eliminación de DEX y REX en Argentina*. Foro de la Cadena Agroindustrial Argentina, Buenos Aries, Argentina.

Organization for Economic Cooperation Development (2012). *Consumption Tax Trends (2004 edition)*. OECD, Paris, France.

Orlin, G. J. (1983). The pricing of call and put options on foreign exchange. *Journal of International Money Finance*, 2(3), 239–253.

Ortega, G. J. and Matesanz, D. (2006). Cross-country hierarchical structure and currency crises. *International Journal of Modern Physics C*, 17(3), 333–341.

Orzechowski, W. and Walker, R. C. (2001). The tax burden on tobacco. Economic Research Service, USDA, Tobacco Briefing Room.

Ostry, J. D., Ghosh, A. R., Chamon, M., and Qureshi, M. S. (2011). Capital controls: When and why? *IMF Economic Review*, 59, 562–580.

Ostry, J. D., Ghosh, A. R., Chamon, M., and Qureshi, M., S. (2012). Tools for managing financial stability risks from capital inflows. *Journal of International Economics*, 88(2), 407–421.

Pastor, M. (1990). Capital flight from Latin America. *World Development*, 18(1), 1–18.

Patrick, F. (1998). How excise tax differentials affect interstate smuggling and cross-border sales of cigarettes in the United States. Background Paper #26. Tax Foundation, Washington, DC.

Phillips, P. C. B. and Durlauf, S. N. (1986). Multiple time series regression with integrated processes. *The Review of Economic Studies*, 53, 473–495.

Pirjetä, A., Ikäheimo, S., and Puttonen, V. (2010). Market pricing of executive stock options and implied risk preferences. *Journal of Empirical Finance*, 17, 394–412.

Pissarides, C. A. (1986). Unemployment and vacancies in Britain. *Economic Policy*, 3, 499–559.

Poincaré, H. (1891). Sur l'integration algébrique des équations différentielles du premier ordre et du premier degré (I and II). *Rend circ mat Palermo*, 5, 161–191.

Poloz, S. S. (1986). Currency substitution and the precautionary demand for money. *Journal of International Money and Finance*, 5, 115–124.

Porto, A. and Lodola, A. (2013). Economic policy and electoral outcomes. *Journal of Applied Economics*, 14(2), 333–356.

Potgieter, D. H. (2009). Fractal asset returns, arbitrage and option pricing. *Chao, Solitions and Fractals*, 42(3), 1792–1795.

Prasad, E. S., Rajan, G. R., and Subramanian, A. (2007). Foreign capital and economic growth. *The Brookings Institution*, 38(1), 153–230.

Qin, X.-Z. and Ying, Y.-R. (2002). Pricing method for contingent claims based on the duality principle of linear programming. *Proceedings of International Conference on Mathematical Programming*, 12, 306–311.

Qin, X.-Z. and Ying, Y.-R. (2004). Asset pricing method based on martingale and entropy theories. *Systems Engineering Theory, Practice and Application*, 13(5), 460–462.

Qin, Y.-X. (1966). On the algebraic limit cycles of second degree of the differential equation $dy/dx = \sum_{0 \le i+j \le 2} a_{ij} x^i y^j \Big/ \sum_{0 \le i+j \le 2} b_{ij} x^i y^j$. *Acta Mathematica Sinica*, 8, 608–619.

Quinn, D. (1997). The correlates of change in international financial regulation. *American Political Science Review*, 91(3), 531–562.

Quinn, D. P. and Toyoda, M. (2008). Does capital account openness lead to growth. *The Review of Financial Studies*, 21(3), 1403–1449.

Qureshi, M. S., Ostry, J. D., Ghosh, A. R., and Chamon, M. (2011). Managing capital inflows: The role of capital controls and prudential policies. National Bureau of Economic Research working paper series, working paper 17363.

Ramsey, F. (1928). A mathematical theory of savings. *Economic Journal*, 38, 543–559.

Reed, W. J. and Jorgensen, M. (2004). The double Pareto-lognormal distribution, A new parametric model for size distributions. *Communications in Statistics—Theory and Methods*, 33(8), 1733–1753.

Richards, G. (2000). The fractal structure of exchange rates: Measurement and forecasting. *Journal of International Financial Market*, 10, 163–180.

Rim, S. (2007). Empirical study on the impacts of major events on inter-market relationships in Asia. *Global Business and Finance Review*, 12(2), 75–87.

Rim, S. (2008). Studies on the financial market integration and financial efficiency: Evidences from Asian markets. *The Business Review*, 10(2), 357–363.

Robert, P. F. and Garber, P. M. (1984). Collapsing exchange-rate regimes. *Journal of International Economics*, 17, 1–13.

Rodriguez, J. C. (2007). Measuring financial contagion: A Copula approach. *Journal of Empirical Finance*, 14(3), 401–423.

Rodrik, D. (2008). The real exchange rate and economic growth. *Brookings Papers on Economic Activity*, pp. 365–412.

Roodman, D. (2009). How to do xtabond2: An introduction to difference and system GMM in Stata. *The Stata Journal*, 9, 86–136.

Rosenberg, N. (1982). *Inside the Black Box: Technology and Economics*. Cambridge University Press, Cambridge, U.K.

Ross, S. A. (2004). Compensation, incentives, and the duality of risk aversion and riskiness. *Journal of Finance*, 59, 207–225.

Roudet, S., Saxegaard, M., and Tsangarides, C. G. (2007). Estimation of equilibrium exchange rates in the WAEMU: A robustness analysis. IMF working paper no. WP/07/94.

Sam, P. (1992). Voters as fiscal conservatives. *Quarterly Journal of Economics*, 107, 327–361.

Sam, P. (1998). Deterministic chaos and fractal attractor as tools for nonparametric dynamical econometric inference with an application to the division monetary aggregates. *Mathematical and Computer Modeling*, 10, 275–296.

Samuelson, P. and Solow, R. (1956). A complete capital model involving heterogeneous capital goods. *Quarterly Journal of Economics*, 70, 537–562.

Sanchirico, J. N. and Springborn, M. (2011). How to get there from here: Ecological and economic dynamics of ecosystem service provision. *Environmental and Resource Economics*, 48, 243–267.

Sarlin, P. and Marghescu, D. (2011). Visual predictions of currency crises using self-organizing maps. *Intelligent Systems in Accounting, Finance and Management*, 18(1), 15–38.

Sarlin P. and Peltonen, T. A. (2013). Mapping the state of financial stability. *Journal of International Financial Markets, Institutions and Money*, 26, 46–76.

Sawada, Y. (2011). Did the financial crisis in Japan affect household welfare seriously? *Journal of Money, Credit and Banking*, 43, 2–3.

Schmitt-Grohé, S. and Uribe, M. (2006). Optimal simple and implementable monetary and fiscal rules: Expanded version. Working paper 12402, National Bureau of Economic Research.

Schneider, F. (2000). Shadow economies: Size, causes, and consequences. *Journal of Economic Literature*, 38, 77–114.

Schneider, F. (2010a). Survey on the shadow economy and undeclared earnings in OECD countries. *German Economic Review*, 11, 109–149.

Schneider, F. (2010b). The influence of public institutions on the shadow economy: An empirical investigation for OECD countries. *Review of Law and Economics*, 6, 113–140.

Schwartz, B. and Yousefi, S. (2003). On complex behavior and exchange rate dynamics. *Chaos Solutions and Fractal*, 18(3), 503–523.

Seidenberg, A. (1968). Reduction of singularities of the differential equation Ady=Bdx. *American Journal of Mathematics*, 90, 248–269.

Shen, C.-H., Lee, C.-C., and Lee, C.-C. (2010). What makes international capital flows promote economic growth? An international cross-country analysis. *Scottish Journal of Political Economy*, 57(5), 515–546.

Shen, M. (2008). Pricing of European option on foreign exchange with time-varying interest rate in fractional market. *Science Technology and Engineering*, 29, 6564–6568.

Shi, S. (1995). Tariffs, unemployment, and the current account: An intertemporal equilibrium model. Manuscript, Queen's University, Kinston, Ont. Canada.

Shi, S. and Wen, Q. (1997). Labor market search and capital accumulation: Some analytical results. *Journal of Economic Dynamics and Control*, 21, 1747–1776.

Shiller, R. J. (1986). The Marsh–Merton model of managers' smoothing of dividends. *American Economic Review*, 76(3), 499–503.

Silva, A. C. (2010). Managerial ability and capital flows. *Journal of Development Economics*, 93, 126–136.

Simos, G. M. and Tsionas, E. (2010). Testing for the generalized normal-Laplace distribution with applications. *Computational Statistics and Data Analysis*, 54(12), 3174–3180.

Sims, C. A. (1980). Macroeconomics and reality. *Econometrica*, 48, 1–48.

Straetmans, S. and Candelon, B. (2013). Long-term asset tail risks in developed and emerging markets. *Journal of Banking and Finance*, 37, 1832–1844.

Steffensen, M. (2010). Optimal consumption and investment under time-varying relative risk aversion. *Journal of Economic Dynamics and Control*, 35(5), 659–667.

Stephen, M. and Hyun S. S. (2004). Coordination risk and the price of debt. *Economic Review*, 48(1), 133–153.

Stix, H. (2007). Impact of Central Bank Intervention during periods of speculative pressure: Evidence from the European Monetary System. *German Economic Review*, 8(3), 399–427.

Stock, J. H. and Watson, M. W. (1993). A simple estimator of co-integrating vectors in higher order integrated systems. *Econometrica*, 61, 783–820.

Stock, J. H., Wright, J. H., and Yogo, M. (2002). A survey of weak instruments and weak identification in generalized method of moments. *Journal of Business and Economic Statistics*, 20(4), 518–529.

Sun, L. (2006). The influence of capital account liberalization on financial system. The postgraduate's paper of Jilin University, Changchun, China.

Sutradhar, B. C. and Jowaheer, V. (2010). Treatment design selection effects on parameter estimation in dynamic logistic models for longitudinal binary data. *Journal of Statistical Computation and Simulation*, 80(9), 1053–1067.

Swallow, S. K. (1990). Depletion of the environmental basis for renewable resources: The economics of interdependent renewable and nonrenewable resources. *Journal of Environmental Economics and Management*, 19, 281–296.

Syaiba, B. A. and Habshah, M. (2010). Robust logistic diagnostics for the identification of high leverage points in logistic regression model. *Journal of Applied Sciences*, 10(23), 3042–3050.

Tangian, A. (2010). Computational application of the mathematical theory of democracy to Arrow's Impossibility Theorem (how dictatorial are Arrow's dictators?). *Social Choice and Welfare*, 35, 129–161.

Ted, O. and Matthew, R. (2006). Optimal sin taxes. *Journal of Public Economics*, 90(10–11), 1825–1849.

Theil, H. (1971). *Principles of Econometrics*, John Wiley and Sons, New York.

Thomas, L. R. (1985). Portfolio theory and currency substitution. *Journal of Money, Credit, and Banking*, 17(3), 105–120.

Tong, T. (2013). The feasibility and routing selection of Chinese carbon financial system construction. Wuhan Finance. No. 04:29–31.

Tong, B. (2014). Risk measurement under heavy tailed environment and its applications. PhD dissertation, Shanghai Jiao Tong University, Shanghai, China.

Tony, C. (2010). What drives monetary policy in post-crisis East Asia? Interest rate or exchange rate monetary policy rules. *Journal of Asian Economics*, 21(5), 456–465.

Tornell, A. and Velasco, A. (1992). The tragedy of the commons and economic growth: Why does capital flow from poor to rich countries? *Journal of Political Economy*, 100, 1208–1231.

Turnovsky, S. J. (1997). *International Macroeconomic Dynamics*. The MIT Press, Cambridge, MA.

Vandewalle, N. and Ausloos, M. (1998). Multi-affine analysis of typical currency exchange rates. *European Physical Journal B*, 4, 257–261.

Veiga, C., and Wystup, U. (2009). Closed formula for options with discrete dividends and its derivatives. *Applied Mathematical Finance*, 16(6), 517–531.

Verdier, G. (2003). The role of capital flows in neoclassical open-economy models with imperfect capital markets. *Mimeo*, 26, 34–46.

Verdier, G. (2008). What drives long-term capital flows? A theoretical and empirical investigation. *Journal of international Economics*, 74, 120–142.

Vesanto, J. and Alhoniemi, E. (2000). Clustering of the self-organizing map. *IEEE Transactions on Neural Networks*, 11(3), 586–600.

Walcher, S. (2000). On the Poincaré problem. *Journal of Differential Equations*, 166, 51–78.

Walker, T. B. and Whiteman, C. H. (2007). Multiple equilibria in a simple asset pricing model. *Economics Letters*, 97(3), 191–196.

Wang, C., Chen, J. N., and Zou. J. (2005). Decomposition of energy-related CO2 emission in China: 1957–2000. *Energy*, 30(1), 73–83.

Wang, X.-T. (2010). Scaling and long-range dependence in option pricing I: Pricing European option with transaction costs under the fractional Black-Scholes model. *Physica A: Statistical Mechanics and Its Applications*, 389(3), 438–444.

Ward, J. H., Jr. (1963). Hierarchical grouping to optimize an objective function. *Journal of the American Statistical Association*, 58, 236–244.

White, H. (1980). A heteroscedasticity-consistent covariance matrix estimator and a direct test for heteroscedasticity. *Econometrica*, 48, 817–838.

Wichitaksorn, N. and Tsurumi, H. (2013). Comparison of MCMC algorithms for the estimation of Tobit model with non-normal error: The case of asymmetric Laplace distribution. *Computational Statistics and Data Analysis*, 67, 226–235.

Wieland, C. (2002). Controlling chaos in higher dimensional maps with constant feedback: An analytical approach. *Physical Review*, 66, 58–75.

William, R., Brian, F., and William, M. P. (2009). Science of analytical reasoning. *Information Visualization*, 8(4), 254–262.

Williamson, J. (1990). What Washington means by policy reform. In: Williamson, J. (ed.), *Latin American Adjustment: How Much Has Happened?* Peterson Institute for International Economics, Washington, DC, pp. 5–20.

Wirl, F. (1996). Pathways to hopfbifurcations in dynamic, continuous time optimization problems. *Journal of Optimization Theory and Application*, 91(2), 299–320.

Wirl, F. (1997). Stability and limit cycles in one-dimensional dynamic optimizations of competitive agents with a market externality. *Journal of Evolutionary Economics*, 7, 73–89.

Wolfgang, S. and Paul, S. (1941). Protection and real wages. *Review of Economic Studies*, 9, 58–73.

WTO. (1997). *Tourism: 2020 Vision, Influences, Directional Flows and Key Trends*, Executive Summary. WTO, Madrid, Spain, p. 31.

Wu, C.-F., Mu, Q.-G., and Wu, W.-F. (2004). The mixed asset pricing model based on industry market-capital market. *Journal of Management Sciences in China*, 7(6), 13–23.

Wu, W.-F., Rui, M., and Chen, G.-M. (2003). The illiquidity compensation of China's stock returns. *World Economy*, 7, 54–60.

Wu, Z. and Wang, G. (2007). A Black-Scholes formula for option pricing with dividends and optimal investment problems under partial information. *Journal of System Science and Mathematical Sciences*, 27(5), 676–683.

Xie, C. and Yang, N. (2008). The chaos of exchange rate behavior and its fractal description. *Journal of Hunan University*, 8, 89–92.

Xu, C. and Gu, E. (2004). Approximation for pricing arithmetic average asian options with fixed strike price. *Communications on Applied Mathematics and Computational Science*, 18(2), 8–12.

Xu, C., Zhou, J., and Ren, X. (2007). Arbitrage analysis of a class of deposit product with option style. *Journal of Tong Ji University (Natural Science)*, 35(7), 994–997.

Yang, N. and Chi, X. (2008). Empirical research on the multiracial characteristics of exchange rate time series. *Journal of Hunan University*, 6, 78–82.

Yang, Z. F. and Jiang, M. M. (2009). Solar energy evaluation for Chinese economy. *Energy Policy*, 38, 875–886.

Ye, W. C. (2009). *Study on Economic Effects of Capital Account Liberalization*. Press of Shanghai University of Finance and Economics, Shanghai, China. [in Chinese]

Ye, Y.-Q. (1986). *Theory of Limit Cycles*. Translations of Math Monographs, 66. American Mathematical Society, Providence, RI.

Ying, Y.-R. and Feng, G. (2012). Stabilization of Michael's model under capital controls. *Proceedings of the 11th International Conference on Information and Management Sciences*, August 2012, pp. 549–554.

Ying, Y.-R., Tong, Y.-Y., and Forrest, J. (2010). Pricing Asian options: Approach of decomposition and estimation. *Advances in Systems Science and Applications*, 10(2), 241–247.

Yinusa, D. O. (2008). Between dollarization and exchange rate volatility: Nigeria's portfolio diversification option. *Journal of Policy Modeling*, 30(5), 811–826.

Yinusa, O. D. (2009). Exchange rate volatility, currency substitution and monetary policy in Nigeria. MPRA Paper, No. 16255, posted 14. July, 12:50.

York, R., Rosa, E. A., Dietz, T. (2003). STIRPAT, IPAT and IMPACT: Analytic tools for unpacking the driving forces of environmental impacts. *Ecological Economics*, 46(3), 351–365.

Zeng, S. (2009). *Study on Financial Fragility Theory*. Chinese Finance Press, Beijing, China.

Zeng, W.-Y. and Liu, F. (2012). An empirical study on asymmetric Laplace distribution of stock index returns in China. *Statistics and Information Forum*, 12, 27–31.

Zhang, J. E. (2001). A semi-analytical method for pricing and hedging continuously sampled arithmetic average rate options. *Journal of Computational Finance*, 5, 59–79.

Zhang, J. E. (2003). Pricing continuously sampled Asian options with perturbation method. *Journal of Futures Markets*, 23, 535–560.

Zhang, W. et al. (2008). Pricing European foreign currency option under jump fractional Brown motion. *Chinese Journal of Management Science*, 16(3), 57–61.

Zhong, L.-M. (2004). Research on the endogenous liquidity risk management of stock market investors. Doctoral dissertation, Shanghai Jiao Tong University, Shanghai, China.

Zhang, M. (2006). An actuarial approach to foreign currency option pricing. *Journal of Hubei University*, 12, 75–78.

Zhang, Y. and Daye, J. (2003). Introduction to fractional Browian motion in finance. *Advanced FE*, 6, 1–4.

Zhong, L.-M., Liu, H.-L., and Wu, C.-F. (2003). Liquidity of China's stock market: Too high or too low—Analysis of international comparative perspective. *Modern Economic Science*, 2, 58–61.

Zhu, Q., Peng, X., Lu, Z., and Yu, J. (2010). Analysis model and empirical study of impact from population and consumption on carbon emissions. *China Population, Resources and Environment*, 20(2), 98–102.

Zoladek, H. (1998). Algebraic invariant curves for the Liénard equation. *Transactions of the American Mathematical Society*, 350, 681–701.

Zou, Y. and Li, H. (2014). Time spans between price maxima and price minima in stock markets. *Physica A: Statistical Mechanics and Its Applications*, 395(1), 303–309.

Index